SLEEP AND BRAIN ACTIVITY

SLEEP AND BRAIN ACTIVITY

Edited by

MARCOS G. FRANK

Amsterdam • Boston • Heidelberg • London • New York • Oxford
Paris • San Diego • San Francisco • Singapore • Sydney • Tokyo
Academic Press is an imprint of Elsevier

Academic Press is an imprint of Elsevier
The Boulevard, Langford Lane, Kidlington, Oxford, OX5 1GB, UK
225 Wyman Street, Waltham, MA 02451, USA

First published 2012

British Library Cataloguing in Publication Data
A catalogue record for this book is available from the British Library

Library of Congress Control Number: 2012940402

ISBN: 978-0-12-384995-3

For information on all Academic Press publications
visit our website at store.elsevier.com

Typeset by MPS Limited, Chennai, India
www.adi-mps.com

Printed and bound in the United States
12 13 14 15 10 9 8 7 6 5 4 3 2 1

CONTENTS

PREFACE

Prior to the middle part of the 20th century, a tome entitled "Brain Activity in Sleep" would likely be greeted with quizzical looks from the scientific community. Although sleep had been shown to be accompanied by slow electroencephalographic (EEG) waves in the 1920s and 30s, the significance of these findings would elude scientists for decades to come. The general consensus was consonant with prescientific thinking, which held that sleep was a time when the brain rested—and, just as the body exhibited reduced muscle activity during sleep, so did the brain. Sleep was thought to involve either a near-complete cessation of brain activity or at best an idling brain state. By the 21st century, it had become abundantly clear that far from a time of brain silence, sleep was characterized by complex patterns of neural and metabolic activity that in some cases exceeded what was observed during wakefulness.

In the late 1940s, a prevailing view was that sleep was essentially a passive brain response to sensory deafferentation. This general idea was often referred to as the "passive theory" of sleep. This passive theory of sleep gave way under overwhelming scientific evidence of brain activity in sleep. REM studies in the 1950s were the first scientific demonstrations that the sleeping brain was crackling with neural activity. Since then, intracellular recordings *in vivo* have revealed that slow oscillations typical of non-REM sleep (and which were originally dismissed as brain idling) reflect the synchronized activity of thousands, if not tens of thousands of neurons. During the cortical "up" state component of the slow oscillation, cortical neurons fire at a rate sometimes exceeding rates measured during wakefulness. REM sleep is additionally accompanied by tonic and rapid firing in thalamic and cortical neurons resembling that of the waking state. REM sleep also shows peculiar, phasic activations of the brainstem, thalamus, and cortex.

Brain activity in sleep has also been shown in imaging studies and molecular approaches. Functional imaging of the human brain has provided stunning pictures of metabolic changes during sleep that are regional and distinct from what is observed during wakefulness. Molecular approaches are now identifying striking changes in gene expression across sleep and wake, as well as the genetic basis of sleep EEG rhythmic activity. What is less clear is to what end all of this brain activity is directed. That is to say, the ultimate functions of brain activity in sleep remain mysterious.

In this book, leading scientists discuss some of the more dramatic developments in our understanding of brain activity in sleep. These include recent findings from measurements of single brain cells *in vivo* and *in vitro* that reveal the network and membrane mechanisms responsible for waking and sleeping brain activity (Chapters 1-2). The discussion of cellular mechanisms continues with treatments of possible roles of glial cells in the sleeping brain (Chapter 3), the molecular basis of sleep EEG rhythms (Chapter 4), and evidence that sleep may be an emergent property of smaller ensembles of neurons (Chapter 5). The remainder of the book is dedicated to perhaps the least understood, but no less important issue of function. We begin this section with an intriguing set of findings indicating that brain activity in songbird sleep contributes to singing (Chapter 6). We then explore the possible roles of REM sleep P-waves in rodent learning (Chapter 7), look inside the sleeping human brain for evidence of memory processing (Chapters 8-9), and consider the possibility that ontogenetic changes in sleep EEG rhythms might shape developing cortical circuits (Chapter 10). I am grateful to all the contributors for their hard work and scholarship.

LIST OF CONTRIBUTORS

Jan Born
University of Lübeck, Department of Neuroendocrinology, Ratzeburger Allee 160, Hs 23a, D-23538 Lübeck

Timothy P. Brawn
Department of Organismal Biology and Anatomy, University of Chicago, Chicago, IL, USA

Vincenzo Crunelli
Neuroscience Division, School of Bioscience, Museum Avenue, Cardiff University, Cardiff CF10 3AX, UK

Subimal Datta
Laboratory of Sleep and Cognitive Neuroscience, Departments of Psychiatry and Neurology and Program in Neuroscience, Boston University School of Medicine, 85 East Newton Street, Suite: M-902 Boston, Massachusetts 02118, USA

Gordon B. Feld
University of Tübingen, Department of Medical Psychology and Behavioral Neurobiology, Gartenstraße 29, 72074 Tübingen, Germany

Ariane Foret
Cyclotron Research Centre (B30), University of Liège, 6, Allée du 8 Août, 4000 Liège, Belgium

Marcos G. Frank
University of Pennsylvania, School of Medicine, Department of Neuroscience, 215 Stemmler Hall, 35th & Hamilton Walk, Philadelphia, PA 19104-6074, USA

Paul Franken
Center for Integrative Genomics, Genopode building, University of Lausanne, CH-1015 Lausanne-Dorigny, Switzerland

Reto Huber
University Children's Hospital, Zurich, Steinwiesstrasse 75, CH-8032 Zurich, Switzerland

Stuart Hughes
Eli Lilly UK, Erl Wood Manor, Windlesham, Surrey, GU20 6PH, UK

Mathieu Jaspar
Cyclotron Research Centre (B30), University of Liège, 6, Allée du 8 Août, 4000 Liège, Belgium

Salome Kurth
Child Development Center, University Children's Hospital Zurich, Steinweisstrasse 75, CH-8032 Zurich, Switzerland and Department of Integrative Physiology, University of Colorado at Boulder, CO, USA

Caroline Kussé
Cyclotron Research Centre (B30), University of Liège, 6, Allée du 8 Août, 4000 Liège, Belgium

Pierre Maquet
Cyclotron Research Centre (B30), University of Liège, 6, Allée du 8 Août, 4000 Liège, Belgium

Daniel Margoliash
University of Chicago, Department of Organismal Biology and Anatomy, 1027 East 57th Street, Chicago, IL 60637, USA

Laura Mascetti
Cyclotron Research Centre (B30), University of Liège, 6, Allée du 8 Août, 4000 Liège, Belgium

Christelle Meyer
Cyclotron Research Centre (B30), University of Liège, 6, Allée du 8 Août, 4000 Liège, Belgium

Vincenzo Muto
Cyclotron Research Centre (B30), University of Liège, 6, Allée du 8 Août, 4000 Liège, Belgium

David Rector
Department of Veterinary Comparative Anatomy, Pharmacology and Physiology, Washington State University, 205 Wegner Hall, Pullman, WA 99164, USA

Igor Timofeev
Department of Psychiatry and Neuroscience, Laval University, Québec, G1V 0A6, Canada.

CHAPTER 1

Neuronal Oscillations in the Thalamocortical System during Sleeping and Waking States

Igor Timofeev
Département de Psychiatrie et de Neurosciences, The Centre de Recherche Université Laval Robert-Giffard (CRULRG), Université Laval, Québec, (QC), G1J 2G3, Canada

All normal brain processes occur over three main states of vigilance: wake, slow-wave sleep, and REM sleep. These states can be subdivided further to passive and active wakefulness, three to four stages of slow-wave sleep, and active and passive REM sleep. There are also abnormal states of the brain like paroxysmal seizure activities or brain activities generated under anesthesia conditions. The states of vigilance by themselves originate via interaction of circadian and homeostatic processes (Achermann & Borbely, 2003; Borbely, Baumann, Brandeis, Strauch, & Lehmann, 1981) with a leading role of interactions between suprachiasmstic nucleus and hypothalamic regions (Fuller, Sherman, Pedersen, Saper, & Lu, 2011; Saper, Scammell, & Lu, 2005). Different brain states are expressed as different forms of global electrographic activities recorded from a brain surface (elecrocorticogram), which are reflected on a head surface and recorded as an electroencephalogram (EEG). These global electrical activities are mediated by synchronous synaptic activities of neurons. Synchronous de- or hyperpolarization of neighboring neurons will generate large amplitude global waves, the asynchronous activities of neurons will not generate the global field potential signals at all, and intermediates neuronal synchrony will produce field potential waves of intermediate amplitude. While the states of vigilance by themselves originate in suprachiasmstic and hypothalamic regions and are transmitted to the other brain structures via ascending activating systems (Steriade & McCarley, 2005), the top level of the brain, primarily the thalamocortical system, generates the electrical activities that are characteristic to different states of vigilance.

NEURONAL SYNCHRONIZATION

Neuronal synchronization requires some form of interactions between neurons. I will only briefly overview this aspect. To understand the

Sleep and Brain Activity
DOI: http://dx.doi.org/10.1016/B978-0-12-384995-3.00001-0

processes of neuronal synchronization, one has to separate local and long-range synchronization. Local synchronization is required to produce simultaneous de- or hyperpolarization of a local group of neurons that mediates generation of local field potentials; and long-range synchronization synchronizes the local field potentials generated at some distance. The amplitude of field potential recordings with one electrode will provide information on levels of local neuronal synchrony. Multisite recordings are used to investigate long-range synchronization.

The most investigated aspect of neuronal interactions are *chemical synaptic interactions* (Eccles, 1964). Action potentials generated in excitatory neurons will propagate to synaptic boutons and release excitatory neurotransmitter (mainly glutamate within the thalamocortical system). The amplitude of single-axon induced excitatory postsynaptic potentials (EPSPs) is small from 0.1 to several millivolts with overall mean less than 2 mV (Thomson, West, Wang, & Bannister, 2002). This neurotransmitter will exert depolarizing action on targets. Because each axon forms multiple contacts with target neurons it can produce sufficient local field effects if the target neurons are located in proximity. If several neurons excite the same group of target cells nearly simultaneously, the overall postsynaptic effect will definitely be larger due to spatial summation. That will produce summated effects, which are sensed at the field potential level. Multiple excitatory connections are formed by long-axon neurons; therefore they are well positioned to mediate long-range synchrony. Activities of inhibitory neurons within the thalamocortical system release mainly GABA, an inhibitory neurotransmitter. All known cortical interneurons have a short-axon (Markram et al., 2004; Somogyi, Tamás, Lujan, & Buhl, 1998), which contacts multiple target neurons in a local network. Therefore, cortical interneurons can contribute to local, but not long-range synchronization. This is not the case for other inhibitory cells. During development, GABAergic neurons of thalamic reticular (RE) nucleus form variable patterns of connectivity from a compact, focal projection to a widespread, diffuse projection encompassing large areas of Ventro-Basal complex (VB) (Cox, Huguenard, & Prince, 1996). Indirect action of reticular thalamic neurons that exerts diffuse projections onto thalamocotical neurons could likely be detected as synchronous field potential events over some large cortical areas when thalamocortical neurons will fire action potentials. Neuronal constellations outside the thalamocortical system may also to cortical synchronization. A recent study (Eschenko, Magri, Panzeri, & Sara, 2012) has demonstrated that in rats, the locus coreleus neurons fire in phase with cortical slow waves and they even preceded onsets of cortical neuronal firing, suggesting a contribution of locus coreleus not only in setting up general cortical excitability, but also in influencing the cortical synchronization processes.

The next mechanism contributing to neuronal synchronization is *electrical coupling* between cells that is mediated by gap junctions. The astrocytic network is tightly connected via gap junctions (Mugnaini, 1986). Gap junctions were also found between multiple groups of cortical interneurons (Galarreta & Hestrin, 1999; Gibson, Beierlein, & Connors, 1999). Electrical coupling was demonstrated between neurons of reticular thalamic nucleus (Fuentealba et al., 2004; Landisman et al., 2002). Dye coupling, presence of spikelets, and modeling experiments suggest the existence of electrical coupling between axons of hippocampal pyramidal cells (Schmitz et al., 2001; Vigmond, Perez Velazquez, Valiante, Bardakjian, & Carlen, 1997). A single study has found spikelets, an accepted signature of electrical coupling, in thalamocortical neurons (Hughes, Blethyn, Cope, & Crunelli, 2002). Indirect data on electrical coupling between thalamocortical neurons and axo-axonal coupling are so far not supported by demonstration of the presence of gap junctions. The gap junctions, mediating electrical coupling, form high resistance contacts between connected cells; therefore, they act as low-pass filters (Galarreta & Hestrin, 2001). Confirmed gap junctions are formed within dendritic arbor of connected cells, therefore, they can contribute to the local synchronization only.

Ephaptic interactions constitute another mechanism of neuronal synchronization. Changes in neuronal membrane potential produce extracellular fields that affect the excitability of neighboring cells (Jefferys, 1995). The extracellular fields produced by a single neuron are weak, however, when a local population of neurons generate nearly simultaneous excitation or inhibition, their summated effects can significantly influence the excitability of neighboring neurons, contributing to local synchronization. External electric field applied with intensities comparable to endogenous fields applied to cortical slices modulated cortical slow oscillation (Frohlich & McCormick, 2010). During seizure activity, the extracellular space reduces and the effects of ephaptic interactions increase (Jefferys, 1995).

Neuronal activities are associated with movement of ions across membrane due to activation of ligand- or voltage-controlled channels. Because extracellular space in the brain is about 20% of total brain volume (Syková, 2004; Sykova & Nicholson, 2008) and an activation of ionic pumps, ionic diffusion, etc., is a time-dependent process, the neuronal activities alter *extracellular concentration of implicated ions* (Somjen, 2002). Changes are temporal and local, but on a short timescale they affect neuronal excitability. For example, during active states of cortical slow oscillation, due to activation of synaptic currents (primarily NMDA receptor-mediated) and Ca^{2+} gated intrinsic channels, the extracellular concentration of Ca^{2+} decreases from 1.2 mM to 1.0 mM (Crochet, Chauvette, Boucetta, & Timofeev, 2005; Massimini & Amzica, 2001). This leads to a dramatic increase in synaptic failure rates (up to 80%) (Crochet, et al., 2005). Dramatic changes of ionic concentrations occur during seizure activity. In these conditions, the extracellular Ca^{2+} drops to below 0.5 mM (Amzica, Massimini, & Manfridi, 2002; Heinemann, Lux, & Gutnick, 1977; Pumain, Kurcewicz, & Louvel, 1983) and extracellular K^+ increases to 7–18 mM (Amzica et al., 2002; Moody, Futamachi, & Prince, 1974; Prince, Lux, & Neher, 1973). This results in an impairment of chemical synaptic transmission and axonal conduction of action potentials is impaired too (Seigneur & Timofeev, 2010). The reversals potential for ionic currents with implicated ions are shifted affecting overall neuronal excitability. Just some little increase in extracellular K^+ concentration and a decrease in extracellular Ca^{2+} and Mg^{2+} concentration may induce slow oscillation in neocortical slices (Reig, Gallego, Nowak, & Sanchez-Vives, 2006). Stronger changes in ion concentration occurring within physiological/pathological range have profound influence on brain network activities.

Therefore, synaptic potentials (primarily excitatory) are in a position to mediate long-range neuronal synchronization. Synaptic activities and all the other mechanisms of neuronal synchronization are local and they contribute to short-range neuronal synchronization. Local synchronous activities of neuronal groups may create propagating or travelling waves of activities that can involve large neuronal territories and can be detected as synchronous activities with shifts in the phase of oscillations (Boucetta, Chauvette, Bazhenov, & Timofeev, 2008; Massimini, Huber, Ferrarelli, Hill, & Tononi, 2004).

BASIC THALAMOCORTICAL STATES

There are two basic states of thalamocortical network: silent and active.

During *silent states* all active conductances (intrinsic and synaptic) are inactive. Basically, during silent states the membrane potential of neurons is mediated by leak current only. This is a theoretical situation and it is unlikely that it occurs in real life. In practice, some inactivating intrinsic currents, for example inward rectifying current, can also be active.

Activities of these currents preset so-called *resting membrane potential* of neurons. The resting membrane potential of cortical neurons is situated somewhere between −70 mV and −80 mV. Given that equilibrium potential for leak current is around −95 mV (McCormick, 1999), an activation of inward rectifying currents appears to play a role in the mediation of the resting membrane potential. The resting membrane potential can be recorded from neurons in nonoscillating slices maintained *in vitro*; it can also be recorded from cortical neurons during silent phases of sleep slow oscillation (see below). Because very few current are active during network silent states, the input resistance of neurons during silent states is higher as compared to active states (Contreras, Timofeev, & Steriade, 1996; Steriade, Timofeev, & Grenier, 2001).

Active states are generated when active coductances either intrinsic or synaptic or both are activated. Active cortical network states are found in wake, REM sleep and active phases of slow oscillation (sleep or anesthesia). During active network states the mean membrane potential of cortical neurons is situated at somewhere around −62 mV (Steriade et al., 2001). Therefore activities of all conductances accompanying active network states depolarize cortical neurons by about 10 to 15 mV. It appears that under anesthesia conditions the excitatory and inhibitory conductances are balanced (Haider, Duque, Hasenstaub, & McCormick, 2006). However, during natural states of vigilance the inhibitory activities dominate the active network states in the majority of neurons and during all states of vigilance except a few tens of milliseconds at the onset of active states during slow-wave activity (Rudolph, Pospischil, Timofeev, & Destexhe, 2007). Although excitatory neurons outnumber inhibitory ones, the spontaneous firing rate of former is higher (Peyrache et al., 2012; Steriade et al., 2001; Timofeev, Grenier, & Steriade, 2001) and all somatic synapses are formed by inhibitory neurons giving overall stronger inhibitory influence on cell somata. The ratio of activation of intrinsic currents versus synaptic currents during active network states is unclear. Because the membrane potential of cortical neurons during active states is situated close to −60 mV, it is reasonable to assume that persistent Na^+ current (Crill, 1996) contributes to the generation of active states. Indeed, the use of QX-314 (use-dependent blocker of Na^+ channels) reduced active states not only in intact brain (Timofeev, Grenier, & Steriade, 2004), but also in neocortical slabs (Timofeev, Grenier, Bazhenov, Sejnowski, & Steriade, 2000). It is reasonable to believe that hyperpolarization activated depolarizing current and some Ca^{2+} currents also contribute to the generation of active states.

Intracellular recordings demonstrated that neuronal firing occurs during active network states only (Contreras & Steriade, 1995; Steriade et al., 2001). Active network states are accompanied with remarkable increase in membrane potential fluctuations (Steriade et al., 2001; Timofeev, Grenier, & Steriade, 2000) that appear to be mediated by synaptic events. Based on these observations it was even proposed that active network states during sleep are functionally very similar to active network states of wake (Destexhe, Rudolph, & Pare, 2003). Several facts suggest that this assumption may not be true: (a) Patch-clamp recordings from slices obtained from human brain tissue (sub-plate zone) at gestation weeks 20–21 demonstrated that a majority of neurons reveal alteration of active and silent states. However, at this time of development, the chemical synaptic connectivity is not functional (Moore, Zhou, Jakovcevski, Zecevic, & Antic, 2011). Therefore, the synaptic potentials not necessary represent an essential component of active network states. (b) As opposed to REM sleep and waking state, during slow-wave sleep the activity of neuromodulatory systems is reduced (Steriade & McCarley, 2005), which has multiple cellular effects. Among others the input resistance is reduced during active phases of slow-wave sleep as compared to the active phases of waking or REM sleep (Steriade et al., 2001). Accordingly the neuronal responsiveness during active phases of slow-wave sleep differs from that of other states of vigilance. (c) The major feature of waking in higher animals is self-awareness. In order to be self-aware some minimal time is required and this time is about 1 sec (Libet, Alberts, Wright, & Feinstein, 1967). During slow-wave sleep or anesthesia, the active network states last less than 1 sec; therefore, the major characteristic of wake is not present.

The duration of active phases of slow oscillation depends on a variety of factors. Our recent study has shown that under ketamine-xylazine anesthesia active states occupy about 80% of the total time, while during slow-wave sleep they occupy around 90% of the total time. The amplitude of slow waves, the long-range synchrony, and the high frequency oscillations are higher during anesthesia as compared to slow-wave sleep (Chauvette, Crochet, Volgushev, & Timofeev, 2011).

Active network states do not need to be depolarizing. Synchronous activities of inhibitory neurons can impose active inhibitory activities on their targets. Synchronous cortical firing during active phases of slow oscillation strongly excite inhibitory reticular thalamic neurons, which induce inhibitory postsynaptic potentials (IPSPs) in target thalamocortical neurons, which may appear in the form of spindles or long-lasting hyperpolarization potentials (Contreras & Steriade, 1995; Fuentealba, Crochet, Timofeev, & Steriade, 2004; Timofeev, Bazhenov, Sejnowski, & Steriade, 2001; Timofeev & Steriade, 1996). In thalamocortical neurons spindles represent active hyperpolarizing states. In associative cortical areas the active inhibitory activities dominate cortical neurons during rapid eye movements of REM sleep (Timofeev, Grenier, et al., 2001).

As was mentioned above, the neuronal firing occurs during active network states. Surprisingly, the extent of spontaneous neuronal firing of individual cortical neurons is not clear yet. Intracellular recordings show that the mean firing rates were in the order of 10–15 Hz with lower rates during slow-wave sleep (Steriade et al., 2001). These numbers likely represent some overestimation, because of membrane leak induced by intracellular penetration. Similar results were obtained with extracellular single-unit recordings in which somewhat higher (Noda & Adey, 1970a, 1970b) or lower firing rates were seen (Evarts, Bental, Bihari, & Huttenlocher, 1962). However, in these studies the bias was to record activities of neurons that fire spikes, and the neurons that had sparse firing or did not fire spikes at all were not included in the database. Patch-clamp, optical, and extracellular unit recordings from superficial neurons demonstrated much lower (generally below 0.5 Hz) firing rates (Greenberg, Houweling, & Kerr, 2008; Peyrache et al., 2012; Waters & Helmchen, 2006). Multiunit recordings from different cortical depth demonstrated that indeed superficially laying neurons fire sparsely, and much higher firing rates can be observed in deep layers (Chauvette, Volgushev, & Timofeev, 2010; Sakata & Harris, 2009). However, in these studies a clear-cut separation of firing of individual neurons is impossible due to technical reasons. A number of recent studies attempt to use timing of multispike activity in order to characterize timing of onset and/or termination of active states (Eschenko et al., 2012; Hasenstaub, Sachdev, & McCormick, 2007; Vyazovskiy et al., 2009). The precision of these measurements is questionable. The averaged delay to the first spike in neurons that fire early during active states occurs usually within tens of milliseconds from the onset of active states and most neurons cease firing hundreds of milliseconds prior to the end of active states (Chauvette et al., 2011).

THALAMOCORTICAL RHYTHMIC ACTIVITIES

Thalamocortical rhythmic activities or oscillations are generated in a wide range of frequencies from 0.01 Hz to hundreds of Hz (Fig. 1.1). The mechanisms of thalamocortical oscillations are different. *Infraslow oscillation* has a

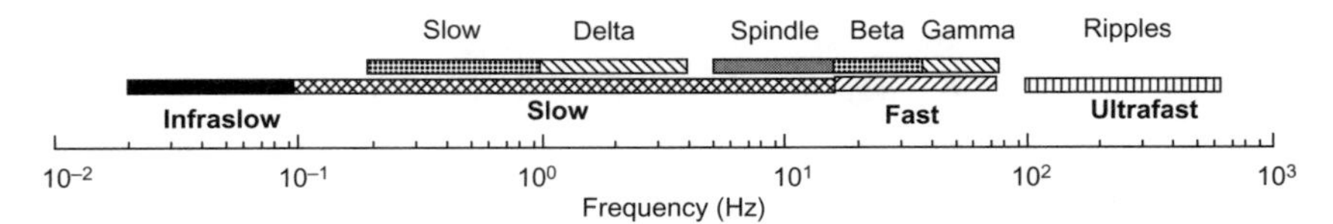

Figure 1.1 Frequency range and groups of thalamocortical oscillations. *(Modified from Bazhenov & Timofeev, 2006)*

period of cycle in the range of seconds to minutes (Aladjalova, 1957, 1962). Infraslow oscillations synchronize faster activities and modulate cortical excitability (Vanhatalo et al., 2004). The mechanisms of infraslow oscillation are unknown. This oscillation likely depends on dynamics of brain metabolism (Aladjalova, 1957; Nita et al., 2004). Multisite recordings indicate that waves of infraslow oscillation can occur nearly simultaneously at different cortical locations (Nita et al., 2004; Vanhatalo et al., 2004) suggesting that long-range synchronizing mechanisms control this type of oscillation.

Slow thalamocortical oscillations are composed of three distinct types of activities: slow oscillation, delta oscillation, and spindle oscillation (Fig. 1.1). During slow-wave sleep and some forms of anesthesia a dominant form of activity within the thalamocortical system is *slow oscillation (0.2–1 Hz)* (Steriade, Nuñez, & Amzica, 1993b; Steriade et al., 2001). During slow oscillation the entire cortical network alternates between silent (hyperpolarizing) and active (depolarizing) states. Depending on cortical area the silent states last between 100 ms and 200 ms and under ketamine-xylazine anesthesia they last between 300 ms and 400 ms (Chauvette et al., 2011). The active states last longer, 80%–90% of a cycle of slow oscillation. During silent cortical states the neurons of both reticular thalamic nucleus and projecting nuclei are also in silent state (Contreras & Steriade, 1995; Contreras, Timofeev, et al., 1996). During active states, cortical firing triggers rhythmic spike bursts of reticular thalamic neurons, which in turn induce rhythmic IPSPs in thalamocortical cells (see Figure 1 in Timofeev & Bazhenov, 2005). Recent experiments on rodents suggest that some thalamocortical neurons can actively participate in the initiation of active states (Crunelli & Hughes, 2010; Slezia, Hangya, Ulbert, & Acsady, 2011; Ushimaru, Ueta, & Kawaguchi, 2012).

Field potential recordings from neocortex in human and animal models during sleep reveal the presence of *delta oscillations (1–4 Hz)*. The delta oscillation likely has two different components, one of which originates in the neocortex and the other in the thalamus. Delta oscillation is distinct from slow oscillation, not only because of some difference in frequencies, but also because the dynamic of the two rhythms during sleep. During sleep the delta activity declines from the first to second episodes of slow-wave sleep, while this was not observed for slow oscillation (Achermann & Borbely, 1997).

Cortical delta activity. Both surgical removal of the thalamus and recordings from neocortical slabs in chronic conditions result in the significant enhancement of neocortical delta activity (Ball, Gloor, & Schaul, 1977; Villablanca & Salinas-Zeballos, 1972). These experiments suggest that cortex itself is able to generate delta activities.

Thalamic delta is a well-known example of rhythmic activity generated intrinsically by thalamic relay neurons as a result of the interplay between their low-threshold Ca^{2+} current (I_T) and hyperpolarization-activated cation current (I_h) (McCormick & Pape, 1990). As such, the delta oscillation may be observed during deep sleep when thalamic relay neurons are hyperpolarized sufficiently to deinactivate I_T. The mechanism of single cell delta activity is the following: a long-lasting hyperpolarization of thalamic relay neuron leads to slow I_h activation that depolarizes the membrane potential and triggers rebound burst, mediated by I_T, which was deinactivated by the hyperpolarization. Both I_h (because of its voltage dependency) and I_T (because of its transient nature) inactivate during burst, so the membrane potential becomes hyperpolarized after the burst termination. This afterhyperpolarization starts in the next cycle of oscillations. Intrinsic delta oscillations in thalamocortical neurons are generated by individual neurons. Thalamocortical neurons are not connected directly. The intrinsic delta oscillation is not synchronized (Timofeev & Steriade, 1996), therefore, it cannot produce field effects at neocortical level.

Sleep spindle oscillations consist of waxing-and-waning field potentials at 7–14 Hz, which last 1–3 seconds and recur every 5–15 seconds. *In vivo*, spindle oscillations are typically observed during stage 2 of sleep or during active phases of slow-wave sleep oscillations.

In vivo, *in vitro*, and *in silico* studies suggest that the minimal substrate accounting for spindle oscillations consists in the interaction between thalamic reticular and relay cells (Steriade & Deschenes, 1984; Steriade, Deschenes, Domich, & Mulle, 1985; von Krosigk, Bal, & McCormick, 1993). Burst firing of reticular thalamic neurons induces inhibitory postsynaptic potentials in thalamocortical neurons. This deinactivates I_T, inducing burst firing in thalamocortical neurons, which, in turn, excite reticular thalamic neurons allowing the cycle to start again. Spontaneous spindle oscillations are synchronized over large cortical areas during natural sleep and barbiturate anesthesia. After complete ipsilateral decortication, however, the long-range synchronization of thalamic spindles changes into disorganized patterns with low spatiotemporal coherence (Contreras, Destexhe, Sejnowski, & Steriade, 1996). Although cortical activities help to initiate spindles (Contreras & Steriade, 1996), persistent cortical firing depolarizes thalamic neurons at the second half of spindles, which contributes to the termination of this thalamic rhythm (Bonjean et al., 2011; Timofeev, Bazhenov, et al., 2001).

Beta (15–30 Hz) and gamma (30–80 Hz) activities occur during active cortical states (Mukovski, Chauvette, Timofeev, & Volgushev, 2007). They dominate cortical oscillations during waking state and REM sleep, and they can be found during active phases of slow oscillation. The power of beta/gamma activities under ketamine-xylazine anesthesia is a double of that recorded during natural sleep (Chauvette et al., 2011). Field potential and cellular recordings demonstrate robust synchronization of fast rhythms between connected areas of cortex and thalamus (Steriade, Contreras, Amzica, & Timofeev, 1996). There can be at least two different sources of origin of fast oscillations in neocortex. The first depends on activities in ascending structures. Indeed, it has been shown that synaptic activities in prethalamic pathways display high frequency oscillations (Castelo-Branco, Neuenschwander, & Singer, 1998; Timofeev & Steriade, 1997), which can be effectively transmitted to cerebral cortex when thalamocortical neurons are depolarized (Timofeev & Steriade, 1997). Another possible source of high frequency oscillations is intracortical. Although the exact mechanism of generation of beta/gamma oscillation is unknown, several mechanisms of their origin have been proposed (Timofeev & Bazhenov, 2005). A subset of cortical neurons called fast-rhythmic bursting cells has an intrinsic ability to generate spike bursts repeated with gamma frequency (Calvin & Sypert, 1976; Gray & McCormick, 1996; Steriade, Timofeev, Dürmüller, & Grenier, 1998). Firing of these neurons may entrain target cells into oscillatory activities creating local field potential oscillations. The second intracortical mechanism of gamma activity generation depends on the activity of inhibitory interneurons and was described both *in vitro* and with computational models (Lytton & Sejnowski, 1991; Traub, Whittington, Stanford, & Jefferys, 1996). Transitions between gamma and beta oscillations were simulated by alternating excitatory coupling pyramidal neurons and by change in K^+-conductances (Kopell, Ermentrout, Whittington, & Traub, 2000; Traub, Whittington, Buhl, Jefferys, & Faulkner, 1999).

Ripples are transient oscillations with frequency range between 100 and 600 Hz (Fig. 1.1). They can be recorded with local field potential electrodes, but not with large EEG electrodes, suggesting that local and not long-range synchronizing mechanisms contribute to their origin. Ripples were described in hippocampus (Buzsaki, Horvath, Urioste, Hetke, & Wise, 1992) and cerebral cortex (Grenier, Timofeev, & Steriade, 2001). Usually, they occur at the beginning of active cortical states (Grenier et al., 2001; Le Van Quyen et al., 2010). In neocortex, ripples with frequency 120–200 Hz occur during normal and paroxysmal activities (Grenier et al., 2001; Grenier, Timofeev, & Steriade, 2003). This is distinct from hippocampus in which paroxysmal activities are associated with high frequency ripples (<300 Hz) (Bragin, Engel, Wilson, Fried, & Mathern, 1999), while in neocortex the high frequency ripples can be recorded in response to sensory stimuli (Jones & Barth, 1999). In addition to active inhibition (Grenier et al., 2001; Ylinen et al., 1995), the electrical coupling mediated by gap junctions contributes to the ripple synchronization (Draguhn, Traub, Schmitz, & Jefferys, 1998; Grenier et al., 2003).

ORIGIN OF SLOW WAVES: CORTEX VERSUS THALAMUS

Slow waves of slow oscillation orchestrate other oscillatory activities (Steriade, 2006). In fact spindle, beta, gamma, and ripple oscillations occur exclusively over active states of slow oscillation. Thus, the other rhythmic events are secondary to slow oscillation. The question is: Where do the slow waves originate? Initial studies pointed to intracortical origin of slow oscillation. This conclusion is based on three main findings: (a) cortical slow oscillation was observed in cats with extensive chemical lesion of thalamus (Steriade, Nuñez, & Amzica, 1993a) or in isolated neocortex (Timofeev, Grenier, Bazhenov, et al., 2000); (b) in one hemisphere of decorticated cats, the slow oscillation was absent in the thalamus on the side of decortication, but it was present in the thalamus of intact hemisphere (Timofeev & Steriade, 1996); and (c) the slow oscillation was obtained in neocortical slices from ferrets (Sanchez-Vives & McCormick, 2000) and cats (Sanchez-Vives, Reig, Winograd, & Descalzo, 2007). These studies conducted on carnivores point to exclusive cortical origin of the slow oscillation. Slow oscillation was also recorded from cortex in multiple experiments conducted on rodents (Clement et al., 2008; Lau et al., 2000; Mohajerani, McVea, Fingas, & Murphy, 2010; Sharma, Wolansky, & Dickson, 2010; Vyazovskiy et al., 2011; Vyazovskiy et al., 2009). Isolated spontaneous active periods were obtained in neocortical slices of rodents (Cossart, Aronov, & Yuste, 2003; MacLean, Watson, Aaron, & Yuste, 2005). However, despite multiple attempts in multiple labs the reliable robust rhythmic slow oscillation was not reported in neocortical slices of rodent cortex (Sanchez-Vives et al., 2007). The absence of reliable slow oscillation in slices obtained from rodent brain can be due to the absence in this order of mammals of so-called patchy intracortical connectivity, which is present in cats (Gilbert & Wiesel, 1983), ferrets (Rockland, 1985) and primates (Lund, Yoshioka, & Levitt, 1993). Therefore, the intracortical connectivity of rodents lacks some essential elements that are likely critical for the generation of slow oscillation.

Slow oscillation is an essential component of brain activities during sleep. Sleep pressure, measured as the power of slow waves, progressively increases during the day, till the onset of sleep (Borbely et al., 1981). *In vitro* experiments on cortical slices demonstrate that the ability of rodent neocortex to generate slow oscillation is small. It is likely that rodents developed adaptive changes enabling other brain structures to contribute to the establishment of this essential rhythm. Recent studies demonstrated that thalamocortical neurons from thalamic slices from rodent brain, subjected to application of 100 μM trans-ACPD (metabotropic glutamate receptor agonist) were able to generate slow oscillation (Hughes, Cope, Blethyn, & Crunelli, 2002; Zhu et al., 2006). This led to the concept of a secondary thalamic oscillator contributing to cortical slow oscillation (Crunelli & Hughes, 2010). *In vivo* experiments on anesthetized rats demonstrated that a large number of thalalmocortical neurons could fire prior to the onset of active phases of cortical slow waves (Slezia et al., 2011; Ushimaru et al., 2012). All this indicates that in rodent brain, the thalamus can play an important, if not leading role in the generation of slow oscillation. This might not be the case in carnivores, in which the slow oscillation is generated in isolated cortical preparations (Ball et al., 1977; Sanchez-Vives & McCormick, 2000; Steriade et al., 1993a; Timofeev, Grenier, Bazhenov, et al., 2000) and in human in whom isolated cortex may even develop paroxysmal discharges (Echlin, Arnet, & Zoll, 1952; Echlin & Battista, 1963).

ORIGIN OF SLOW WAVES: HORIZONTAL AND VERTICAL PROPAGATION

Every cycle of slow oscillation can originate from any part of the cortex and propagate toward other areas. However, there are preferential sites of origin of slow waves. In adult humans the slow waves originate most commonly in frontal areas and from these areas they propagate to involve other cortical regions (Carrier et al., 2011; Massimini et al., 2004). The sites of preferential origin of slow waves are not fixed. In young human (2–8 years) the slow waves start preferentially in occipital areas, than the strongest power of slow waves moves to parietal areas (8–14 years) to become highest in frontal areas after the age of 14 years (Kurth et al., 2010). In cats, like in humans, each active state of slow wave has a preferential site of origin and propagates toward other areas (Volgushev, Chauvette, Mukovski, & Timofeev, 2006). Most commonly active states in cats originate in parietal cortex at the border of area 5 and 7 (Volgushev et al., 2006), where the neurons show the longest silent states and the slow waves are of highest amplitude (Chauvette et al., 2011). Propagation of active state onsets was also found in mice (Mohajerani et al., 2010; Ruiz-Mejias, Ciria-Suarez, Mattia, & Sanchez-Vives, 2011). Interestingly, the onset of silent network states occurred more synchronously than the onset of active states (Volgushev et al., 2006) suggesting implication of a long-range synaptic mechanism in termination of active network states during slow oscillation. Pathological processes shift the preferential sites of origin of sleep slow waves. In epileptic patients, most of the slow waves are local, preferentially originate in medial prefrontal cortex, and if they propagate, they invade the medial temporal lobe and hippocampus (Nir et al., 2011). Seemingly, in cats undergoing trauma-induced epileptogenesis, the slow wave activities start around traumatized cortex (Nita, Cissé, Timofeev, & Steriade, 2007; Topolnik, Steriade, & Timofeev, 2003).

How and where exactly do the cortical active states start when all the neurons are silent? There should be a first neuron that is depolarized to the firing threshold and the firing of this neuron engages target cells into the active state. There are two main possibilities for the neuron to fire. One possibility is that a hyperpolarization achieved during silent state activates I_h, which depolarizes the first neuron to the firing threshold. Indeed, in neocortical slices from ferret visual cortex the layer 5 pyramidal neurons reveal depolarizing sag (suggesting the presence of h current) that bring these neurons to the firing (Sanchez-Vives & McCormick, 2000). However, the I_h in neocortical neurons is weak (Timofeev, Bazhenov, Sejnowski, & Steriade, 2002) and even in conditions of bath solution that increase neuronal excitability it triggers slow oscillation with a period longer than 3 sec (Sanchez-Vives & McCormick, 2000). Thus, it is unlikely that the I_h-based mechanism is implicated in the generation of slow oscillation in intact cortex. Another possibility is that spike-independent neurotransmitter release (minis (Katz, 1969)) leads to an occasional summation of depolarizing events, activating some inward current (i.e., persistent sodium current) that depolarizes neurons to the threshold of action potential generation (Chauvette et al., 2010; Timofeev, Grenier, Bazhenov, et al., 2000). Because of the stochastic nature of spontaneous neurotransmitter release, this mechanism should be more efficient in neurons possessing larger number of synapses. Indeed, layer 5 neurons are by far the biggest cortical neurons with 50,000–60,000 synapses (DeFelipe & Farinas, 1992). Due to multiple mechanisms of dendritic democracy (Häusser, 2001; Rumsey & Abbott, 2006), even remote synapses can efficiently depolarize the somatic compartment of a neuron. In addition, a large number of layer 5 neurons are intrinsically bursting (Connors & Gutnick, 1990), thus, they are in a position to efficiently excite their targets (Lisman, 1997; Timofeev, Grenier, Bazhenov, et al., 2000). Extra- and intracellular recordings from rat and cat neocortex demonstrated that indeed the deep layer (presumably layer 5) neurons are the first to be depolarized at the onset of new active state, they are the first to fire action potentials, and they generate a larger number of action potentials during active periods of slow waves (Chauvette et al., 2010; Sakata & Harris, 2009). Surprisingly, intracortical recordings from human epileptic patients demonstrate that active states start predominantly from superficial cortical layers (Cash et al., 2009; Csercsa et al., 2010). Does it suggest that in terms of slow wave sleep the neocortex of human is fundamentally different from animals or that epileptic brain is different from normal brain? Our current experiments on cats with cortical trauma show that during epileptogenesis lasting several months, the most likely site of origin of slow waves moves from deep layers to the more superficial layers (Avramescu, Timofeev, unpublished). This change in the origin of slow waves can be explained by a progressive loss of deeply laying neurons occurring on a background of increased connectivity among remaining neurons (Avramescu, Nita, & Timofeev, 2009; Avramescu & Timofeev, 2008). This suggests that in healthy human subjects, the slow waves of sleep can originate in deep cortical layers.

ACKNOWLEDGEMENTS

Currently my laboratory is supported by CHIR, NSERC, FRSQ and NIH. I deeply thank all current and former members of my laboratory as well as collaborators for their contribution in advancement of knowledge on the physiology of normal and pathological thalamocortical oscillations.

REFERENCES

Achermann, P., & Borbely, A. A. (1997). Low-frequency (<1 Hz) oscillations in the human sleep electroencephalogram. *Neuroscience*, *81*(1), 213–222.

Achermann, P., & Borbely, A. A. (2003). Mathematical models of sleep regulation. *Frontiers in Bioscience*, *8*, s683–693.

Aladjalova, N. A. (1957). Infra-slow rhythmic oscillations of the steady potential of the cerebral cortex. *Nature*, *4567*(11), 957–959.

Aladjalova, N. A. (1962). *Slow electrical processes in the brain*. Moscow: Acad Sci USSA.

Amzica, F., Massimini, M., & Manfridi, A. (2002). Spatial buffering during slow and paroxysmal sleep oscillations in cortical networks of glial cells in vivo. *Journal of Neuroscience*, *22*(3), 1042–1053.

Avramescu, S., Nita, D., & Timofeev, I. (2009). Neocortical post-traumatic epileptogenesis is associated with the loss of GABAergic neurons. *Journal of Neurotrauma*, *26*, 799–812.

Avramescu, S., & Timofeev, I. (2008). Synaptic strength modulation following cortical trauma: A role in epileptogenesis. *Journal of Neuroscience*, *28*(27), 6760–6772.

Ball, G. J., Gloor, P., & Schaul, N. (1977). The cortical electromicrophysiology of pathological delta waves in the electroencephalogram of cats. *Electroencephalography and Clinical Neurophysiology*, *43*(3), 346–361.

Bazhenov, M., & Timofeev, I. (2006). Thalamocortical oscillations. http://www.scholarpedia.org/article/Thalamocortical_Oscillation.

Bonjean, M., Baker, T., Lemieux, M., Timofeev, I., Sejnowski, T., & Bazhenov, M. (2011). Corticothalamic feedback controls sleep spindle duration in vivo. *Journal of Neuroscience*, *31*(25), 9124–9134. doi:10.1523/JNEUROSCI.0077-11.2011.

Borbely, A. A., Baumann, F., Brandeis, D., Strauch, I., & Lehmann, D. (1981). Sleep deprivation: Effect on sleep stages and EEG power density in man. *Electroencephalography and Clinical Neurophysiology*, *51*(5), 483–495.

Boucetta, S., Chauvette, S., Bazhenov, M., & Timofeev, I. (2008). Focal generation of paroxysmal fast runs during electrographic seizures. *Epilepsia*, *49*(11), 1925–1940.

Bragin, A., Engel, J., Jr., Wilson, C. L., Fried, I., & Mathern, G. W. (1999). Hippocampal and entorhinal cortex high-frequency oscillations (100–500 Hz) in human epileptic brain and in kainic acid-treated rats with chronic seizures. *Epilepsia*, *40*(2), 127–137.

Buzsaki, G., Horvath, Z., Urioste, R., Hetke, J., & Wise, K. (1992). High-frequency network oscillation in the hippocampus. *Science*, *256*(5059), 1025–1027. (Research Support, Non-U.S. Gov't Research Support, U.S. Gov't, P.H.S.)

Calvin, W. H., & Sypert, G. W. (1976). Fast and slow pyramidal tract neurons: An intracellular analysis of their contrasting repetitive firing properties in the cat. *Journal of Neurophysiology*, *39*(2), 420–434.

Carrier, J., Viens, I., Poirier, G., Robillard, R., Lafortune, M., Vandewalle, G., et al. (2011). Sleep slow wave changes during the middle years of life. *European Journal of Neuroscience*, *33*(4), 758–766. doi:10.1111/j.1460-9568.2010.07543.x.

Cash, S. S., Halgren, E., Dehghani, N., Rossetti, A. O., Thesen, T., Wang, C., et al. (2009). The human K-complex represents an isolated cortical down-state. *Science*, *324*(5930), 1084–1087. doi:10.1126/science.1169626.

Castelo-Branco, M., Neuenschwander, S., & Singer, W. (1998). Synchronization of visual responses between the cortex, lateral geniculate nucleus, and retina in the anesthetized cat. *Journal of Neuroscience*, *18*(16), 6395–6410.

Chauvette, S., Crochet, S., Volgushev, M., & Timofeev, I. (2011). Properties of slow oscillation during slow-wave sleep and anesthesia in cats. *Journal of Neuroscience*, *31*(42), 14998–15008. doi:10.1523/jneurosci.2339-11.2011.

Chauvette, S., Volgushev, M., & Timofeev, I. (2010). Origin of active states in local neocortical networks during slow sleep oscillation. *Cerebral Cortex*, *20*, 2660–2674. (doi: 10.1093/cercor/bhq009).

Clement, E. A., Richard, A., Thwaites, M., Ailon, J., Peters, S., & Dickson, C. T. (2008). Cyclic and sleep-like spontaneous alternations of brain state under urethane anaesthesia. *PLoS ONE, 3*(4), e2004.

Connors, B. W., & Gutnick, M. J. (1990). Intrinsic firing patterns of divers neocortical neurons. *Trends in Neurosciences, 13*(3), 99–104.

Contreras, D., Destexhe, A., Sejnowski, T. J., & Steriade, M. (1996). Control of spatiotemporal coherence of a thalamic oscillation by corticothalamic feedback. *Science, 274*(5288), 771–774.

Contreras, D., & Steriade, M. (1995). Cellular basis of EEG slow rhythms: A study of dynamic corticothalamic relationships. *Journal of Neuroscience, 15*(1), 604–622.

Contreras, D., & Steriade, M. (1996). Spindle oscillation in cats: The role of corticothalamic feedback in a thalamically generated rhythm. *The Journal of Physiology, 490*(Pt 1), 159–179.

Contreras, D., Timofeev, I., & Steriade, M. (1996). Mechanisms of long-lasting hyperpolarizations underlying slow sleep oscillations in cat corticothalamic networks. *The Journal of Physiology, 494*(Pt 1), 251–264.

Cossart, R., Aronov, D., & Yuste, R. (2003). Attractor dynamics of network UP states in the neocortex. *Nature, 423*(6937), 283–288.

Cox, C. L., Huguenard, J. R., & Prince, D. A. (1996). Heterogeneous axonal arborizations of rat thalamic reticular neurons in the ventrobasal nucleus. *The Journal of Comparative Neurology, 366*(3), 416–430.

Crill, W. E. (1996). Persistent sodium current in mammalian central neurons. *Annual Review of Physiology, 58*, 349–362.

Crochet, S., Chauvette, S., Boucetta, S., & Timofeev, I. (2005). Modulation of synaptic transmission in neocortex by network activities. *The European Journal of Neuroscience, 21*(4), 1030–1044.

Crunelli, V., & Hughes, S. W. (2010). The slow (<1 Hz) rhythm of non-REM sleep: A dialogue between three cardinal oscillators. *Nature Neuroscience, 13*(1), 9–17.

Csercsa, R., Dombovari, B., Fabo, D., Wittner, L., Eross, L., Entz, L., et al. (2010). Laminar analysis of slow wave activity in humans. *Brain, 133*(9), 2814–2829. (doi: 10.1093/brain/awq169).

DeFelipe, J., & Farinas, I. (1992). The pyramidal neuron of the cerebral cortex: Morphological and chemical characteristics of the synaptic inputs. *Progress in Neurobiology, 39*(6), 563–607.

Destexhe, A., Rudolph, M., & Pare, D. (2003). The high-conductance state of neocortical neurons in vivo. *Nature Reviews Neuroscience, 4*(9), 739–751.

Draguhn, A., Traub, R. D., Schmitz, D., & Jefferys, J. G. (1998). Electrical coupling underlies high-frequency oscillations in the hippocampus in vitro. *Nature, 394*(6689), 189–192.

Eccles, J. C. (1964). *The physiology of synapses*. Berlin: Springer.

Echlin, F. A., Arnet, V., & Zoll, J. (1952). Paroxysmal high voltage discharges from isolated and partially isolated human and animal cerebral cortex. *Electroencephalography and Clinical Neurophsysiology, 4*, 147–164.

Echlin, F. A., & Battista, A. (1963). Epileptiform seizures from chronic isolated cortex. *Archives of Neurology, 168*, 154–170.

Eschenko, O., Magri, C., Panzeri, S., & Sara, S. J. (2012). Noradrenergic Neurons of the Locus Coeruleus Are Phase Locked to Cortical Up-Down States during Sleep. *Cerebral Cortex, 22*(2), 426–435. doi:10.1093/cercor/bhr121.

Evarts, E. V., Bental, E., Bihari, B., & Huttenlocher, P. R. (1962). Spontaneous discharge of single neurons during sleep and waking. *Science, 135*(3505), 726–728.

Frohlich, F., & McCormick, D. A. (2010). Endogenous electric fields may guide neocortical network activity. *Neuron, 67*(1), 129–143. (doi: 10.1016/j.neuron.2010.06.005).

Fuentealba, P., Crochet, S., Timofeev, I., Bazhenov, M., Sejnowski, T. J., & Steriade, M. (2004). Experimental evidence and modeling studies support a synchronizing role for

electrical coupling in the cat thalamic reticular neurons in vivo. *The European Journal of Neuroscience*, *20*(1), 111–119.
Fuentealba, P., Crochet, S., Timofeev, I., & Steriade, M. (2004). Synaptic interactions between thalamic and cortical inputs onto cortical neurons in vivo. *Journal of Neurophysiology*, *91*(5), 1990–1998.
Fuller, P., Sherman, D., Pedersen, N. P., Saper, C. B., & Lu, J. (2011). Reassessment of the structural basis of the ascending arousal system. *The Journal of Comparative Neurology*, *519*(5), 933–956. doi:10.1002/cne.22559.
Galarreta, M., & Hestrin, S. (1999). A network of fast-spiking cells in the neocortex connected by electrical synapses. *Nature*, *402*(6757), 72–75.
Galarreta, M., & Hestrin, S. (2001). Electrical synapses between GABA-releasing interneurons. *Nature Reviews Neuroscience*, *2*(6), 425–433.
Gibson, J. R., Beierlein, M., & Connors, B. W. (1999). Two networks of electrically coupled inhibitory neurons in neocortex. *Nature*, *402*(6757), 75–79.
Gilbert, C., & Wiesel, T. (1983). Clustered intrinsic connections in cat visual cortex. *The Journal of Neuroscience*, *3*(5), 1116–1133.
Gray, C. M., & McCormick, D. A. (1996). Chattering cells: Superficial pyramidal neurons contributing to the generation of synchronous oscillations in the visual cortex. *Science*, *274*(4 October), 109–113.
Greenberg, D. S., Houweling, A. R., & Kerr, J. N. D. (2008). Population imaging of ongoing neuronal activity in the visual cortex of awake rats. *Nature Neuroscience*, *11*(7), 749–751.
Grenier, F., Timofeev, I., & Steriade, M. (2001). Focal synchronization of ripples (80–200 Hz) in neocortex and their neuronal correlates. *Journal of Neurophysiology*, *86*(4), 1884–1898.
Grenier, F., Timofeev, I., & Steriade, M. (2003). Neocortical very fast oscillations (ripples, 80–200 Hz) during seizures: Intracellular correlates. *Journal of Neurophysiology*, *89*(2), 841–852.
Haider, B., Duque, A., Hasenstaub, A. R., & McCormick, D. A. (2006). Neocortical network activity in vivo is generated through a dynamic balance of excitation and inhibition. *Journal of Neuroscience*, *26*(17), 4535–4545.
Hasenstaub, A., Sachdev, R. N. S., & McCormick, D. A. (2007). State changes rapidly modulate cortical neuronal responsiveness. *Journal of Neuroscience*, *27*(36), 9607–9622. doi:10.1523/jneurosci.2184-07.2007.
Häusser, M. (2001). Synaptic function: Dendritic democracy. *Current Biology*, *11*(1), R10–R12.
Heinemann, U., Lux, H. D., & Gutnick, M. J. (1977). Extracellular free calcium and potassium during paroxysmal activity in the cerebral cortex of the cat. *Experimental Brain Research*, *27*(3–4), 237–243.
Hughes, S. W., Blethyn, K. L., Cope, D. W., & Crunelli, V. (2002). Properties and origin of spikelets in thalamocortical neurones in vitro. *Neuroscience*, *110*(3), 395–401.
Hughes, S. W., Cope, D. W., Blethyn, K. L., & Crunelli, V. (2002). Cellular mechanisms of the slow (<1 Hz) oscillation in thalamocortical neurons in vitro. *Neuron*, *33*(6), 947–958.
Jefferys, J. G. (1995). Nonsynaptic modulation of neuronal activity in the brain: Electric currents and extracellular ions. *Physiological Reviews*, *75*(4), 689–723.
Jones, M. S., & Barth, D. S. (1999). Spatiotemporal organization of fast (>200 Hz) electrical oscillations in rat vibrissa/barrel cortex. *Journal of Neurophysiology*, *82*(3), 1599–1609.
Katz, B. (1969). *The release of neuronal transmitter substances*. Springfield, Illinois: Thomas.
Kopell, N., Ermentrout, G. B., Whittington, M. A., & Traub, R. D. (2000). Gamma rhythms and beta rhythms have different synchronization properties. *Proceedings of the National Academy of Sciences of the United States of America*, *97*(4), 1867–1872.
Kurth, S., Ringli, M., Geiger, A., LeBourgeois, M., Jenni, O. G., & Huber, R. (2010). Mapping of cortical activity in the first two decades of life: A high-density sleep electroencephalogram study. *Journal of Neuroscience*, *30*(40), 13211–13219. (doi: 10.1523/JNEUROSCI.2532-10.2010).

Landisman, C. E., Long, M. A., Beierlein, M., Deans, M. R., Paul, D. L., & Connors, B.W. (2002). Electrical synapses in the thalamic reticular nucleus. *Journal of Neuroscience*, *22*(3), 1002–1009.

Lau, D., de Miera, E.V., Contreras, D., Ozaita, A., Harvey, M., Chow, A., et al. (2000). Impaired fast-spiking, suppressed cortical inhibition, and increased susceptibility to seizures in mice lacking kv3.2 K+ channel proteins. *Journal of Neuroscience*, *20*(24), 9071–9085.

Le Van Quyen, M., Staba, R., Bragin, A., Dickson, C., Valderrama, M., Fried, I., et al. (2010). Large-scale microelectrode recordings of High-Frequency gamma oscillations in human cortex during sleep. *The Journal of Neuroscience*, *30*(23), 7770–7782. doi:10.1523/jneurosci.5049-09.2010.

Libet, B., Alberts, W. W., Wright, E. W., Jr., & Feinstein, B. (1967). Responses of human somatosensory cortex to stimuli below threshold for conscious sensation. *Science*, *158*(808), 1597–1600.

Lisman, J. E. (1997). Bursts as a unit of neural information: Making unreliable synapses reliable. *Trends in Neurosciences*, *20*(1), 38–43.

Lund, J. S., Yoshioka, T., & Levitt, J. B. (1993). Comparison of intrinsic connectivity in different areas of macaque monkey cerebral cortex. *Cerebral Cortex*, *3*(2), 148–162.

Lytton, W. W., & Sejnowski, T. J. (1991). Simulations of cortical pyramidal neurons synchronized by inhibitory interneurons. *Journal of Neurophysiology*, *66*(3), 1059–1079.

MacLean, J. N., Watson, B. O., Aaron, G. B., & Yuste, R. (2005). Internal dynamics determine the cortical response to thalamic stimulation. *Neuron*, *48*(5), 811–823.

Markram, H., Toledo-Rodriguez, M., Wang, Y., Gupta, A., Silberberg, G., & Wu, C. (2004). Interneurons of the neocortical inhibitory system. *Nature Reviews Neuroscience*, *5*(10), 793–807.

Massimini, M., & Amzica, F. (2001). Extracellular calcium fluctuations and intracellular potentials in the cortex during the slow sleep oscillation. *Journal of Neurophysiology*, *85*(3), 1346–1350.

Massimini, M., Huber, R., Ferrarelli, F., Hill, S., & Tononi, G. (2004). The sleep slow oscillation as a traveling wave. *Journal of Neuroscience*, *24*(31), 6862–6870.

McCormick, D. A. (1999). Membrane potential and action potential. In M. J. Zigmond, F. E. Bloom, S. C. Landis, J. L. Roberts, & L. R. Squire (Eds.), *Fundamental neuroscience* (pp. 129–154). San Diego, London, Boston, New York, Sydney, Tokyo, Toronto: Academic Press.

McCormick, D. A., & Pape, H. C. (1990). Properties of a hyperpolarization-activated cation current and its role in rhythmic oscillation in thalamic relay neurons. *The Journal of Physiology*, *431*, 291–318.

Mohajerani, M. H., McVea, D. A., Fingas, M., & Murphy, T. H. (2010). Mirrored bilateral Slow-Wave cortical activity within local circuits revealed by fast bihemispheric Voltage-Sensitive dye imaging in anesthetized and awake mice. *The Journal of Neuroscience*, *30*(10), 3745–3751. doi:10.1523/jneurosci.6437-09.2010.

Moody, W. J., Futamachi, K. J., & Prince, D. A. (1974). Extracellular potassium activity during epileptogenesis. *Experimental Neurology*, *42*(2), 248–263.

Moore, A. R., Zhou, W. -L., Jakovcevski, I., Zecevic, N., & Antic, S. D. (2011). Spontaneous electrical activity in the human fetal cortex in vitro. *Journal of Neuroscience*, *31*(7), 2391–2398. doi:10.1523/jneurosci.3886-10.2011.

Mugnaini, E. (1986). Cell junctions of astrocytes, ependyma, and related cells in the mammalian central nervous system, with emphasis on the hypothesis of a generalized functional syncytium of supporting cells. In S. Fedoroff & A. Vernadakis (Eds.), *Astrocytes* (Vol. 1, pp. 329–371). New York: Academic.

Mukovski, M., Chauvette, S., Timofeev, I., & Volgushev, M. (2007). Detection of active and silent states in neocortical neurons from the field potential signal during slow-wave sleep. *Cerebral Cortex*, *17*(2), 400–414.

Nir, Y., Staba, R. J., Andrillon, T., Vyazovskiy, V. V., Cirelli, C., Fried, I., et al. (2011). Regional slow waves and spindles in human sleep. *Neuron, 70*(1), 153–169.

Nita, D., Cissé, Y., Timofeev, I., & Steriade, M. (2007). Waking-sleep modulation of paroxysmal activities induced by partial cortical deafferentation. *Cerebral Cortex, 17*(2), 272–283. doi:10.1093/cercor/bhj1145.

Nita, D. A., Vanhatalo, S., Lafortune, F. D., Voipio, J., Kaila, K., & Amzica, F. (2004). Nonneuronal origin of CO2-related DC EEG shifts: An in vivo study in the cat. *Journal of Neurophysiology, 92*(2), 1011–1022.

Noda, H., & Adey, W. R. (1970). Changes in neuronal activity in association cortex of the cat in relation to sleep and wakefulness. *Brain Research, 19*(2), 263–275.

Noda, H., & Adey, W. R. (1970). Firing variability in cat association cortex during sleep and wakefulness. *Brain Research, 18*(3), 513–526.

Peyrache, A., Dehghani, N., Eskandar, E. N., Madsen, J. R., Anderson, W. S., Donoghue, J. A., et al. (2012). Spatiotemporal dynamics of neocortical excitation and inhibition during human sleep. *Proceedings of the National Academy of Sciences, 109*(5), 1731–1736. doi:10.1073/pnas.1109895109.

Prince, D. A., Lux, H. D., & Neher, E. (1973). Measurement of extracellular potassium activity in cat cortex. *Brain Research, 50*(2), 489–495.

Pumain, R., Kurcewicz, I., & Louvel, J. (1983). Fast extracellular calcium transients: Involvement in epileptic processes. *Science, 222*(4620), 177–179.

Reig, R., Gallego, R., Nowak, L. G., & Sanchez-Vives, M. V. (2006). Impact of cortical network activity on short-term synaptic depression. *Cerebral Cortex, 16*(5), 688–695.

Rockland, K. S. (1985). Anatomical organization of primary visual cortex (area 17) in the ferret. *The Journal of Comparative Neurology, 241*(2), 225–236. doi:10.1002/cne.902410209.

Rudolph, M., Pospischil, M., Timofeev, I., & Destexhe, A. (2007). Inhibition determines membrane potential dynamics and controls action potential generation in awake and sleeping cat cortex. *Journal of Neuroscience, 27*(20), 5280–5290.

Ruiz-Mejias, M., Ciria-Suarez, L., Mattia, M., & Sanchez-Vives, M. V. (2011). Slow and fast rhythms generated in the cerebral cortex of the anesthetized mouse. *Journal of Neurophysiology, 106*(6), 2910–2921. doi:10.1152/jn.00440.2011.

Rumsey, C. C., & Abbott, L. F. (2006). Synaptic democracy in active dendrites. *Journal of Neurophysiology, 96*(5), 2307–2318. (doi: 10.1152/jn.00149.2006).

Sakata, S., & Harris, K. D. (2009). Laminar structure of spontaneous and Sensory-Evoked population activity in auditory cortex. *Neuron, 64*(3), 404–418.

Sanchez-Vives, M. V., & McCormick, D. A. (2000). Cellular and network mechanisms of rhythmic recurrent activity in neocortex. *Nature Neuroscience, 3*(10), 1027–1034.

Sanchez-Vives, M. V., Reig, R., Winograd, M., & Descalzo, V. F. (2007). An active cortical network in vitro. In I. Timofeev (Ed.), *Mechanisms of spontaneous actrive states in neocortex* (pp. 23–44). Kerala, India: Research Signpost.

Saper, C. B., Scammell, T. E., & Lu, J. (2005). Hypothalamic regulation of sleep and circadian rhythms. *Nature, 437*(7063), 1257–1263.

Schmitz, D., Schuchmann, S., Fisahn, A., Draguhn, A., Buhl, E. H., Petrasch-Parwez, E., et al. (2001). Axo-axonal coupling a novel mechanism for ultrafast neuronal communication. *Neuron, 31*(5), 831–840.

Seigneur, J., & Timofeev, I. (2010). Synaptic impairment induced by paroxysmal ionic conditions in neocortex. *Epilepsia, 52*(1), 132–139. doi:10.1111/j.1528-1167.2010.02784.x.

Sharma, A. V., Wolansky, T., & Dickson, C. T. (2010). A Comparison of sleeplike slow oscillations in the hippocampus under ketamine and urethane anesthesia. *Journal of Neurophysiology, 104*(2), 932–939. doi:10.1152/jn.01065.2009.

Slezia, A., Hangya, B., Ulbert, I., & Acsady, L. (2011). Phase advancement and nucleus-specific timing of thalamocortical activity during slow cortical oscillation. *Journal of*

Neuroscience, 31(2), 607–617. doi:10.1523/JNEUROSCI.3375-10.2011. (Research Support, Non-U.S. Gov't).
Somjen, G. G. (2002). Ion regulation in the brain: Implications for pathophysiology. *Neuroscientist, 8*(3), 254–267.
Somogyi, P., Tamás, G., Lujan, R., & Buhl, E. H. (1998). Salient features of synaptic organisation in the cerebral cortex. *Brain Research Reviews, 26*(2–3), 113–135.
Steriade, M. (2006). Grouping of brain rhythms in corticothalamic systems. *Neuroscience, 137*(4), 1087–1106.
Steriade, M., Contreras, D., Amzica, F., & Timofeev, I. (1996). Synchronization of fast (30–40 Hz) spontaneous oscillations in intrathalamic and thalamocortical networks. *Journal of Neuroscience, 16*(8), 2788–2808.
Steriade, M., & Deschenes, M. (1984). The thalamus as a neuronal oscillator. *Brain Research Reviews, 8*, 1–63.
Steriade, M., Deschenes, M., Domich, L., & Mulle, C. (1985). Abolition of spindle oscillations in thalamic neurons disconnected from nucleus reticularis thalami. *Journal of Neurophysiology, 54*(6), 1473–1497.
Steriade, M., & McCarley, R. W. (2005). *Brainstem control of wakefulness and sleep*. New York: Plenum.
Steriade, M., Nuñez, A., & Amzica, F. (1993). Intracellular analysis of relations between the slow (<1 Hz) neocortical oscillations and other sleep rhythms of electroencephalogram. *Journal of Neuroscience, 13*, 3266–3283.
Steriade, M., Nuñez, A., & Amzica, F. (1993). A novel slow (<1 Hz) oscillation of neocortical neurons *in vivo*: Depolarizing and hyperpolarizing components. *Journal of Neuroscience, 13*, 3252–3265.
Steriade, M., Timofeev, I., Dürmüller, N., & Grenier, F. (1998). Dynamic properties of corticothalamic neurons and local cortical interneurons generating fast rhythmic (30–40 Hz) spike bursts. *Journal of Neurophysiology, 79*(1), 483–490.
Steriade, M., Timofeev, I., & Grenier, F. (2001). Natural waking and sleep states: A view from inside neocortical neurons. *Journal of Neurophysiology, 85*(5), 1969–1985.
Syková, E. (2004). Extrasynaptic volume transmission and diffusion parameters of the extracellular space. *Neuroscience, 129*(4), 861–876.
Sykova, E., & Nicholson, C. (2008). Diffusion in brain extracellular space. *Physiological Reviews, 88*(4), 1277–1340. doi:10.1152/physrev.00027.2007.
Thomson, A. M., West, D. C., Wang, Y., & Bannister, A. P. (2002). Synaptic connections and small circuits involving excitatory and inhibitory neurons in layers 2–5 of adult rat and cat neocortex: Triple intracellular recordings and biocytin labelling in vitro. *Cerebral Cortex, 12*(9), 936–953.
Timofeev, I., & Bazhenov, M. (2005). Mechanisms and biological role of thalamocortical oscillations. In F. Columbus (Ed.), *Trends in chronobiology research* (pp. 1–47). New York: Nova Science Publishers.
Timofeev, I., Bazhenov, M., Sejnowski, T., & Steriade, M. (2001). Contribution of intrinsic and synaptic factors in the desynchronization of thalamic oscillatory activity. *Thalamus & Related Systems, 1*(1), 53–69.
Timofeev, I., Bazhenov, M., Sejnowski, T., & Steriade, M. (2002). Cortical hyperpolarization-activated depolarizing current takes part in the generation of focal paroxysmal activities. *Proceedings of the National Academy of Sciences of the United States of America, 99*(14), 9533–9537.
Timofeev, I., Grenier, F., Bazhenov, M., Sejnowski, T. J., & Steriade, M. (2000). Origin of slow cortical oscillations in deafferentiated cortical slabs. *Cerebral Cortex, 10*(12), 1185–1199.
Timofeev, I., Grenier, F., & Steriade, M. (2000). Impact of intrinsic properties and synaptic factors on the activity of neocortical networks in vivo. *The Journal of Physiology (Paris), 94*(5–6), 343–355.

Timofeev, I., Grenier, F., & Steriade, M. (2001). Disfacilitation and active inhibition in the neocortex during the natural sleep-wake cycle: An intracellular study. *Proceedings of the National Academy of Sciences of the United States of America, 98*(4), 1924–1929.

Timofeev, I., Grenier, F., & Steriade, M. (2004). Contribution of intrinsic neuronal factors in the generation of cortically driven electrographic seizures. *Journal of Neurophysiology, 92*(2), 1133–1143.

Timofeev, I., & Steriade, M. (1996). Low-frequency rhythms in the thalamus of intact-cortex and decorticated cats. *Journal of Neurophysiology, 76*(6), 4152–4168.

Timofeev, I., & Steriade, M. (1997). Fast (mainly 30–100 Hz) oscillations in the cat cerebellothalamic pathway and their synchronization with cortical potentials. *The Journal of Physiology, 504*(1), 153–168.

Topolnik, L., Steriade, M., & Timofeev, I. (2003). Partial cortical deafferentation promotes development of paroxysmal activity. *Cerebral Cortex, 13*(8), 883–893.

Traub, R. D., Whittington, M. A., Buhl, E. H., Jefferys, J. G., & Faulkner, H. J. (1999). On the mechanism of the gamma -> beta frequency shift in neuronal oscillations induced in rat hippocampal slices by tetanic stimulation. *Journal of Neuroscience, 19*(3), 1088–1105.

Traub, R. D., Whittington, M. A., Stanford, I. M., & Jefferys, J. G. (1996). A mechanism for generation of long-range synchronous fast oscillations in the cortex. *Nature, 383*(6601), 621–624.

Ushimaru, M., Ueta, Y., & Kawaguchi, Y. (2012). Differentiated participation of thalamocortical subnetworks in Slow/Spindle waves and desynchronization. *The Journal of Neuroscience, 32*(5), 1730–1746. doi:10.1523/jneurosci.4883-11.2012.

Vanhatalo, S., Palva, J. M., Holmes, M. D., Miller, J. W., Voipio, J., & Kaila, K. (2004). Infraslow oscillations modulate excitability and interictal epileptic activity in the human cortex during sleep. *PNAS, 101*(14), 5053–5057.

Vigmond, E. J., Perez Velazquez, J. L., Valiante, T. A., Bardakjian, B. L., & Carlen, P. L. (1997). Mechanisms of electrical coupling between pyramidal cells. *Journal of Neurophysiology, 78*(6), 3107–3116.

Villablanca, J., & Salinas-Zeballos, M. E. (1972). Sleep-wakefulness, EEG and behavioral studies of chronic cats without the thalamus: The 'athalamic' cat. *Archives Italiennes de Biologie, 110*, 383–411.

Volgushev, M., Chauvette, S., Mukovski, M., & Timofeev, I. (2006). Precise long-range synchronization of activity and silence in neocortical neurons during slow-wave sleep. *Journal of Neuroscience, 26*(21), 5665–5672.

von Krosigk, M., Bal, T., & McCormick, D. A. (1993). Cellular mechanisms of a synchronized oscillation in the thalamus. *Science, 261*(5119), 361–364.

Vyazovskiy, V. V., Olcese, U., Hanlon, E. C., Nir, Y., Cirelli, C., & Tononi, G. (2011). Local sleep in awake rats. *Nature, 472*(7344), 443–447. doi:10.1038/nature10009.

Vyazovskiy, V. V., Olcese, U., Lazimy, Y. M., Faraguna, U., Esser, S. K., Williams, J. C., et al. (2009). Cortical Firing and Sleep Homeostasis. *Neuron, 63*(6), 865–878.

Waters, J., & Helmchen, F. (2006). Background synaptic activity is sparse in neocortex. *Journal of Neuroscience, 26*(32), 8267–8277.

Ylinen, A., Bragin, A., Nadasdy, Z., Jando, G., Szabo, I., Sik, A., et al. (1995). Sharp wave-associated high-frequency oscillation (200 Hz) in the intact hippocampus: Network and intracellular mechanisms. *Journal of Neuroscience, 15*(1 Pt 1), 30–46.

Zhu, L., Blethyn, K. L., Cope, D. W., Tsomaia, V., Crunelli, V., & Hughes, S. W. (2006). Nucleus- and species-specific properties of the slow (<1 Hz) sleep oscillation in thalamocortical neurons. *Neuroscience, 141*(2), 621–636.

CHAPTER 2

Corticothalamic Rhythms during States of Reduced Vigilance

Vincenzo Crunelli and Stuart Hughes*
Neuroscience Division, School of Bioscience, Cardiff University, Cardiff, UK
*Present address: Lilly UK, Windlesham, Surrey, UK

Since the discovery of the first rhythmic activity in the human electroencephalogram (EEG) by Hans Berger in the late 1920s (Berger, 1929), neuroscientists and laymen alike have been fascinated by the variety and complexity of the electrical waves that the human brain is able to express during clearly defined behavioral states (Brismar, 2007; Dijk, 2009; Krueger, Rector, Roy, Van Dongen, Belenky, & Panksepp, 2008; Monto, Palva, Voipio, & Palva, 2008; Siegel, 2008; Tallon-Baudry, 2009), from the attentive states of perception and decision-making to the unconscious states of non-rapid-eye-movement (NREM) sleep, anesthesia, and coma. This has been particularly true in the last 20 years with the advent of full band and high density EEG recordings (Vanhatalo, Voipio, & Kaila, 2005; Monto et al., 2008). At the same time, our understanding of the physiological and pathological significance of these EEG biomarkers recorded from the human scalp has benefited greatly from the ability to correlate them with noninvasive functional imaging of deep brain structures (Dang-Vu, Schabus, Desseilles, Sterpenich, Bonjean, & Maquet, 2010; Desseilles, Dang-Vu, Schabus, Sterpenich, Maquet, & Schwartz, 2008; Salek-Haddadi, Friston, Lemieux, & Fish, 2003). Our knowledge of the molecular, cellular, synaptic, and network processes underlying these waves and rhythms has also greatly progressed, though at a relatively slower speed, such that more than 80 years after Berger's pioneering observations we still lack a fully comprehensive mechanistic picture of some of these EEG activities and thus refer to them using long-standing, frequency band-based classifications.

As far as sleep rhythms are concerned, however, two notable exceptions need to be stressed. First is the discovery of powerful reciprocal connections between neocortex and thalamus, both with respect to the sensory thalamus and primary cortices as well as to intralaminar thalamic nuclei and association cortices (Jones, 2008; Sherman & Guillery, 1996): this

Sleep and Brain Activity
DOI: http://dx.doi.org/10.1016/B978-0-12-384995-3.00002-2

has focused relevant research not only to the neocortex but also to the thalamus when searching for potential generator(s) of sleep-related EEG rhythms. In this respect, particular emphasis has been given to the corticothalamic synapses on thalamocortical (TC) neurons in sensory thalamic nuclei, which well outnumber those of afferents from the periphery and represent more than 50% of the overall synaptic count onto these neurons (Godwin, Van Horn, Sesma, Romano, & Sherman, 1996; Sherman & Guillery, 1996). Second is the identification of a key modulatory role for the ascending cholinergic and monoaminergic systems, which are the main determinants of the membrane potential of cortical and thalamic neurons (McCormick, 1992). In particular, during the progression from the waking attentive state to deep NREM sleep there is an overall net decrease in the tone of these afferents, which leads to a progressively more negative membrane potential in almost all types of cortical and thalamic neurons (Hirsch, Fourment, & Marc, 1983; Crunelli & Hughes, 2010; Steriade, Amzica, & Nuñez, 1993a).

In this chapter, we summarize current views on the cellular and network mechanisms of three EEG rhythms: the classical posterior alpha rhythm, the theta waves of light NREM sleep, and the slow (<1 Hz) sleep oscillation. The emerging picture in all these activities is one that highlights a smooth interplay between cortical and thalamic networks as the essential element that underlies the full expression of each of these brain waves in the EEG: this finding, we believe, is key in achieving a comprehensive knowledge of brain function in health and disease states. The other important point that arises from these studies is that the now classical textbook dichotomy of TC neuron electrophysiology (Sherman & Guillery, 1996; Llinas & Steriade, 2006) as expressing two firing modes, i.e., tonic single action potential firing in wakefulness and T-type Ca^{2+} channel-mediated high frequency burst firing in sleep, appears today to be highly restrictive and often an impediment for a full understanding of thalamic physiology: as illustrated in detail throughout this chapter, TC neurons express a much richer repertoire of intrinsic and synaptically driven firing activities than previously thought (Fig. 2.1).

THE ALPHA RHYTHM OF RELAXED WAKEFULNESS AND THE THETA WAVES OF EARLY NREM SLEEP

A section dealing with the classical posterior alpha rhythm may seem out of place in a book dedicated to brain activities during sleep. However,

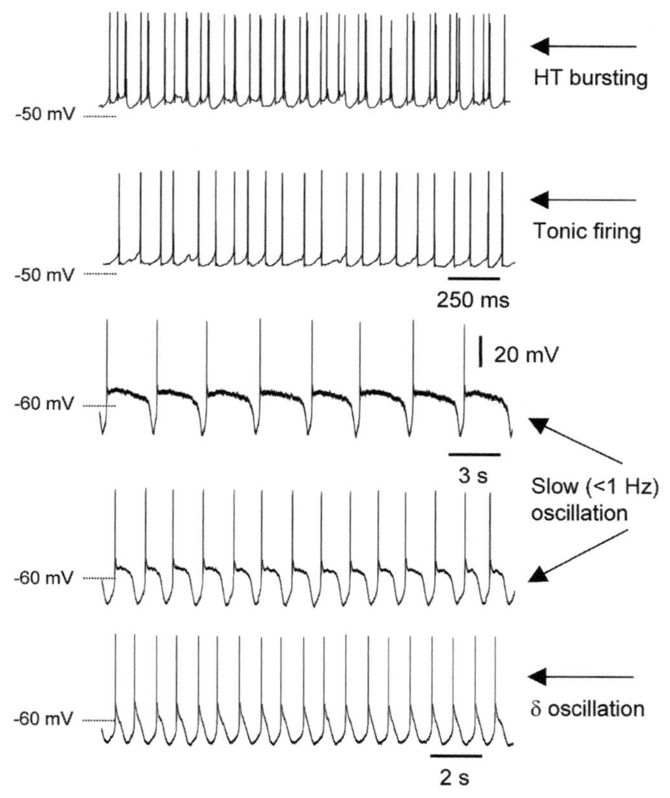

Figure 2.1 ***Repertoire of intrinsic activities of TC neurons.*** Intracellular recordings from a single TC neuron in the presence of a metabotrobic glutamate receptor agonist (t-ACPD, 50 μM) *in vitro* show the variety of intrinsic activities that can be elicited by TC neurons with different levels of increasing steady hyperpolarization (from top to bottom) mimicking the decrease in depolarizing tone that occurs during progression from relaxed wakefulness, when high threshold (HT) bursting is observed at the cellular level, to deep NREM sleep, where slow (<1 Hz) and delta oscillations are present. The addition of the glutamate metabotropic receptor agonist is used to mimic the effect of the cortical input on TC neurons. Note, however, that identical activities can be obtained *in vitro* in the absence of a glutamate metabotropic receptor agonist by electrically stimulating the cortical afferent present in the slice (for further details, see Blethyn et al., 2006; Crunelli & Hughes, 2010; Hughes et al., 2002b, 2004; Zhu et al., 2006).

as will be become evident later, the reason for this inclusion stems from the discovery that the alpha rhythm of relaxed wakefulness and the theta waves that are observed in the EEG during the early stages of NREM sleep rely on a very similar cellular mechanism (Hughes, Lőrincz, Cope, Blethyn, Kekesi, Parri, Juhasz, & Crunelli, 2004). Indeed, a discussion of these rhythms provides the opportunity to highlight the full spectrum of intrinsic activities that TC neurons can generate at different membrane potentials, i.e., from the tonic action potential firing and high threshold (HT) bursts at depolarized potentials, to the intrinsically generated slow (<1 Hz) sleep oscillation and delta oscillation at progressively more hyperpolarized potentials (Fig. 2.1).

In contrast to other brain rhythms, for which extensive investigations have been carried out both in cortical and thalamic territories, the analysis of the mechanisms underlying the alpha rhythm has mainly focused, in part for historical reasons, on the thalamus (Hughes & Crunelli, 2005). Thus, following attempts to investigate the alpha rhythm mechanism by the use of an inappropriate model, i.e., the barbiturate-induced spindles (Andersen & Andersson, 1968), the work of Buser (Buser & Rougeul-Buser, 1995), Lopes Da Silva (Lopes da Silva, van Lierop, Schrijer, & van Leeuwen, 1973; Lopes da Silva, Vos, Mooibroek, & Van Rotterdam, 1980), and others (Basar E, Schurmann M, Basar-Eroglu C, & Karakas S. 1997; Bouyer, Tilquin, & Rougeul, 1983; Feige, Scheffler, Esposito, Di Salle, Hennig, & Seifritz, 2005; Isaichev, Derevyankin, Koptelov, & Sokolov, 2001), together with more recent noninvasive imaging investigations (Danos, Guich, Abel, & Buchsbaum, 2001; Goldman, Stern, Engel, & Cohen, 2002; Larson, Davidson, Abercrombie, Ward, Schaefer, Jackson, Holden, & Perlman, 1998; Lindgren, Larson, Schaefer, Abercrombie, Ward, Oakes, Holden, Perlman, Benca, & Davidson, 1999), have pointed to this subcortical structure as the potential rhythm generator for the alpha rhythm that is observed in the primary visual cortex of both humans and higher mammals during periods of relaxed wakefulness. Nevertheless, in freely behaving cats the alpha rhythm is clearly driven by a novel firing pattern, i.e., HT bursting, that is present in a subset (about 30%) of TC neurons in the visual thalamus, i.e., the dorsal lateral geniculate nucleus (LGN) (Figs. 2.1 and 2.2A) (Hughes et al., 2004; Lőrincz, Kékesi, Juhász, Crunelli, & Hughes, 2009). These HT bursts are very different from the classical high frequency bursts of action potentials that are elicited on the crest of T-type Ca^{2+} channel-mediated, low threshold Ca^{2+} potentials (LTCPs) (Figs. 2.1 and 2.2) (Hughes et al., 2004; Lőrincz et al., 2009). First, HT bursts occur when the membrane potential is in the range of −50 to −55 mV (Fig. 2.2D), whereas LTCP-mediated bursts are evoked only from membrane potentials <−60 mV. Second, the interspike intervals of HT bursts do not change as the burst progresses (Fig. 2.2C), whereas those of LTCP-mediated bursts show a characteristic increase within a

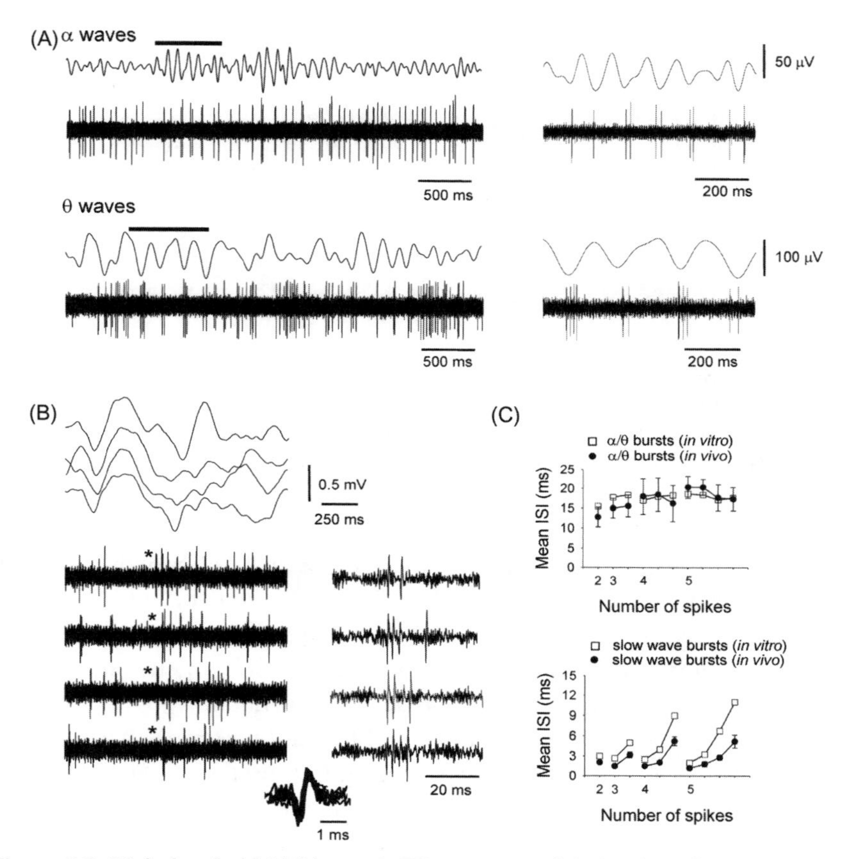

Figure 2.2 ***High threshold (HT) bursts in TC neurons and their role in the generation of alpha rhythm and sleep theta waves.*** A. Simultaneous field and single-unit recording in the LGN of a freely behaving animal showing correlated activity with both α (upper traces) and θ (lower traces) field oscillations during a period of relaxed wakefulness and in the early stage of NREM sleep, respectively. Marked sections are enlarged on the right. B. Four consecutive slow waves recorded from the LGN of a freely behaving animal during deep NREM sleep *in vivo* (upper 4 traces) and the corresponding single unit activity (lower 4 traces). Note that the high frequency burst of action potentials with interspike intervals that increase as the burst progresses. The bursts marked by asterisks are enlarged on the right, and the bottom inset is an overlay of all spikes in these bursts. C. Average interspike intervals (ISI) for action potentials in HT bursts correlated to α and θ activity *in vivo* (•) and *in vitro* (□) (top plot). Average interspike interval (ISI) for action potentials correlated to slow wave activity *in vivo* (•) and *in vitro* (□) (bottom plot) (note the contrasting pattern to that occurring during α and θ rhythms shown in the top plot). *In vitro* data were obtained in the presence of a glutamate metabotropic receptor agonist (t-ACPD).

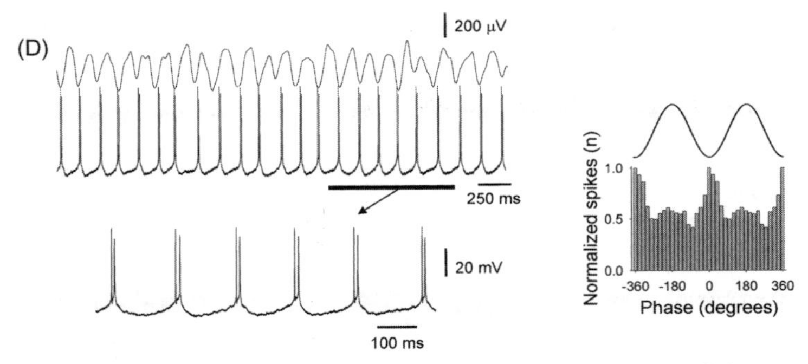

Figure 2.2 (*Continued*) D. Local field potential (top trace) and intracellular recordings (bottom trace) showing the waveform of HT bursts during alpha waves recorded *in vitro*. Spike timing histogram of the intracellular activity is shown to the right (for further details, see Hughes et al., 2004; Hughes & Crunelli, 2005).

burst (Fig. 2.2D). Third, the interspike intervals of HT bursts are rarely less than ~10 msec, whereas those at the start of an LTCP-mediated burst are typically in the range 2–5 ms (Fig. 2.2C, D). Fourth, HT bursts are partially sensitive to both high- and low-voltage activated Ca^{2+} channel blockers, whereas LTCP-mediated events are exclusively generated by the activation of low-voltage activated T-type Ca^{2+} channels.

Importantly, the frequency of HT bursting progressively increases with increasing membrane hyperpolarization (Hughes et al., 2004). It is therefore not surprising that when an animal transitions from a state of relaxed wakefulness (with prominent EEG alpha activity) to the early stage of sleep the occasional sequences of theta waves that are recorded from the occipital cortex in this behavioral condition are also correlated with HT bursts, which now show a larger number of action potentials per burst (generally 3 to 7) than during the alpha rhythm (Figs. 2.1 and 2.2A) (Hughes et al., 2004).

HT bursts are locally synchronized within the thalamus, since HT bursting TC neurons are coupled by gap junctions (Hughes, Blethyn, Cope, & Crunelli, 2002a; Hughes, Lőrincz, Blethyn, Kékesi, Juhász, Turmaine, Parnavelas, & Crunelli, 2011), leading to a powerful thalamic output to cortex, as indicated by the presence of a large local field potential (LFP) in the alpha frequency band in slices of the LGN following activation of either metabotropic glutamate receptors (mGluRs) or muscarinic acetylcholine receptors (mAchRs) (Hughes et al., 2004; Lőrincz, Crunelli, & Hughes, 2008b; Lőrincz et al., 2009). Indeed, blocking either mGluRs or mAchRs as well as gap-junction coupling by reverse

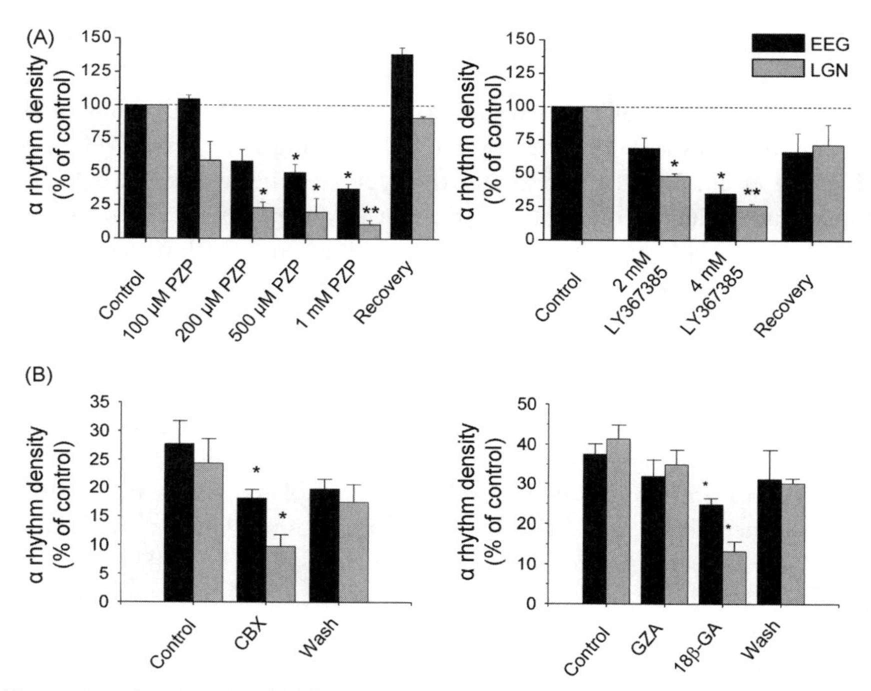

Figure 2.3 ***Gap-junction blockers and metabotropic cholinergic and glutamatergic receptor antagonists block alpha rhythm.*** A. Histograms summarizing the effect of bilateral thalamic injection by reverse microdialysis of pirenzipine (mAchR antagonist) (left) and LY367385 (mGluR1a antagonist) (right) on EEG (black bars) and LGN (gray bars) alpha rhythm density. B. Histograms summarizing the effect of bilateral thalamic injection by reverse microdialysis of two gap-junction blockers, carbenoxolone (CBX) (left plot) and 18β-glycyrrhetinic acid (18β-GA) (right plot) on EEG (black bars) and LGN (gray bars) alpha rhythm density. Note the lack of action of glycyrrhizic acid (GZA) (right plot), a glycyrrhetinic acid derivative that is inactive at gap-junction blockers (for further details, see Hughes et al., 2011; Lőrincz et al., 2009).

microdialyis application of appropriate antagonists directly in the LGN of freely moving animals almost abolishes the alpha rhythm locally and markedly decreases the alpha waves recorded from the occipital cortex (Fig. 2.3) (Hughes et al., 2011; Lőrincz et al., 2009). Moreover, in LGN slices transient elimination of HT bursts in a single TC neuron by membrane hyperpolarization abolishes alpha rhythm-related firing in closely surrounding neurons (Hughes et al., 2004) and decreases the amplitude of the thalamic LFP measured in the vicinity of the recorded neuron (Lõrincz et al., 2008b). Conversely, inhibition of firing in a single TC neuron that does not fire HT bursts leaves the output activity of the surrounding neurons and the thalamic LFP unchanged (Lőrincz et al., 2008, 2009).

During both alpha and theta waves in freely moving animals, the activity of the local GABAergic interneurons in the LGN and that of the remaining 70% of TC neurons that do not fire HT bursts but single action potentials is determined by the HT bursting TC neurons (Fig. 2.4) (Lőrincz et al., 2009). Interestingly, the type of firing expressed by LGN interneurons during the alpha rhythm strictly depends on their level of membrane polarization; thus, interneurons that are in a less depolarized state elicit single action potentials while those in a more depolarized state fire in a peculiar manner that consists of a single action potential followed some 20–50 msec later by a short burst of 3–5 action potentials (Fig. 2.4). These two interneuron firing modes in turn lead to two distinct patterns of $GABA_A$ receptor-mediated inhibitory postsynaptic potentials (IPSPs) in tonic firing TC neurons, thus splitting the tonic firing TC neuron population into two groups: one where a few isolated IPSPs are in phase with the negative peak of the LFP and another one where a short burst of IPSPs is in an anti-phase relation with the LFP (Fig. 2.4). These two patterns of phasic inhibition tightly control the output of the tonic firing TC neurons, giving rise to a short period of suppression of firing that is either in an in-phase or an anti-phase relationship with the negative peak of the thalamic LFP (Fig. 2.4) (Lőrincz et al., 2009). Importantly, the GABAergic neurons in the visual sector of the nucleus reticularis thalami (NRT), the other main GABAergic input to TC neurons in the LGN, are either silent during alpha rhythm episodes in freely moving animals or their occasional firing is not correlated to the ongoing thalamic or EEG alpha waves, indicating that NRT neurons are unlikely to contribute to the generation of this rhythm of relaxed wakefulness (Lőrincz et al., 2009).

Thus, although phasic inhibition onto TC neurons is also a key component in the generation of another sleep rhythm, i.e., sleep spindles (Andersen & Andersson, 1968; von Krosigk et al., 1993), fundamental differences exist between this activity of early NREM sleep and the alpha rhythm. Most notably, phasic inhibition during alpha activity is derived from probably one or very few LGN interneurons leading to small amplitude IPSPs (~0.5–1 mV) that delicately control the single action potential output of tonic firing TC neurons (Fig. 2.4) (Lőrincz et al., 2009), whereas inhibition during spindles originates from widespread LTCP-mediated bursting in the perigeniculate nucleus (the visual sector of the NRT) leading to large amplitude IPSPs (~10–15 mV) in the entire population of TC neurons (von Krosigk et al., 1993). Moreover, during the alpha rhythm

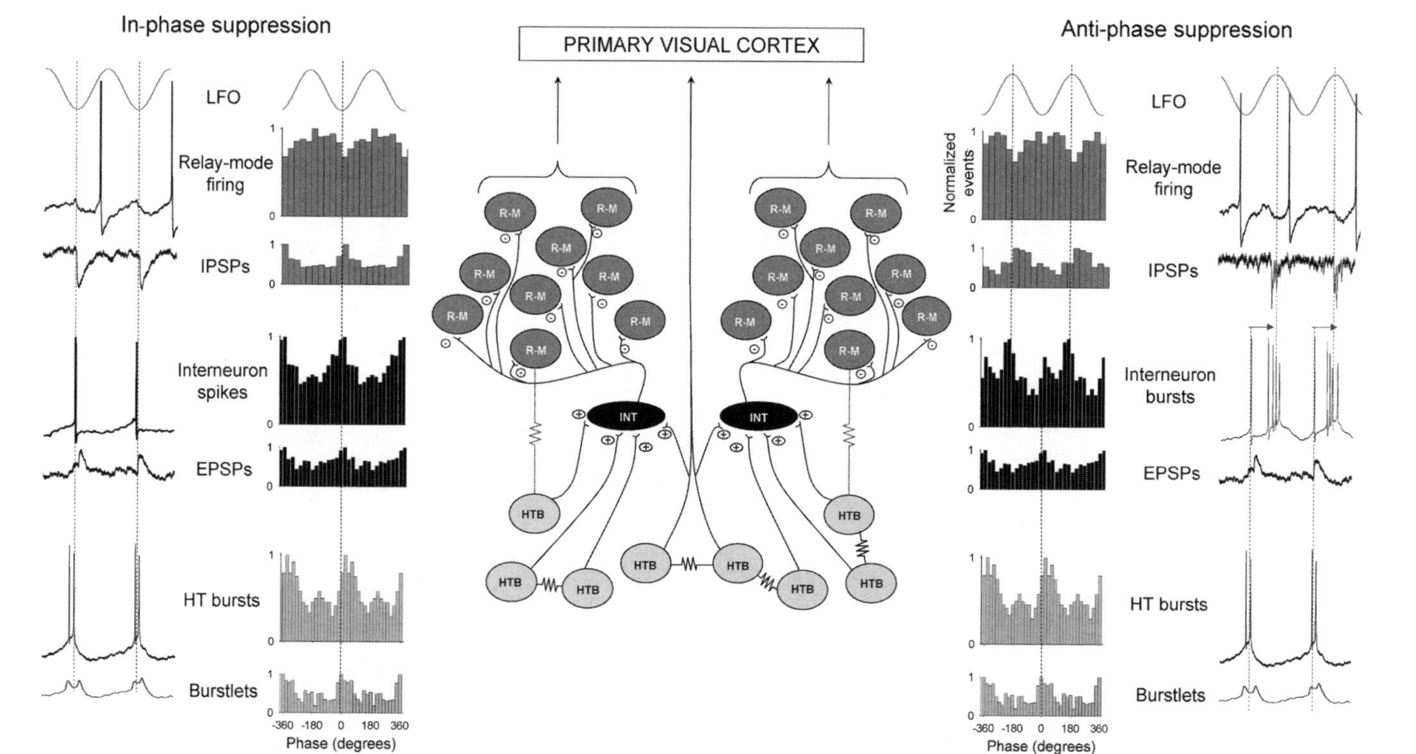

Figure 2.4 ***Cellular and network mechanism of alpha and theta wave generation.*** HT bursting TC neurons (light gray filled circles, HTB) form a gap-junction coupled network in the LGN that generates synchronized oscillations at alpha/theta frequencies. These cells may also couple via weak gap junctions (thin dotted lines) to a subset of TC neurons that exhibit conventional tonic single action potential firing, i.e., relay mode firing (dark gray filled circles, R-M). The main sequence of alpha/theta rhythm-related events that shape relay-mode firing, however, is as follows. First, HT bursting TC neurons provide a convergent excitation of local GABAergic interneurons (black filled circles, INT), probably via axon collaterals. This convergent excitation rhythmically drives interneuron firing, which can consist of either primarily single spike output (left side) or bursting (right), with bursting occurring when interneurons occupy a more depolarized state. During single spike output, action potentials in interneurons occur predominantly close to the negative LFP peak. In contrast, during bursting, a large interval between the first and second spikes in individual bursts means that action potentials are translated to occur mainly close to the positive LFP peak. These two forms of interneuron firing are then reflected in relay-mode TC neurons as mainly single IPSPs that occur near the negative LFP peak or IPSP bursts that occur near the positive LFP peak, respectively. Ultimately, these two forms of inhibition lead to a differential temporal framing of output from relay-mode TC neurons (for further details, see Lőrincz et al., 2009).

TC neurons are relatively depolarized and can thus only generate either HT bursts or single action potentials (Lőrincz et al., 2009), whereas during sleep spindles TC neurons are hyperpolarized and produce action potentials as LTCP-mediated high frequency bursts (von Krosigk et al., 1993). These differences are clearly commensurate with the fact that spindle waves are mostly pervasive brain oscillations that occur during sleep (and certain types of anesthesia) whereas alpha waves are a more localized, responsive rhythm of the awake, albeit "relaxed" brain (Hughes & Crunelli, 2005).

Another important feature of alpha waves is that HT bursting TC neurons often spontaneously switch between an in-phase and anti-phase association with the LFP (Hughes et al,, 2004; Hughes & Crunelli, 2005). This dynamic phase relationship contrasts markedly with that which occurs in other brain areas, such as the hippocampus, where the firing patterns of neurons of a similar type exhibit a reproducible monophasic relationship with respect to the ongoing field oscillations (Klausberger & Somogyi, 2004). The dynamic phase shifting in the thalamus during alpha waves dictates that at any point in time the local network state essentially reflects the destructive interference between two populations of neurons that oscillate in an essentially anti-phase relationship to each other (Hughes et al., 2004; Hughes & Crunelli, 2005). Because phase shifting appears to be a response to increasing depolarization, one prediction of this scheme is that the reduction of alpha waves that occurs in the intact brain with increased attention, and particularly upon eye opening, may not necessarily reflect a decrease in cellular alpha activity but, at least at the thalamic level, a new dynamic condition with increased contribution of anti-phase HT bursts (Hughes et al., 2004; Hughes & Crunelli, 2005).

In summary, the alpha rhythm of relaxed wakefulness and the theta waves of early NREM sleep are mainly driven by thalamic pacemaker units comprising a small number of HT bursting TC neurons that are connected by gap junctions and entrain local GABAergic interneurons and the remaining TC neurons that fire in a classical relay mode, i.e., with a tonic single action potential output (Fig. 2.4). It remains unclear whether the failure of pharmacological inactivation of the thalamic HT burst pacemaker units to fully abolish alpha and theta waves in the EEG results from the inability of this procedure to affect the entire visual thalamus or whether the primary visual cortex is also capable of its own independent output in these frequency bands. In the latter case, it will be important to establish whether the underlying mechanism of any putative alpha rhythm pacemaker unit in primary visual cortex is mainly driven by cortical neurons intrinsically generating alpha and sleep theta waves or by synaptic interactions across local networks.

THE SLOW (<1 HZ) EEG RHYTHM OF NREM SLEEP

The pioneering work of Steriade's group in the early 1990s identified the slow (<1 Hz) rhythm as one of the key EEG waves of NREM sleep and demonstrated that it is underpinned by the rhythmic sequences of depolarizations (associated with neuronal firing, i.e., UP states) and hyperpolarizations (associated with neuronal quiescence, i.e., the DOWN states) that occur in a quasi-synchronous manner in all cortical and thalamic neurons during this behavioral state (Fig. 2.5) (Steriade et al., 1993a; Steriade, Nuñez, & Amzica, 1993b; Steriade, Contreras, Curró Dossi, & Nuñez, 1993c; Steriade, Timofeev, & Grenier, 2001). It is now well established that this slow (<1 Hz) EEG rhythm, and the underlying cellular activity, i.e., the slow (<1 Hz) sleep oscillation, permeates all stages of natural NREM sleep in humans (Achermann & Borbély, 1997; Cash, Halgren, Dehghani, Rossetti, Thesen, Wang, Devinsky, Kuzniecky, Doyle, Madsen, Bromfield, Eross, Halász, Karmos, Csercsa, Wittner, & Ulbert, 2009; Clemens, Mölle, Eross, Barsi, Halász, & Born, 2007; Dang-Vu et al., 2010; Simon, Manshanden, & Lopes da Silva, 2000) and other mammals (Destexhe, Contreras, & Steriade, 1999). Thus, the K-complex (of stage 2 sleep) represents the EEG manifestation of a single cycle of the slow oscillation (Cash et al., 2009; Amzica & Steriade, 1997, 2002), but as sleep deepens, the frequency of the slow oscillation, and thus that of K-complexes, increases until a proper slow rhythm develops, occupying an increasingly larger component of the EEG signal (Amzica & Steriade, 1997, 1998, 2002). Thus, in contrast to the original observation in ketamine-xylazine anesthetized animals, the frequency of the slow rhythm during sleep progression is not set at a given frequency but changes from 0.03 Hz in the early sleep stages to about 0.6–1 Hz in deep NREM sleep. In contrast, during anesthesia the frequency of the slow sleep rhythm is determined by the anesthetic agent and the depth of anesthesia; for example, the ketamine-xylazine combination elicits a slow oscillation with a higher frequency (0.6–1 Hz) than that observed under urethane (0.3–0.4 Hz) (Steriade et al., 1993a), and isoflurane drastically lowers the slow oscillation frequency (Doi, Mizuno, Katafuchi, Furue, Koga, & Yoshimura, 2007), whereas barbiturates abolish the characteristic UP and DOWN state fluctuations

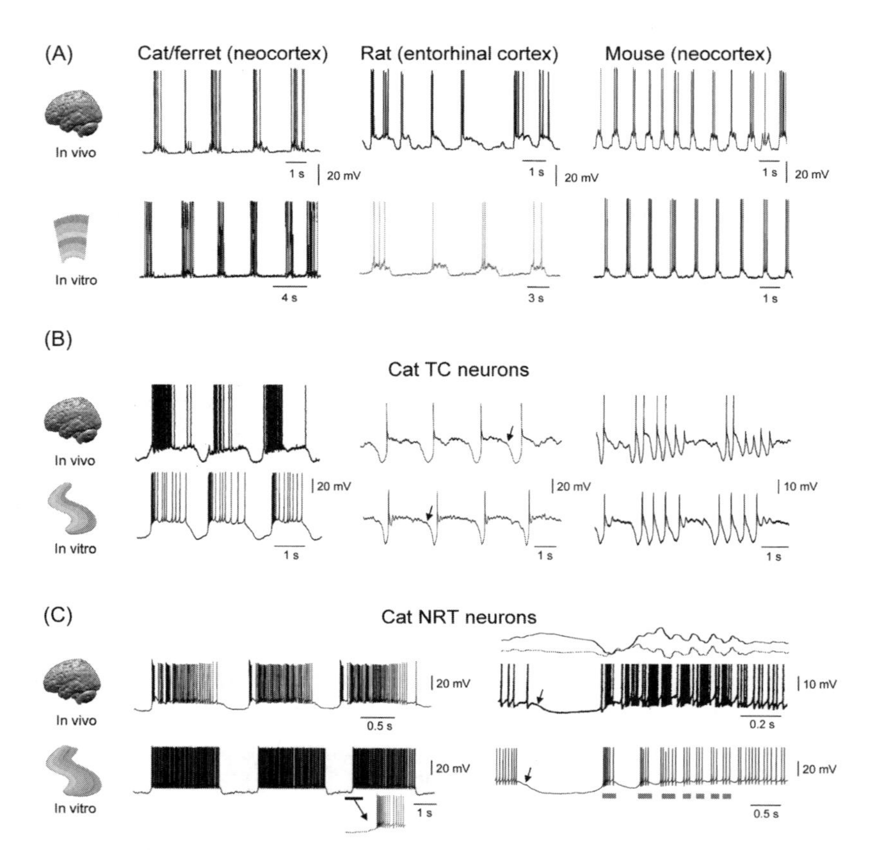

Figure 2.5 ***The slow (<1 Hz) sleep oscillation in cortical and thalamic neurons.*** A. Comparison of the slow (<1 Hz) sleep oscillation recorded in the indicated species and cortical areas *in vivo* during anesthesia (top traces) and in slices (bottom traces). B. Comparison of the slow oscillation recorded *in vivo* in three TC neurons of anesthetized cats (top traces) with that observed in slices of the cat dorsal lateral geniculate nucleus (LGN) *in vitro* in the presence of a metabotropic glutamate agonist (t-ACPD) (bottom traces). Arrows mark inflection points in the membrane potential at the transition from the UP to the DOWN state. C. Comparison of the slow oscillation recorded *in vivo* in two NRT neurons of anesthetized cats (top traces) with that observed in NRT neurons from cat LGN-perigeniculate nucleus slices *in vitro* in the presence of an mGluR agonist (bottom traces). Note the presence of sleep spindle activity in the top two traces on the right (surface and depth EEG records) (for further details, see Crunelli & Hughes, 2010). *Reproduced with permission from Crunelli & Hughes, 2010.*

(Steriade et al., 1993a; Destexhe et al., 1999). Another important feature is that both in anesthesia and during natural sleep the frequency of the slow oscillation is higher in lower mammals (reaching up to 2 Hz in mice) than in humans (Crunelli, Lőrincz, Errington, & Hughes, 2011; Fellin, Halassa,

Terunuma, Succol, Takano, Frank, Moss, & Haydon, 2009; Isomura, Sirota, Ozen, Montgomery, Mizuseki, Henze, & Buzsáki, 2006; Ruiz-Mejias, Ciria-Suarez, Mattia, & Sanchez-Vives, 2011; Sirota, Csicsvari, Buhl, & Buzsáki, 2003).

Despite many claims to the contrary, the most parsimonious view of the mechanism of the slow (<1 Hz) sleep oscillation is one that considers this activity as resulting from the intricate interplay of cortical and thalamic neurons (Crunelli & Hughes, 2010). In fact, the original view of this sleep rhythm as being exclusively of cortical origin has been strongly challenged by the results of recent *in vivo* and *in vitro* experiments showing that both an isolated cortex as well as an isolated thalamus are unable to express the variety, complexity, and periodicity of this brain wave as it is observed in the intact brain. Thus, in a cortical slab (i.e., a small block of cortical tissue in anesthetized animals where thalamic and other cortical inputs are severed) or in an intact cortex following pharmacological inactivation of a limited thalamic territory the slow oscillation is abolished in some neurons while in others its periodicity is drastically compromised (Lemieux & Timofeev, 2010; Timofeev, Grenier, Bazhenov, Sejnowski, & Steriade, 2000) (see also Chapter 1, I. Timofeev's chapter in this book). Similarly, neither *in vitro* nor *in vivo* recordings from thalamic neurons that lack their cortical afferents are able to show the slow sleep oscillation (Hughes et al., 2002b; Steriade et al., 1993b; Timofeev & Steriade, 1996). On the other hand, both an isolated cortex maintained under certain *in vitro* conditions (Beierlein, Fall, Rinzel, & Yuste, 2002; Cossart, Aronov, & Yuste, 2003; Sanchez-Vives &, McCormick, 2000; Shu, Hasenstaub &, McCormick, 2003) as well as an isolated thalamus where the impact of missing cortical afferents is compensated for by reactivation of mGluRs (Hughes, Cope, Blethyn, & Crunelli, 2002b; Blethyn, Hughes, Tóth, Cope, & Crunelli, 2006; Zhu, Blethyn, Cope, Tsomaia, Crunelli, & Hughes, 2006) can both generate slow oscillations in their respective cellular types that are almost identical to those observed *in vivo*. More specifically, in cortical slices modifications of the Ca^{2+} concentration of the perfusing solution (Cossart et al., 2003; Sanchez-Vives & McCormick, 2000) or addition of the cholinergic agonist, carbachol (Lőrincz, Bao, Crunelli, & Hughes, 2007; Lőrincz et al., 2008a), are required for any type of cortical neuron to elicit an activity that closely reproduces the slow oscillation of NREM sleep and anesthesia, while activation of mGluRs is essential for both TC and NRT neurons to express the slow sleep oscillation because it is these receptors that are targeted by corticothalamic fibers in the intact brain (McCormick & von Krosigk, 1992; Sherman & Guillery, 1996; Turner & Salt, 2000).

These studies have also demonstrated that the slow oscillation expressed by cortical neurons mainly results from the regular recurrence of intense, but balanced, intracortical excitatory and inhibitory synaptic barrages, which generate the UP state, and their absence, that constitutes the DOWN state (Cossart et al., 2003; Haider, Duque, Hasenstaub, & McCormick 2006; Sanchez-Vives & McCormick, 2000; Shu et al., 2003). There are, however, pyramidal neurons in layers 2/3 and 5, and Martinotti cells in layer 5 that can intrinsically generate UP and DOWN states (Le Bon-Jego & Yuste, 2007). In the latter case the persistent Na^+ current (I_{NaP}) appears to be particularly important for generating UP states whereas in a general sense $I_{K(Ca)}$, $I_{K(Na)}$, and $I_{K(ATP)}$ (Ca^{2+}-, Na^+-, and ATP dependent, K^+ currents, respectively) play an essential role in bringing about the DOWN state (Compte, Sanchez-Vives, McCormick, & Wang, 2003; Cunningham, Pervouchine, Racca, Kopell, Davies, Jones, Traub, & Whittington, 2006; Le Bon-Jego & Yuste, 2007). Importantly, although complex interactions among intracortical neuronal ensembles within a cortical slice allow potential multiple foci (layer 2/3, 4, and/or 5 in different studies) of initiation for the slow oscillation (MacLean, Watson, Aaron & Yuste 2005; Sanchez-Vives & McCormick, 2000), many investigations both *in vitro* and *in vivo* have consistently demonstrated that stimulation of thalamic afferents more efficiently and reliably triggers UP and DOWN state dynamics in cortical networks than intracortical stimulation (Cossart et al., 2003; Rigas & Castro-Alamancos, 2007; Shu et al., 2003; Steriade, Timofeev, Grenier, & Dürmüller, 1998).

The most striking difference in the slow oscillation between cortical and thalamic neurons is that in thalamic neurons recorded *in vivo* and *in vitro* this activity shows a highly stereotypical appearance and a conserved waveform from cycle to cycle (Fig. 2.5B, C) (Contreras & Steriade, 1996; Steriade et al., 1993c; Rudolph, Pospischil, Timofeev, & Destexhe, 2007; Hughes et al., 2002; Blethyn et al., 2006; Crunelli et al., 2011), properties which are clearly suggestive of an underlying intrinsic mechanism. In particular, there is a large (15–20 mV) and constant voltage difference between the UP and DOWN states, and the evolution of the membrane potential during the slow oscillation is characterized by two unmistakable signatures at the transition between states. Thus, the transition from DOWN to UP state is punctuated by an LTCP and associated high frequency (150–300 Hz) burst of action potentials (Fig. 2.5B, C), whereas the transition between the UP and DOWN state is marked by a clear inflection point in the membrane potential (arrows in Fig. 2.5B). Moreover, the slow oscillation recorded from single TC and NRT neurons *in vitro* is not blocked by tetrodotoxin, $GABA_A$ and $GABA_B$ antagonists, and ionotropic glutamate receptor blockers, and its frequency is strictly governed by the amount of steady intracellular current injection (Hughes et al., 2002b; Blethyn et al., 2006; Zhu et al., 2006), confirming that it is intrinsically generated as a pacemaker activity. In fact, it results from the membrane potential bistability (Hughes et al., 1999; Tóth, Hughes, & Crunelli, 1998; Williams, Turner, Tóth, Hughes, & Crunelli, 1997) that is generated by the interplay of I_{Leak} and the window component of the low threshold Ca^{2+} current I_T ($I_{Twindow}$), such that the UP state essentially corresponds to the condition when $I_{Twindow}$ is active and the DOWN state to when $I_{Twindow}$ is inactive (Figs. 2.6 and 2.7) (for a detailed description of the biophysics underlying thalamic neuron bistability, see Crunelli, Tóth, Cope, Blethyn, & Hughes, 2005). Other membrane currents that are essential for

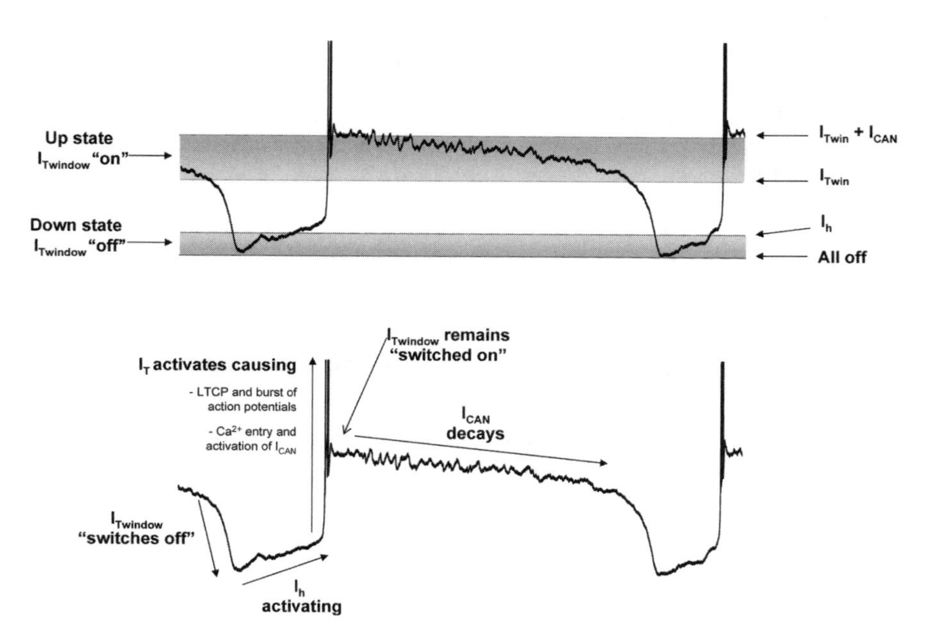

Figure 2.6 ***Cellular mechanism of slow (<1 Hz) sleep oscillation in TC neurons.*** The upper diagram illustrates the different ionic currents that are active at various points during the slow sleep oscillation in a TC neuron, whereas the lower diagram depicts the sequence of ionic events that shape this oscillation. The presence of UP and DOWN states is primarily due to the presence and absence of $I_{Twindow}$, respectively (for further details, see Hughes et al., 2002; Crunelli et al., 2005, 2006; Crunelli & Hughes, 2010).

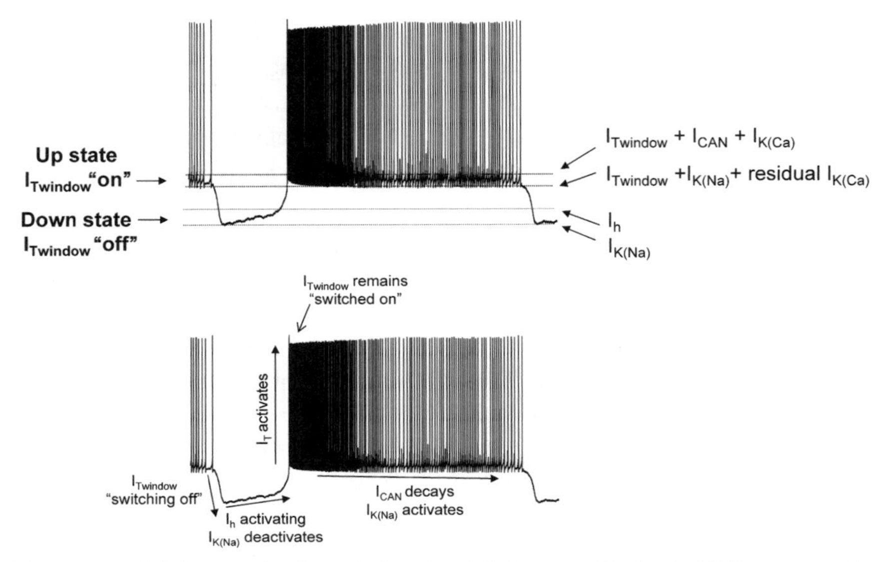

Figure 2.7 ***Cellular mechanism of slow (<1 Hz) sleep oscillation in NRT neurons.*** The upper diagram illustrates the different ionic currents that are active at various points during the slow (<1 Hz) sleep oscillation in an NRT neuron whereas the lower diagram depicts the sequence of ionic events that shape this oscillation. The presence of UP and DOWN states is primarily due to the presence and absence of $I_{Twindow}$, respectively (for further details, see Blethyn et al., 2006; Crunelli et al., 2005, 2006; Crunelli & Hughes, 2010).

the expression of the slow oscillation in TC and NRT neurons include the Ca^{2+}-activated nonselective cation current, I_{CAN}, and the hyperpolarization-activated mixed cation current, I_h (Hughes et al., 2002b; Blethyn et al., 2006; Crunelli et al., 2005; Crunelli, Cope, & Hughes, 2006), which in the absence of synaptic inputs are the main determinants of the duration of the UP and DOWN states, respectively. In NRT neurons, the slow oscillation is also shaped by $I_{K(Ca)}$ and $I_{K(Na)}$ (Fig. 2.7) (Blethyn et al., 2006). Importantly, in a thalamic slice a slow oscillation can also be recorded as a large LFP (see Fig. 1B in Hughes et al., 2004), indicating the ability of the thalamus to produce a synchronized output even in the absence of the cortical input. Whether the synchronization underlying these thalamic LFPs depends on TC neuron axon collaterals onto other TC neurons or interneurons (Cox, Reichova, & Sherman, 2003; Soltesz & Crunelli, 1992; Lorincz et al., 2009) and/or on gap junction-based electrical synapses among TC neurons (Hughes et al., 2002b; Lorincz et al., 2009; Hughes et al., 2011) remains to be determined. In summary, a <1 Hz neocortical rhythm is not necessary for the expression of the slow oscillation in either TC or NRT neurons, which instead can be generated by each of these neuronal types operating as "conditional oscillators," i.e., by the dynamic interplay of their intrinsic voltage-dependent membrane currents, with sustained mGluRs (subtype 1) providing the necessary "condition" for these oscillators to function (Crunelli & Hughes, 2010).

Thus, contrary to the pervading corticocentric view, we believe that the EEG slow (<1 Hz) rhythm of NREM sleep is an emergent property of corticothalamic networks, which originates from the dynamic interplay of three cardinal oscillators: the mainly, but not necessarily exclusively, synaptically based cortical oscillator (with a layer 4 thalamofugal input and a layer 5/6 corticofugal output), and two intrinsic, conditional thalamic oscillators, i.e., TC and NRT neurons (Fig. 2.8) (Crunelli & Hughes, 2010). Although each of these three oscillators is capable of producing its own slow oscillation, the full EEG manifestation of the slow sleep rhythm requires the essential dynamic tuning provided by their complex synaptic interactions. Since two of the oscillators (i.e., the cortical network and the TC neuron population) are capable of an independent synchronized output when isolated from each other, the start of a new UP state within the intact brain will depend on the relative strength and timing of both the TC neuron and the cortical network oscillator. Although this question has not been directly addressed *in vivo*, the available indirect *in vivo* and *in vitro* evidence strongly points to the TC neuron output (i.e., the LTCP-mediated burst of action potentials that is invariably present at the onset of the TC neuron UP states) as being at least as frequent and effective a signal for eliciting the start of a new cortical UP state in a given thalamocortical module as an intracortical input (for a full discussion of the evidence supporting this point, see Crunelli and Hughes, 2010). This three-oscillator model that we have proposed does not diminish the important contribution played by astrocytic networks in modulating the strength and periodicity of the slow sleep oscillation, as recently shown using transgenic mice with impaired gliotransmission (Fellin et al., 2009) (see Chapter 3 in this book).

A key issue that originates from the above studies and which we believe has not received due attention by researchers in the field is that the frequency of the slow sleep oscillation clearly overlaps with the lower portion of the frequency band of the delta waves. This raises the questions of whether the classical definition of delta waves (as encompassing rhythms between 0.5 and 4 Hz) needs to be revised since it may be possible

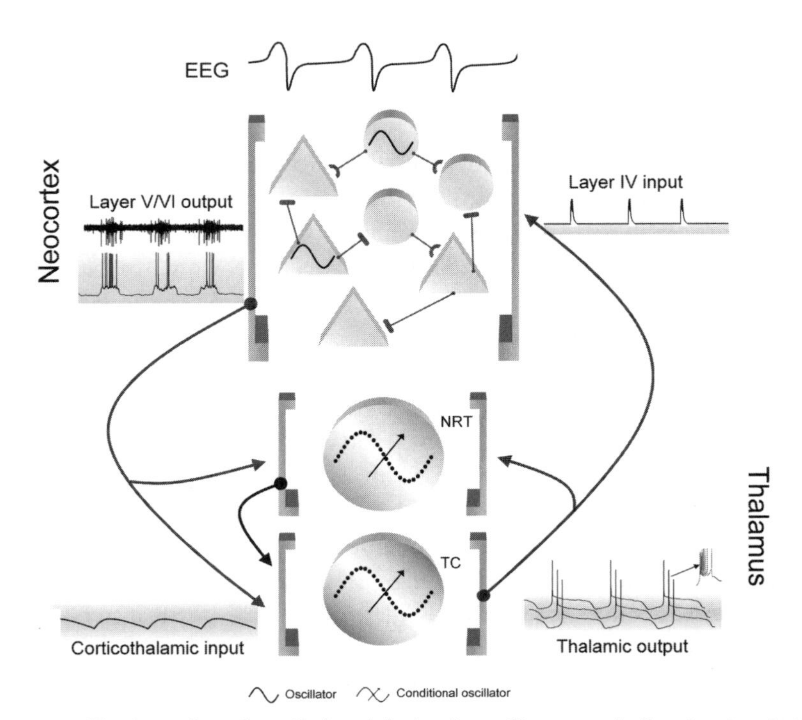

Figure 2.8 ***The interplay of cortical and thalamic oscillators underlies the slow (<1 Hz) sleep rhythm.*** The firing during the UP states of the slow oscillation in layer 5/6 cortical neurons lead to long-lasting corticothalamic EPSPs in TC and NRT neurons. These slow EPSPs bring about the mGluR-induced reduction in I_{Leak}, which is the necessary "condition" for thalamic neurons to exhibit the slow oscillation (for further details, see Crunelli et al., 2005). The LTCP-mediated high-frequency burst that is invariably present at the start of each UP state of the TC neuron slow oscillation leads to highly effective bursts of thalamocortical EPSPs that initiate a new UP state in NRT and layer 4 neurons. The overall UP and DOWN state dynamics of a cortical region, however, are maintained by synaptically generated barrages of excitation and inhibition from other cortical neurons as well as being potentially fine-tuned by inputs from intrinsically oscillating cortical neurons in layers 2/3 and 5. Additional synchronizing inputs (not shown) are provided by short- and long-distance intracortical connections and by intralaminar thalamic afferents that are not restricted to layer 4 (not shown). The symbol ~, without and with a crossing arrow, indicates neurons that function as oscillators or conditional oscillators, respectively (for further details, see Crunelli & Hughes, 2010). *Reproduced with permission from Crunelli & Hughes, 2010.*

that EEG delta waves between 0.5 and 1 Hz in humans and felines and between 0.5 and 2 Hz in rodents may in reality represent the slow sleep oscillation. Indeed, the mechanisms that generate delta waves, particularly in the neocortex, have not been systematically investigated. Moreover, whereas thalamic delta waves, i.e., the rhythmic occurrence of LTCPs at delta frequency in TC and NRT neurons, can be recorded undisturbed for hours in thalamic slices (see the bottom trace of Fig. 2.1) (Hughes et al., 2002b; Blethyn et al., 2006), they are mostly present in the intact brain only transiently (see Fig. 1 in Steriade, 1993).

CONCLUDING REMARKS

The EEG waves of relaxed wakefulness and NREM sleep, as well as sleep spindles, require the fine dynamic interplay of both intrinsic and synaptic activities across cortical and thalamic territories. The more we understand of these brain rhythms at the cellular, synaptic, and network level the more it becomes manifest that the cortex is built to preferentially generate network-driven rhythms and oscillations whereas thalamic neurons are exquisitely equipped for the expression of highly diverse and powerful, intrinsic oscillations (Fig. 2.1).

ACKNOWLEDGMENTS

Our work in this area is supported by The Wellcome Trust.

REFERENCES

Achermann, P., & Borbély, A. (1997). Low-frequency (<1 Hz) oscillations in the human sleep EEG. *Neuroscience*, *81*, 213–222.

Amzica, F., & Steriade, M. (1997). The K-complex: Its slow rhythmicity and relation to delta waves. *Neurology*, *49*, 952–959.

Amzica, F., & Steriade, M. (1998). Cellular substrates and laminar profile of sleep K-complex. *Neuroscience*, *82*, 671–686.

Amzica, F., & Steriade, M. (2002). The functional significance of K-complexes. *Sleep Medicine Reviews*, *6*, 139–149.

Andersen, P., & Andersson, S. A. (1968). *Physiological basis of the alpha rhythm*. New York: Century Crofts.

Basar, E., Schurmann, M., Basar-Eroglu, C., & Karakas, S. (1997). Alpha oscillations in brain functioning: An integrative theory. *International Journal of Psychophysiology*, *26*(1–3), 5–29.

Beierlein, M., Fall, C. P., Rinzel, J., & Yuste, R. (2002). Thalamocortical bursts trigger recurrent activity in neocortical networks: Layer 4 as a frequency-dependent gate. *Journal of Neuroscience*, *22*, 9885–9894.

Berger, H. (1929). Über das elektroenkephalogramm des menschen. *Archives Psychiatry*, *87*, 527–570.

Blethyn, K. L., Hughes, S. W., Tóth, T. I., Cope, D. W., & Crunelli, V. (2006). Neuronal basis of the slow (<1 Hz) oscillation in neurons of the nucleus reticularis thalami *in vitro*. *Journal of Neuroscience*, *26*, 2474–2486.

Bouyer, J. J., Tilquin, C., & Rougeul, A. (1983). Thalamic rhythms in cat during quiet wakefulness and immobility. *Electroencephalography and Clinical Neurophysiology*, *55*(2), 180–187.

Brismar, T. (2007). The human EEG: Physiological and clinical studies. *Physiology and Behaviour, 92*(1–2), 141–147.
Buser, P., & Rougeul-Buser, A. (1995). Do cortical and thalamic bioelectric oscillations have a functional role? A brief survey and discussion. *Journal of Physiology (Paris), 89*(4–6), 249–254.
Cash, S. S., Halgren, E., Dehghani, N., Rossetti, A. O., Thesen, T., Wang, C., et al. (2009). The human K-complex represents an isolated cortical down-state. *Science, 324*, 1084–1087.
Clemens, Z., Mölle, M., Eross, L., Barsi, P., Halász, P., & Born, J. (2007). Temporal coupling of parahippocampal ripples, sleep spindles and slow oscillations in humans. *Brain, 130*, 2868–2878.
Compte, A., Sanchez-Vives, M. V., McCormick, D. A., & Wang, X-J. (2003). Cellular and network mechanisms of slow oscillatory activity (<1 Hz) and wave propagations in a cortical network model. *Journal of Neurosphysiology, 89*, 2707–2725.
Contreras, D., & Steriade, M. (1996). Spindle oscillation in cats: The role of corticothalamic feedback in a thalamically generated rhythm. *Journal of Physiology, 490*, 159–179.
Cossart, R., Aronov, D., & Yuste, R. (2003). Attractor dynamics of network UP states in the neocortex. *Nature, 423*, 283–288.
Cox, C. L., Reichova, I., & Sherman, S. M. (2003). Functional synaptic contacts by intranuclear axon collaterals of thalamic relay neurons. *Journal of Neuroscience, 23*, 7642–7646.
Cunningham, M. O., Pervouchine, D. D., Racca, C., Kopell, N. J., Davies, C. H., Jones, R. S., et al. (2006). Neuronal metabolism governs cortical network response state. *Proceedings of the National Academy of Sciences (USA), 103*, 5597–5601.
Crunelli, V., Tóth, T. I., Cope, D. W., Blethyn, K., & Hughes, S. W. (2005). The 'window' T-type calcium current in brain dynamics of different behavioural states. *Journal of Physiology, 562*, 121–129.
Crunelli, V., Cope, D. W., & Hughes, S. W. (2006). Thalamic T-type Ca^{2+} channels and NREM sleep. *Cell Calcium, 40*, 175–190.
Crunelli, V., & Hughes, S. W. (2010). The slow (<1 Hz) sleep rhythm: A dialogue of three cardinal oscillators. *Nature Neuroscience, 13*, 9–17.
Crunelli, V., Lőrincz, M. L., Errington, A. C., & Hughes, S. W. (2011). Activity of cortical and thalamic neurons during the slow (<1 Hz) rhythm in the mouse in vivo. *Pflugers Archives* (PMID:21892727)
Dang-Vu, T. T., Schabus, M., Desseilles, M., Sterpenich, V., Bonjean, M., & Maquet, P. (2010). Functional neuroimaging insights into the physiology of human sleep. *Sleep, 33*(12), 1589–1603.
Danos, P., Guich, S., Abel, L., & Buchsbaum, M. S. (2001). EEG alpha rhythm and glucose metabolic rate in the thalamus in schizophrenia. *Neuropsychobiology, 43*, 265–272.
Desseilles, M., Dang-Vu, T. T., Schabus, M., Sterpenich, V., Maquet, P., & Schwartz, S. (2008). Neuroimaging insights into the pathophysiology of sleep disorders. *Sleep, 31*(6), 777–794.
Destexhe, A., Contreras, D., & Steriade, M. (1999). Spatiotemporal analysis of local field potentials and unit discharges in cat cerebral cortex during natural wake and sleep states. *Journal of Neuroscience, 19*, 4595–4608.
Dijk, D. J. (2009). Regulation and functional correlates of slow wave sleep. *Journal of Clinical Sleep Medicine, 5*(2 Suppl.), S6–15.
Doi, A., Mizuno, M., Katafuchi, T., Furue, H., Koga, K., & Yoshimura, M. (2007). Slow oscillation of membrane currents mediated by glutamatergic inputs of rat somatosensory cortical neurons: *In vivo* patch-clamp analysis. *European Journal of Neuroscience, 26*, 2565–2575.
Feige, B., Scheffler, K., Esposito, F., Di Salle, F., Hennig, J., & Seifritz, E. (2005). Cortical and subcortical correlates of electroencephalographic alpha rhythm modulation. *Journal of Neurophysiology, 93*(5), 2864–2872.

Fellin, T., Halassa, M. M., Terunuma, M., Succol, F., Takano, H., Frank, M., et al. (2009). Endogenous non-neuronal modulators of synaptic transmission control cortical slow oscillations in vivo. *Proceedings of the National Academy of Sciences (USA), 106*(35), 15037–15042.

Godwin, D. W., Van Horn, S. C., Sesma, M., Romano, C., & Sherman, S. M. (1996). Ultrastructural localization suggests that retinal and cortical inputs access different metabotropic glutamate receptors in the lateral geniculate nucleus. *Journal of Neuroscience, 16*, 8181–8192.

Goldman, R. I., Stern, J. M., Engel, J., Jr., & Cohen, M. S. (2002). Simultaneous EEG and fMRI of the alpha rhythm. *Neuroreport, 13*, 2487–2492.

Haider, B., Duque, A., Hasenstaub, A. R., & McCormick, D. A. (2006). Neocortical network activity in vivo is generated through a dynamic balance of excitation and inhibition. *Journal of Neuroscience, 26*, 4535–4545.

Hirsch, J. C., Fourment, A., & Marc, M. E. (1983). Sleep-related variations of membrane potential in the lateral geniculate body relay neurons of the cat. *Brain Research, 259*(2), 308–312.

Hughes, S. W., & Crunelli, V. (2005). Thalamic mechanisms of EEG alpha rhythms and their pathological implications. *Neuroscientist, 11*(4), 357–372.

Hughes, S. W., Blethyn, K. L., Cope, D. W., & Crunelli, V. (2002). Properties and origin of spikelets in thalamocortical neurones in vitro. *Neuroscience, 110*, 395–401.

Hughes, S. W., Cope, D. W., Blethyn, K. L., & Crunelli, V. (2002). Cellular mechanisms of the slow (<1 Hz) oscillation in thalamocortical neurons *in vitro*. *Neuron, 33*, 947–958.

Hughes, S. W., Cope, D. W., Tóth, T. I., Williams, S. R., & Crunelli, V. (1999). All thalamocortical neurones possess a T-type Ca^{2+} 'window' current that enables the expression of bistability-mediated activities. *Journal of Physiology, 517*, 805–815.

Hughes, S. W., Lorincz, M., Cope, D. W., Blethyn, K. L., Kekesi, K. A., Parri, H. R., et al. (2004). Synchronized oscillations at alpha and theta frequencies in the lateral geniculate nucleus. *Neuron, 42*, 253–268.

Hughes, S. W., Lőrincz, M. L., Blethyn, K., Kékesi, K. A., Juhász, G., Turmaine, M., et al. (2011). Thalamic gap junctions control local neuronal synchrony and influence macroscopic oscillation amplitude during EEG alpha rhythms. *Frontiers in Psychology, 2* (article 193).

Isaichev, S. A., Derevyankin, V. T., Koptelov, Yu. M., & Sokolov, E. N. (2001). Rhythmic alpha-activity generators in the human EEG. *Neuroscience Behaviour Physiology, 31*(1), 49–53.

Isomura, Y., Sirota, A., Ozen, S., Montgomery, S., Mizuseki, K., Henze, D. A., et al. (2006). Integration and segregation of activity in entorhinal-hippocampal subregions by neocortical slow oscillations. *Neuron, 52*, 871–882.

Jones, E. G. (2008). *The thalamus*. New York: Plenum Press.

Klausberger, T., & Somogyi, P. (2004). Neuronal diversity and temporal dynamics: The unit of hippocampal circuit dynamics. *Science, 321*, 53–57.

Krueger, J. M., Rector, D. M., Roy, S., Van Dongen, H. P., Belenky, G., & Panksepp, J. (2008). Sleep as a fundamental property of neuronal assemblies. *Nature Review Neuroscience, 9*, 910–919.

Larson, C. L., Davidson, R. J., Abercrombie, H. C., Ward, R. T., Schaefer, S. M., Jackson, D. C., et al. (1998). Relations between PET-derived measures of thalamic glucose metabolism and EEG alpha power. *Psychophysiology, 35*, 162–169.

Le Bon-Jego, M., & Yuste, R. (2007). Persistently active, pacemaker-like neurons in neocortex. *Frontiers in Neuroscience, 1*, 123–129.

Lemieux, M., & Timofeev, I. (2010). Impact of thalamic inactivation on slow oscillation in the suprasylvian gyrus of cat. *Society for Neuroscience Abstracts, 36* (552.16).

Lindgren, K. A., Larson, C. L., Schaefer, S. M., Abercrombie, H. C., Ward, R. T., Oakes, T. R., et al. (1999). Thalamic metabolic rate predicts EEG alpha power in healthy control subjects but not in depressed patients. *Biological Psychiatry, 45*, 943–952.

Llinás, R. R., & Steriade, M. (2006). Bursting of thalamic neurons and states of vigilance. *Journal of Neurophysiology*, *95*, 3297–3308.

Lopes da Silva, F. H., van Lierop, T. H., Schrijer, C. F., & van Leeuwen, W. S. (1973). Organization of thalamic and cortical alpha rhythms: Spectra and coherences. *Electroencephalography and Clinical Neurophysiology*, *35*(6), 627–639.

Lopes da Silva, F. H., Vos, J. E., Mooibroek, J., & Van Rotterdam, A. (1980). Relative contributions of intracortical and thalamo-cortical processes in the generation of alpha rhythms, revealed by partial coherence analysis. *Electroencephalography and Clinical Neurophysiology*, *50*(5–6), 449–456.

Lőrincz, M. L., Bao, Y., Crunelli, V., & Hughes, S. W. (2007). Contribution of intrinsic membrane potential bistability to neocortical UP and DOWN states. *Society for Neuroscience Abstracts*, *33* (881.19).

Lőrincz, M. L., Bao, Y., Crunelli, V., & Hughes, S. W. (2008a). Cholinergic activation determines the dynamics of neocortical UP states. *Society for Neuroscience Abstracts*, *34* (41.6).

Lőrincz, M. L., Crunelli, V., & Hughes, S. W. (2008b). Cellular dynamics of cholinergically-induced alpha (8–13 Hz) rhythms in sensory thalamic nuclei in vitro. *Journal of Neuroscience*, *28*, 660–671.

Lőrincz, M. L., Kékesi, K. A., Juhász, G., Crunelli, V., & Hughes, S. W. (2009). Temporal framing of thalamic relay-mode firing by phasic inhibition during the alpha rhythm. *Neuron*, *63*, 683–696.

MacLean, J. N., Watson, B. O., Aaron, G. B., & Yuste, R. (2005). Internal dynamics determine the cortical response to thalamic stimulation. *Neuron*, *48*, 811–823.

McCormick, D. A. (1992). Neurotransmitter actions in the thalamus and cerebral cortex and their role in neuromodulation of thalamocortical activity. *Progress in Neurobiology*, *39*, 337–388.

McCormick, D. A., & von Krosigk, M. (1992). Corticothalamic activation modulates thalamic firing through glutamate "metabotropic" receptors. *Proceedings of the National Academy of Sciences (USA)*, *89*, 2774–2778.

Monto, S., Palva, S., Voipio, J., & Palva, J. M. (2008). Very slow EEG fluctuations predict the dynamics of stimulus detection and oscillation amplitudes in humans. *Journal of Neuroscience*, *28*(33), 8268–8272.

Rigas, P., & Castro-Alamancos, M. A. (2007). Thalamocortical Up states: Different effects of intrinsic and extrinsic cortical inputs on persistent activity. *Journal of Neuroscience*, *27*, 4261–4272.

Rudolph, M., Pospischil, M., Timofeev, I., & Destexhe, A. (2007). Inhibition determines membrane potential dynamics and controls action potential generation in awake and sleeping cat cortex. *Journal of Neuroscience*, *27*, 5280–5290.

Ruiz-Mejias, M., Ciria-Suarez, L., Mattia, M., & Sanchez-Vives, M. V. (2011). Slow and fast rhythms generated in the cerebral cortex of the anesthetized mouse. *Journal of Neurophysiology* (PMID: 21880935).

Salek-Haddadi, A., Friston, K. J., Lemieux, L., & Fish, D. R. (2003). Studying spontaneous EEG activity with fMRI. *Brain Research, Brain Research Reviews*, *43*(1), 110–133.

Sanchez-Vives, M. V., & McCormick, D. A. (2000). Cellular and network mechanisms of rhythmic recurrent activity in neocortex. *Nature Neuroscience*, *3*, 1027–1034.

Sherman, S. M., & Guillery, R. W. (1996). Functional organization of thalamocortical relays. *Journal of Neurophysiology*, *76*, 1367–1395.

Shu, Y., Hasenstaub, A., & McCormick, D. A. (2003). Turning on and off recurrent balanced cortical activity. *Nature*, *423*, 288–293.

Siegel, J. M. (2008). Do all animals sleep? *Trends in Neuroscience*, *31*, 208–213.

Simon, N. R., Manshanden, I., & Lopes da Silva, F. H. (2000). A MEG study of sleep. *Brain Research*, *860*, 64–76.

Sirota, A., Csicsvari, J., Buhl, D., & Buzsáki, G. (2003). Communication between neocortex and hippocampus during sleep in rodents. *Proceedings of the National Academy of Sciences (USA), 100*, 2065–2069.

Soltesz, I., & Crunelli, V. (1992). A role for low-frequency, rhythmic synaptic potentials in the synchronization of cat thalamocortical cells. *Journal of Physiology, 457*, 257–276.

Steriade, M. (1993). Oscillations in interacting thalamic and neocortical neurons. *News in Physiological Sciences, 8*, 111–116.

Steriade, M., Amzica, F., & Nuñez, A. (1993). Cholinergic and noradrenergic modulation of the slow (approximately 0.3 Hz) oscillation in neocortical cells. *Journal of Neurophysiology, 70*, 1385–1400.

Steriade, M., Contreras, D., Curró Dossi, R., & Nuñez, A. (1993). The slow (<1 Hz) oscillation in reticular thalamic and thalamocortical neurons: Scenario of sleep rhythm generation in interacting thalamic and neocortical networks. *Journal of Neuroscience, 13*, 3284–3299.

Steriade, M., Nuñez, A., & Amzica, F. (1993). A novel slow (<1 Hz) oscillation of neocortical neurons *in vivo*: Depolarizing and hyperpolarizing components. *Journal of Neuroscience, 13*, 3253–3265.

Steriade, M., Timofeev, I., & Grenier, F. (2001). Natural waking and sleep states: A view from inside neocortical neurons. *Journal of Neurophysiology, 85*, 1969–1985.

Steriade, M., Timofeev, I., Grenier, F., & Dürmüller, N. (1998). Role of thalamic and cortical neurons in augmenting responses and self-sustained activity: Dual intracellular recordings in vivo. *Journal of Neuroscience, 18*, 6425–6443.

Tallon-Baudry, C. (2009). The roles of gamma-band oscillatory synchrony in human visual cognition. *Frontiers in Bioscience, 14*, 321–332.

Timofeev, I., Grenier, F., Bazhenov, M., Sejnowski, T. J., & Steriade, M. (2000). Origin of slow cortical oscillations in deafferented cortical slabs. *Cerebral Cortex, 10*, 1185–1199.

Timofeev, I., & Steriade, M. (1996). Low-frequency rhythms in the thalamus of intact-cortex and decorticated cats. *Journal of Neurophysiology, 76*, 4152–4168.

Tóth, T. I., Hughes, S. W., & Crunelli, V. (1998). Analysis and biophysical interpretation of bistable behaviour in thalamocortical neurons. *Neuroscience, 87*, 519–523.

Turner, J. P., & Salt, T. E. (2000). Synaptic activation of the group I metabotropic glutamate receptor mGlu1 on the thalamocortical neurons of the rat dorsal lateral geniculate nucleus *in vitro*. *Neuroscience, 100*, 493–505.

Vanhatalo, S., Voipio, J., & Kaila, K. (2005). Full-band EEG (FbEEG): An emerging standard in electroencephalography. *Clinical Neurophysiology, 116*, 1–8.

von Krosigk, M., Bal, T., & McCormick, D. A. (1993). Cellular mechanisms of a synchronized oscillation in the thalamus. *Science, 261*, 361–364.

Williams, S. R., Turner, J. P., Tóth, T. I., Hughes, S. W., & Crunelli, V. (1997). The 'window' component of the low threshold Ca^{2+} current produces input signal amplification and bistability in cat and rat thalamocortical neurones. *Journal of Physiology, 505*, 689–705.

Zhu, L., Blethyn, K. L., Cope, D. W., Tsomaia, V., Crunelli, V., & Hughes, S. W. (2006). Nucleus- and species-specific properties of the slow (<1 Hz) sleep oscillation in thalamocortical neurons. *Neuroscience, 141*, 621–636.

CHAPTER 3

Glial Modulation of Sleep and Electroencephalographic Rhythms

Marcos G. Frank
Department of Neuroscience, University of Pennsylvania School of Medicine

INTRODUCTION

The three principal classes of glia in the mature brain are astrocytes, microglia, and oligodendrocytes. Astrocytes perform a number of "housekeeping" functions in the brain, including buffering ions, recycling neurotransmitter, and regulating metabolism. Microglia are the immune cells of the central nervous system and play critical roles in the response to neural injury and cellular stress. Oligodendrocytes produce myelin and enwrap axons and dysfunction in these cells leads to demyelination and disease. They are considered nonexcitable, as they do not produce action potentials and respond linearly to current injections. For these reasons, they have historically been viewed as supportive brain cells with no special or direct roles in brain activity or behavior. However, at least two classes of glial cells (astrocytes and microglia) are known to secrete substances that can alter the activity of surrounding neurons. This provides a means of influencing behavior and brain activity. In astrocytes this process is known as gliotransmission and is now recognized as an important means by which glial and neurons exchange chemical signals (Fiacco, Agulhon, & McCarthy, 2009; Halassa & Haydon, 2010; Hamilton & Atwell, 2010). Therefore, in conjunction with neuronal circuits and neuromodulatory nuclei, glia likely play important roles in brain activity patterns during sleep.

EARLY CONCEPTS OF GLIAL ROLES IN SLEEP

In 1895, Ramon y Cajal hypothesized that astrocytes might control the expression of sleep and wakefulness (García-Marín, García-López, & Freire, 2007). He hypothesized that by either contracting or expanding their processes between synapses, astrocytes could modify the flow

Sleep and Brain Activity
DOI: http://dx.doi.org/10.1016/B978-0-12-384995-3.00003-4

of "nervous current" between neurons. The withdrawal of this physical barrier would promote wakefulness, while its expansion would promote sleep. Early empirical support for a role for glia in sleep came from a study of metabolic enzymes in the brainstem of sleeping rabbits (Hyden & Lange, 1965). Hyden and Lange found sleep-wake rhythms in succinooxidase (a key enzymatic step in the Krebs cycle) activity in neurons and glia (type not specified), with glia showing reduced activity and neurons heightened activity during sleep (Hyden & Lange, 1965). The authors suggested that "neuron and glia form a functional unit" and that sleep and wakefulness resulted in an exchange of signaling molecules between these cell types.

GLIA, NEURAL METABOLISM, AND SLEEP

A more modern twist on the ideas of Hyden and Lange was proposed in 1995 by Joel Benington and Craig Heller. According to the Benington-Heller hypothesis, astrocytic glycogen—which acts as a reserve glucose store for neurons—is depleted during wakefulness and restored during NREM sleep. The depletion of glycogen is mediated by the heightened release of excitatory neurotransmitters during wake, which through enzymatic mechanisms convert astrocytic glycogen into glucose. The utilization of this glucose by neurons (and the subsequent hydrolysis of ATP to AMP) leads to an increase in neuronal adenosine production—which diffuses across the cell membrane and by acting on A1 receptors reduces neuronal excitability. The restoration of glycogen is then favored by states with reduced release of excitatory neurotransmitters, such as NREM sleep (Benington & Heller, 1995).

The Benington-Heller theory was very appealing because it connected the regulation of sleep to a function with clear adaptive value (cerebral metabolism) and elegantly incorporated previous work demonstrating a central role for adenosine in sleep regulation. The evidence for this theory, however, is equivocal. Sleep deprivation increases the activity of glycogen synthase (Petit, Tobler, Allaman, Borbely, & Magistretti, 2002), and one study showed decreases in brain glycogen content following sleep deprivation and increases following recovery sleep (Kong et al., 2002). These latter findings, however, have not been replicated by other labs (Franken, Gip, Hagiwara, Ruby, & Heller, 2003; Gip, Hagiwara, Ruby, & Heller, 2002; Gip et al., 2004; Zimmerman et al., 2004). Therefore, while it remains possible that astrocyte-neuronal metabolic interactions influence sleep (Magistretti, 2006; Scharf, Naidoo, Zimmerman, & Pack, 2008), the specific role posited for astrocytes in the Benington-Heller hypothesis is not supported by current data.

ION BUFFERING AND EEG ACTIVITY

Astrocytes through passive and active transport buffer ions in the extracellular space. This buffering may facilitate synchronized neuronal activity necessary for slow, cortical oscillations (slow-wave activity [SWA]) typical of NREM sleep (Crunelli et al., 2002). For example, astrocytes in thalamic slices (*in situ*) exhibit spontaneous intracellular calcium oscillations that fall within the slower EEG bands of slow-wave activity (<0.1 Hz). These oscillations can propagate within the slice and elicit NMDA currents in neighboring neurons. This particular coupling between neurons and astrocytes does not appear to be related to the EEG rhythms of sleep, as it predominates at ages when thalamocortical and intracortical EEG activity typical of adult sleep is absent (Crunelli & Hughes, 2010). More compelling evidence that astrocytes contribute to slow, EEG rhythms of sleep come from studies using dual intracellular recording in cortical astrocytes and neurons in the adult cat (Amzica, 2002; Amzica & Massimini, 2002; Amzica & Neckelmann, 1999). These investigators showed that astrocyte membrane polarization and capacitance oscillate in phase with slow EEG waves during natural NREM sleep, suggesting that the cation buffering by these cells is a critical component of neuronal "up" and "down" states.

GLIAL SECRETION OF SOMNOGENIC SUBSTANCES: ASTROCYTES

Cultured astrocytes secrete and/or exocytose a variety of molecules that when injected either systemically or into the brain can increase sleep time or NREM SWA. For example, the cytokine IL-1 derived from cultured mouse astrocytes increases NREM sleep in rats when administered into the ventricles. (Tobler, Borbély, Schwyzer, & Fontana, 1984). Cultured astrocytes also secrete neurotrophins (e.g., brain-derived neurotrophin factor [BDNF]), prostaglandins (PGD2) and the cytokine TNFα that increase sleep time or intensity (e.g.,. NREM SWA) when injected intraventricularly and/or infused and applied to the neocortex (Faraguna, Vyazovskiy, Nelson, Tononi, & Cirelli, 2008; Hayaishi, 2002; Huang, Urade, & Hayaishi, 2007; Krueger, 2008; Kushikata, Fang, & Krueger, 1999). Astrocytes release some of these substances in response to neuronal signals, including ATP acting at astrocytic PP2 receptors (Krueger, 2008). These findings support a mechanism by which neuronal activation of glia via purinergic receptors during wake leads to the release of substances that can increase sleep amounts and indices of sleep intensity (SWA) (Krueger, 2008).

Whether or not such mechanisms actually exist *in vivo* is unclear. Cultured cells can show very different properties than cells *in vivo*, or in brain slice preparations (Inagaki & Wada, 1994; Yamamoto, Miwa, Ueno, & Hayaishi, 1988). In addition, although mutant mice lacking the TNFα, IL-1, and PP2 receptors have sleep phenotypes consistent with this general hypothesis (i.e., reduced NREM sleep amounts and intensity), it is not known if this reflects neuronal or astrocytic influences, as neurons release and respond to many cytokines as well (Baracchi & Opp, 2008; Kapas et al., 2008; Krueger et al., 2010b). A recent study showed increased IL-β1 immunoreactivity in astrocytes following 2 hours of whisker stimulation in awake rodents (Hallett, Churchill, Taishi, De, & Krueger, 2010). While this latter finding is quite intriguing, it is not known if similar stimulation also produces IL-β1 secretion in astrocytes and subsequent sleep (Hallett et al., 2010). Therefore, determining the precise role of astrocyte cytokine signaling in sleep must await more selective manipulations of these signaling pathways *in vivo.*

In addition to neurotrophins, cytokines, and prostaglandins, there is evidence that astrocytes exocytose additional chemical transmitters that modulate neuronal excitability, including ATP, which is hydrolyzed to adenosine in the extracellular space (Halassa, Fellin, & Haydon, 2009; Pascual et al., 2005). Although the precise mechanisms of gliotransmission are unclear (Fiacco et al., 2009; Hamilton & Atwell, 2010), for some gliotransmitters, exocytosis may depend on the formation of a SNARE complex between vesicles and the target membrane (Scales, Bock, & Scheller, 2000). Conditional astrocyte-selective expression of the SNARE domain of the protein synaptobrevin II prevents both tonic and activity-dependent extracellular accumulation of adenosine that acts on A1 receptors *in situ* (Pascual et al., 2005). Fortunately, the role of this signaling pathway in sleep can be investigated *in vivo* using transgenic approaches in mice (Pascual et al., 2005).

We investigated this possibility (Halassa, Florian et al., 2009) using the tet-off system (Morozov, Kellendonk, Simpson, & Tronche, 2003) to allow conditional expression of a dnSNARE transgene selectively in astrocytes (Pascual et al., 2005). Astrocyte specificity of transgene expression is achieved by using the astrocyte-specific Glial Fibrillary Acidic Protein (GFAP) promoter to drive the expression of tetracycline transactivator (tTA) only in this subset of glia. GFAP.tTA mice were crossed with tetO.dnSNARE mice. The tet-operator (tet.O) drives the expression of dnSNARE and the EGFP reporter. Thus, in bigenic offspring of this mating transgenes are only expressed in GFAP-positive astrocytes (Pascual et al., 2005). Conditional suppression of transgene expression is achieved by including doxycycline (Dox) in the diet. Dox binds to tTA preventing it from activating the tet.O promoter, and by maintaining all bigenic mice on DOX throughout gestation and early development, the transgene was only expressed, when desired, in adult mice (Halassa, Florian, et al., 2009).

Suppressing gliotransmission *in vivo* had surprisingly little effect on baseline sleep-wake architecture, except for a reduction in the normal accumulation of NREM EEG slow-wave activity—a classic index of sleep pressure in mammals (Halassa, Florian, et al., 2009). This suggested that in the absence of gliotransmission, the accumulation of sleep need was reduced. This was confirmed by examining compensatory responses to sleep deprivation in dnSNARE mutant mice and wild-type controls. While wild-type mice showed normal compensatory increases in NREM sleep time, bout duration, and SWA after 6 hours of sleep deprivation, these changes did not occur (or were greatly attenuated) in the mutant mice. Interestingly, the cognitive effects of sleep deprivation were also absent in the mutant mice. Subsequent investigations showed that this phenotype could be copied in wild-type mice by antagonists to the A1 (but not A2) receptor, demonstrating that the gliotransmitter of interest was ATP (Halassa, Florian, et al., 2009). Manipulations of adenosine receptors in wild-type mice duplicated most of the effects of the dnSNARE mutation; however, subsequent investigations showed that gliotransmission of D-serine is also necessary for SWA (Fellin et al., 2009). These findings represent the first direct demonstration that astrocytes *in vivo* influence mammalian sleep.

Further studies *in situ* and *in vivo* in the dnSNARE mouse demonstrated that gliotransmission also modulates sleep EEG rhythms (Fellin et al., 2009). Mutant mice normally show reduced levels of NREM SWA under baseline conditions, and after sleep deprivation—which suggests impairment intracortically in the ability to generate slow, neuronal oscillations (Fellin et al., 2009; Halassa, Florian, et al., 2009). This was confirmed using a combination of extracellular local field potential (LFP) and patch-clamp recordings *in vivo* from the somatosensory cortex of urethane anesthetized dnSNARE animals and wild-type littermates. Attenuation of gliotransmission in transgenic animals significantly decreased the power of slow oscillations (<1 Hz), pyramidal neurons from dnSNARE animals have a significantly lower probability of being at the depolarized

Table 3.1 Glial Substances Linked to Sleep

Substance	Effects on Sleep	Evidence That Glial Secretion *in vivo* Regulates Sleep?
ATP>> adenosine	Mediates NREM sleep homeostasis (EEG slow-wave activity (SWA) and state architecture) (Halassa, Florian et al., 2009)	Yes
D-serine	Necessary for normal SWA (Fellin et al., 2009)	Yes
TNFα	Intracranial infusions increase sleep time and SWA (Krueger, 2008)	No
IL-1	Intracranial infusions increase sleep time and SWA (Tobler et al., 1984)	No
PDG2	Intracranial infusions increase sleep time and SWA (Huang et al., 2007)	No
BDNF	Intracranial infusions increases sleep time and SWA (Faraguna et al., 2008; Kushikata et al., 1999)	No

"Intracranial" infusions include intraventricular injections, topical application to the cortex, and intracortical infusions.

potential (up-state probability) compared to controls, and up-state transitions occurred at lower frequency in the mutants (Fellin et al., 2009). Subsequent studies *in situ* and *in vivo* demonstrated that the attenuation of slow, EEG rhythms in sleep was due to the absence of two gliotransmitters: D-serine and ATP (hydrolyzed to adenosine) (Fellin et al., 2009).

GLIAL SECRETION OF SOMNOGENIC SUBSTANCES: MICROGLIA AND OLIGODENDROCYTES

Cerebral microglia and oligodendrocytes cells secrete a number of substances *in vitro* known to influence sleep or brain activity in sleep (e.g., cytokines, prostaglandins, and nitric oxide, see Table 3.1) (Matsui et al., 2010). Because sleep deprivation is associated with an increase in markers of cellular stress, it has been proposed that substances secreted by microglia may play a central role in sleep regulation (Wisor, Clegern, & Schmidt, 2011; Wisor, Schmidt, & Clegern, 2011). For example, attenuation of microglia reactivity with minocycline reduces the normal compensatory increases in NREM SWA in mice following sleep deprivation (Wisor & Clegern, 2011). Interestingly, microglia contain membrane-bound purinergic receptors, providing a means of interaction with astrocyte-derived ATP and adenosine

(Gyoneva, Orr, & Traynelis, 2009; Haynes et al., 2006). On the other hand, a putative transducer of microglial-mediated effects (the toll-like receptor 4 (TLR4)) does not appear to play a central role in sleep brain activity. Constitutive deletion of TLR4 minimally impacts NREM EEG SWA under baseline conditions or after sleep deprivation) (Wisor, Clegern, et al., 2011). Oligodendrocytes are a source of prostaglandin D2 in the mature brain, which has been shown to be a potent sleep-inducing substance when applied exogenously (Urade & Hayaishi, 2011). In addition to increasing behavioral indices of sleep, stimulation of D2 receptors increases NREM SWA in a physiological manner(Urade & Hayaishi, 2011). However, in contrast to astrocytes, much less is known about exocytosis and secretion in microglia and oligodendrocytes. It is also unknown what signals normally trigger the release of these substances across the sleep-wake cycle.

UNANSWERED QUESTIONS AND FUTURE DIRECTIONS

An important future area of investigation is to determine where glial cells exert their effects on sleep and/or brain activity. Glia are dispersed widely in subcortical and cortical brain areas (Zhang & Haydon, 2005) including regions known to trigger sleep and wakefulness (Halassa, Florian, et al., 2009). Therefore, they may regulate sleep and brain activity by acting within specific neocortical circuits or via modulation of basal forebrain and hypothalamic sleep and arousal centers (Benington & Heller, 1995; Krueger et al., 2008; Strecker et al., 2000; Szymusiak, Gvilia, & McGinty, 2007). It is presently unknown which of these two models is valid. Addressing this issue is complicated by several factors. First, it is not yet clear if glia in different brain regions secrete the same or different sets of somnogenic substances. As discussed above, cultured cells appear to secrete a variety of neuroactive molecules, but evidence for similar secretion *in vivo* is very sparse, and regional patterns of secretion within the intact brain are poorly understood. Second, cultured glia express numerous neurotransmitter receptors (including glutamate NMDA and metabotropic, neurotrophin (TRKb), and purinergic receptors), raising the possibility that substances secreted by astrocytes also feedback onto their sites of release. These potential feedback loops add yet another layer of complexity to any proposed model of neuronal-glial interactions (Fiacco et al., 2009; Halassa & Haydon, 2010; Hamilton & Atwell, 2010).

A related important question to address is the relative role of different glial-secreted substances in the sleeping brain. Astrocytic adenosine (acting at A1 receptors) is likely to be a key mediator of sleep behavior and brain activity. This is because (a) gliotransmission of ATP provides a large amount of extracellular adenosine in the brain (Halassa, Florian, et al., 2009; Pascual et al., 2005), (b) activation of gliotransmission in cultured astrocytes increases adenosine release (Figueiredo et al., 2010), (c) similar activation *in vivo* leads to a suppression of surrounding neuronal activity consistent with activation of A1 receptors (Gradinaru, Mogri, Thompson, Henderson, & Deisseroth, 2009), and (d) adenosine is widely recognized as an endogenous sleep-inducing substance (Bjorness & Greene, 2009; Strecker et al., 2000). However, other signaling pathways and secreted substances may play complementary roles (Frank, 2011). For example, glial-derived adenosine might activate A2 receptors in sleep-promoting hypothalamic neurons (Szymusiak & McGinty, 2008) and glial-derived glutamate might excite sleep-promoting neurons in the forebrain and brainstem (Kaushik, Kumar, & Mallick, 2010; Luppi et al., 2007). This may act in concert with A1-mediated inhibition of wake-promoting neurons to increase sleep drive. There may also be important roles for purinergic receptors activated directly by ATP (Krueger et al., 2010a) and other substances secreted by glia (e.g., cytokines (Krueger, 2008; Opp, 2005)).

REFERENCES

Amzica, F. (2002). In vivo electrophysiological evidences for cortical neuron-glia interactions during slow (<1 Hz) and paroxysmal sleep oscillations. *Journal of Physiology-Paris*, *96*(3–4), 209–219.

Amzica, F., & Massimini, M. (2002). Glial and neuronal interactions during slow wave and paroxysmal activities in the neocortex. *Cerebral Cortex*, *12*(10), 1101–1113.

Amzica, F., & Neckelmann, D. (1999). Membrane capacitance of cortical neurons and glia during sleep oscillations and spike-wave seizures. *Journal of Neurophysiology*, *82*(5), 2731–2746.

Baracchi, F., & Opp, M. R. (2008). Sleep-wake behavior and responses to sleep deprivation of mice lacking both interleukin-1[beta] receptor 1 and tumor necrosis factor-[alpha] receptor 1. *Brain, Behavior, and Immunity*, *22*(6), 982–993. doi:10.1016/j.bbi.2008.02.001.

Benington, J., & Heller, H. C. (1995). Restoration of brain energy metabolism as the function of sleep. *Progress in Neurobiology*, *45*(4), 347–360.

Bjorness, T. E., & Greene, R. W. (2009). Adenosine and sleep. *Current Neuropharmacology*, *7*(3), 238–245.

Crunelli, V., Blethyn, K. L., Cope, D. W., Hughes, S. W., Parri, H. R., Turner, J. P., et al. (2002). Novel neuronal and astrocytic mechanisms in thalamocortical loop dynamics. *Philosophical Transactions of the Royal Society of London Series B, Biological Sciences*, *357*(1428), 1675–1693.

Crunelli, V., & Hughes, S. W. (2010). The slow (<1 Hz) rhythm of non-REM sleep: A dialogue between three cardinal oscillators. *Nature Neuroscience*, *13*(1), 9–17.

Faraguna, U., Vyazovskiy, V. V., Nelson, A. B., Tononi, G., & Cirelli, C. (2008). A causal role for brain-derived neurotrophic factor in the homeostatic regulation of sleep. *The Journal of Neuroscience*, *28*(15), 4088–4095.

Fellin, T., Halassa, M. M., Terunuma, M., Succol, F., Takano, H., Frank, M., et al. (2009). Endogenous nonneuronal modulators of synaptic transmission control cortical slow oscillations in vivo. *Proceedings of the National Academy of Sciences of the United States of America*, *106*(35), 15037–15042.

Fiacco, T. A., Agulhon, C., & McCarthy, K. D. (2009). Sorting out astrocyte physiology from pharmacology. *Annual Review of Pharmacology and Toxicology*, *49*, 151–174.

Figueiredo, M., Lane, S., Tang, F., Liu, B. H., Hewinson, J., Marina, N., et al. (2010). Optogenetic experimentation on astrocytes. *Experimental Physiology*, *96*(1), 40–50.

Frank, M. G. (2011). Beyond the neuron: Astroglial regulation of mammalian sleep. *Current Topics in Medicinal Chemistry*, *11*(19), 2452–2456.

Franken, P., Gip, P., Hagiwara, G., Ruby, N. F., & Heller, H. C. (2003). Changes in brain glycogen after sleep deprivation vary with genotype. *American Journal of Physiology Regulatory, Integrative and Comparative Physiology*, *285*(2), R413–419.

García-Marín, V., García-López, P., & Freire, M. (2007). Cajal's contributions to glia research. *Trends in Neurosciences*, *30*(9), 479–487.

Gip, P., Hagiwara, G., Ruby, N. F., & Heller, H. C. (2002). Sleep deprivation decreases glycogen in the cerebellum but not in the cortex of young rats. *American Journal of Physiology Regulatory, Integrative and Comparative Physiology*, *283*(1), R54–59.

Gip, P., Hagiwara, G., Sapolsky, R. M., Cao, V. H., Heller, H. C., & Ruby, N. F. (2004). Glucocorticoids influence brain glycogen levels during sleep deprivation. *American Journal of Physiology Regulatory, Integrative and Comparative Physiology*, *286*(6), R1057–1062.

Gradinaru, V., Mogri, M., Thompson, K. R., Henderson, J. M., & Deisseroth, K. (2009). Optical deconstruction of parkinsonian neural circuitry. *Science*, *324*(5925), 354–359.

Gyoneva, S., Orr, A. G., & Traynelis, S. F. (2009). Differential regulation of microglial motility by ATP/ADP and adenosine. *Parkinsonism & Related Disorders*, *15*(*Supplement 3*), S195–S199.

Halassa, M. M., Fellin, T., & Haydon, P. G. (2009). Tripartite synapses: Roles for astrocytic purines in the control of synaptic physiology and behavior. *Neuropharmacology*, *57*(4), 343–346.

Halassa, M. M., Florian, C., Fellin, T., Munoz, J. R., Lee, S.Y., Abel, T., et al. (2009). Astrocytic modulation of sleep homeostasis and cognitive consequences of sleep loss. *Neuron*, *61*(2), 213–219.

Halassa, M. M., & Haydon, P. G. (2010). Integrated brain circuits: Astrocytic networks modulate neuronal activity and behavior. *Annual Review of Physiology*, *72*, 335–355.

Hallett, H., Churchill, L., Taishi, P., De, A., & Krueger, J. M. (2010). Whisker stimulation increases expression of nerve growth factor- and interleukin-1[beta]-immunoreactivity in the rat somatosensory cortex. *Brain Research* (*In Press, Accepted Manuscript*).

Hamilton, N. B., & Atwell, D. (2010). Do astrocytes really exocytose neurotransmitters? *Nature Reviews Neuroscience*, *11*, 227–238.

Hayaishi, O. (2002). Functional genomics of sleep and circadian rhythm: Invited review: Molecular genetic studies on sleep-wake regulation, with special emphasis on the prostaglandin D2 system. *Journal of Applied Physiology*, *92*(2), 863–868.

Haynes, S. E., Hollopeter, G., Yang, G., Kurpius, D., Dailey, M. E., Gan, W. B., et al. (2006). The P2Y12 receptor regulates microglial activation by extracellular nucleotides. *Nature Neuroscience*, *9*(12), 1512–1519.

Huang, Z.-L., Urade, Y., & Hayaishi, O. (2007). Prostaglandins and adenosine in the regulation of sleep and wakefulness. *Current Opinion in Pharmacology*, 7(1), 33–38.

Hyden, H., & Lange, P. W. (1965). Rhythmic enzyme changes in neurons and glia during sleep. *Science*, *149*(3684), 654–656.

Inagaki, N., & Wada, H. (1994). Histamine and prostanoid receptors on glial cells. *Glia*, *11*, 102–109.

Kapas, L., Bohnet, S. G., Traynor, T. R., Majde, J. A., Szentirmai, E., Magrath, P., et al. (2008). Spontaneous and influenza virus-induced sleep are altered in TNF-α double-receptor deficient mice. *Journal of Applied Physiology*, *105*(4), 1187–1198.

Kaushik, M. K., Kumar, V. M., & Mallick, H. N. (2010). Glutamate microinjection at the medial preoptic area enhances slow wave sleep in rats. *Behavioural Brain Research*, *217*(1), 240–243.

Kong, J., Shepel, P. N., Holden, C. P., Mackiewicz, M., Pack, A. I., & Geiger, J. D. (2002). Brain glycogen decreases with increased periods of wakefulness: Implications for homeostatic drive to sleep. *The Journal of Neuroscience*, *22*(13), 5581–5587.

Krueger, J. M. (2008). The role of cytokines in sleep regulation. *Current Pharmaceutical Design*, *14*(32), 3408–3416.

Krueger, J. M., Rector, D. M., Roy, S., Van Dongen, H. P., Belenky, G., & Panksepp, J. (2008). Sleep as a fundamental property of neuronal assemblies. *Nature Reviews Neuroscience*, *9*(12), 910–919.

Krueger, J. M., Taishi, P., De, A., Davis, C. J., Winters, B. D., Clinton, J., et al. (2010a). ATP and the purine type 2 X 7 receptor affect sleep. *Journal of Applied Physiology*, *109*(5), 1318–1327.

Krueger, J. M., Taishi, P., De, A., Davis, C. J., Winters, B. D., Clinton, J., et al. (2010b). ATP and the purine type 2 X 7 receptor affect sleep. *Journal of Applied Physiology* (japplphysiol.00586.02010).

Kushikata, T., Fang, J., & Krueger, J. M. (1999). Brain-derived neurotrophic factor enhances spontaneous sleep in rats and rabbits. *AJP – Regulatory, Integrative and Comparative Physiology*, *276*(5), R1334–1338.

Luppi, P. H., Gervasoni, D., Verret, L., Goutagny, R., Peyron, C., Salvert, D., et al. (2007). Paradoxical (REM) sleep genesis: The switch from an aminergic-cholinergic to a GABAergic-glutamatergic hypothesis. *Journal of Physiology, Paris*.

Magistretti, P. J. (2006). Neuron-glia metabolic coupling and plasticity. *The Journal of Experimental Biology*, *209*(12), 2304–2311.

Matsui, T., Svensson, C. I., Hirata, Y., Mizobata, K., Hua, X.-Y., & Yaksh, T. L. (2010). Release of prostaglandin E2 and nitric oxide from spinal microglia is dependent on activation of p38 mitogen-activated protein kinase. *Anesthesia & Analgesia*, *111*(2), 554–560.

Morozov, A., Kellendonk, C., Simpson, E., & Tronche, F. (2003). Using conditional mutagenesis to study the brain. *Biological Psychiatry*, *54*(11), 1125–1133.

Opp, M. R. (2005). Cytokines and sleep. *Sleep Medicine Reviews*, *9*(5), 355–364.

Pascual, O., Casper, K. B., Kubera, C., Zhang, J., Revilla-Sanchez, R., Sul, J. Y., et al. (2005). Astrocytic purinergic signaling coordinates synaptic networks. *Science*, *310*(5745), 113–116.

Petit, J. M., Tobler, I., Allaman, I., Borbely, A. A., & Magistretti, P. J. (2002). Sleep deprivation modulates brain mRNAs encoding genes of glycogen metabolism. *The European Journal of Neuroscience*, *16*(6), 1163–1167.

Scales, S. J., Bock, J. B., & Scheller, R. H. (2000). The specifics of membrane fusion. *Nature*, *407*(6801), 144–146.

Scharf, M. T., Naidoo, N., Zimmerman, J. E., & Pack, A. I. (2008). The energy hypothesis of sleep revisited. *Progress in Neurobiology*, *86*(3), 264–280.

Strecker, R. E., Morairty, S., Thakkar, M. M., Porkka-Heiskanen, T., Basheer, R., Dauphin, L. J., et al. (2000). Adenosinergic modulation of basal forebrain and preoptic/anterior hypothalamic neuronal activity in the control of behavioral state. *Behavioural Brain Research*, *115*(2), 183–204.

Szymusiak, R., Gvilia, I., & McGinty, D. (2007). Hypothalamic control of sleep. *Sleep Medicine Advances in Sleep Medicine*, *8*(4), 291–301.

Szymusiak, R., & McGinty, D. (2008). Hypothalamic regulation of sleep and arousal. *Annals of the New York Academy of Sciences*, *1129*(1), 275–286.

Tobler, I., Borbély, A. A., Schwyzer, M., & Fontana, A. (1984). Interleukin-1 derived from astrocytes enhances slow wave activity in sleep EEG of the rat. *European Journal of Pharmacology*, *104*(1–2), 191–192.

Urade, Y., & Hayaishi, O. (2011). Prostaglandin D2 and sleep/wake regulation. *Sleep Medicine Reviews*, *15*(6), 411–418.

Wisor, J. P., & Clegern, W. C. (2011). Quantification of short-term slow wave sleep homeostasis and its disruption by minocycline in the laboratory mouse. *Neuroscience Letters*, *490*(3), 165–169.

Wisor, J. P., Clegern, W. C., & Schmidt, M. A. (2011). Toll-like receptor 4 is a regulator of monocyte and electroencephalographic responses to sleep loss. *Sleep*, *34*(10), 1335–1345.
Wisor, J. P., Schmidt, M. A., & Clegern, W. C. (2011). Evidence for neuroinflammatory and microglial changes in the cerebral response to sleep loss. *Sleep*, *34*(3), 261–272.
Yamamoto, K., Miwa, T., Ueno, R., & Hayaishi, O. (1988). Muramyl dipeptide-elicited production of PGD2 from astrocytes in culture. *Biochemical and Biophysical Research Communications*, *156*(2), 882–888.
Zhang, Q., & Haydon, P. G. (2005). Roles for gliotransmission in the nervous system. *Journal of Neural Transmission*, *112*(1), 121–125. doi:10.1007/s00702-004-0119-x.
Zimmerman, J. E., Mackiewicz, M., Galante, R. J., Zhang, L., Cater, J., Zoh, C., et al. (2004). Glycogen in the brain of *Drosophila melanogaster*: Diurnal rhythm and the effect of rest deprivation. *Journal of Neurochemistry*, *88*(1), 32–40.

CHAPTER 4

Genetic Mechanisms Underlying Rhythmic EEG Activity during Sleep

Paul Franken
Center for Integrative Genomics, University of Lausanne, Switzerland

INTRODUCTION

To unambiguously determine whether a mammal or bird is asleep or not and if so, in which sleep state, electroencephalographic (EEG) recordings are indispensable. The EEG captures the summed activity of millions of neurons. In humans, activity of neurons of the cerebral cortex are thought to contribute the most to the EEG signal while activity from deeper sources contribute less to currents measured with scalp electrodes. In laboratory animals, such as rats and mice, epidural electrodes are used and, since cortical thickness in these species is greatly reduced as compared to humans, activity of deeper structures such as the hippocampus importantly contribute to the EEG patterns observed. The EEG signal shows oscillations of specific frequency ranges representing synchronized and coherent activity over large networks of neuronal assemblies. For instance, thalamocortical and corticocortical networks underlie the slow oscillation (<1 Hz), delta oscillations (1–4 Hz), and spindle oscillation (11–15 Hz; i.e., the sigma frequency range), characteristic of non-rapid-eye-movement (NREM) sleep (McCormick & Bal, 1997; Steriade, 2003), while the septo-hippocampal/entorhinal-cortex system underlies the theta oscillations (5–9 Hz) that dominate the EEG during rapid-eye-movement (REM) sleep in rodents (Chrobak, Lorincz, & Buzsaki, 2000; Vertes & Kocsis, 1997). Based on the visual inspection of these state-specific EEG patterns, usually in combinations with the electromyogram (EMG) to quantify muscle tone, the states waking, NREM, and REM sleep can be readily discerned (Fig. 4.1) and measures such as the daily amount of sleep, its distribution over the day, the period of the NREM-REM sleep alternation, sleep continuity or quality, sleep-onset latency, and the changes in these variables in response to experimental interventions, can all be quantified (Mang & Franken, 2012).

Sleep and Brain Activity
DOI: http://dx.doi.org/10.1016/B978-0-12-384995-3.00004-6

Besides its use in classifying sleep-wake state, the EEG signal itself contains a wealth of information not only important for the study of the neuronal processes that shape the EEG but also for gaining inside into the regulation, physiology, and pharmacology of sleep. The frequency components that contribute to the EEG within each state can be quantified using spectral analyses algorithms such as the widely used fast Fourier transform (FFT) (Borbely, Baumann, Brandeis, Strauch, & Lehmann, 1981) (Fig. 4.1). Such algorithms allow for the quantification of the activity or power over a broad frequency range or spectrum (e.g., 0.5–100 Hz) or in specific frequency bands such as those introduced above. Besides the pronounced state-dependent changes in the spectral composition of the EEG on which state classification is based, within each state EEG spectra vary as a function of circadian time and time-spent-awake (or asleep) (Dijk & Czeisler, 1995). The latter changes are especially clearly reflected by the sleep-wake driven or homeostatic changes in EEG delta power (Daan, Beersma, & Borbely, 1984; Franken, Chollet, & Tafti, 2001), a measure capturing both the amplitude and prevalence of EEG delta oscillations. The spectral changes in the sleep and waking EEG after drug administration are specific to the class of drug resulting in a recognizable "EEG fingerprint" (Hasan et al., 2009; Tobler, Kopp, Deboer, & Rudolph, 2001; Winsky-Sommerer, 2009). The term EEG fingerprint has also been used to describe the surprisingly stable and trait-like spectral composition of the sleep EEG within an individual over time contrasting the marked EEG differences that can be observed among individuals (Buckelmuller, Landolt, Stassen, & Achermann, 2006; De Gennaro, Ferrara, Vecchio, Curcio, & Bertini, 2005; Finelli, Achermann, & Borbely, 2001). Identifying the molecular basis underlying such EEG fingerprints and the homeostatic EEG changes has led to new insights and hypotheses (Krueger et al., 2008; Tononi & Cirelli, 2006). Since to date no systematic genetic studies have been performed in humans to identify the genetic variants contributing to the rhythmic EEG activity, this chapter will have to rely largely on available mouse data. In the context of this book, a further focus is that only the genetics of the EEG during physiological sleep will be considered and not the genetics of the manifestation of sleep *per se* (i.e., its duration, distribution, and so forth) or the genetics of sleep disorders that have been reviewed elsewhere (e.g., Franken & Tafti, 2003; Sehgal & Mignot, 2011).

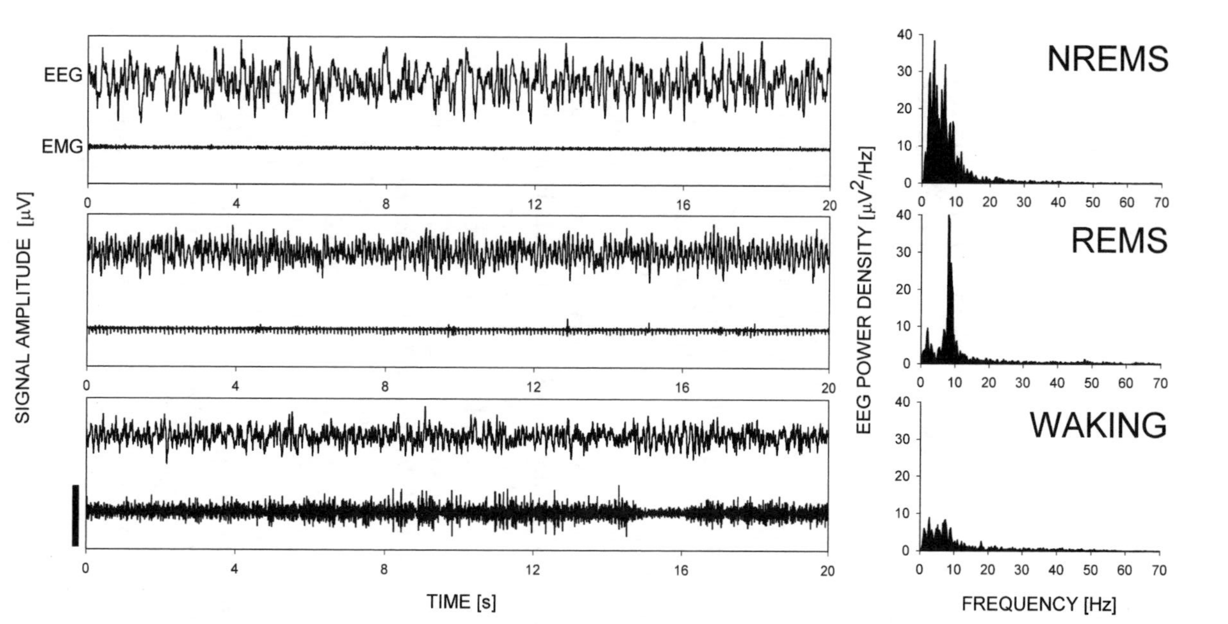

Figure 4.1 ***Sleep-wake state determination is based on EEG and EMG patterns.*** *Left panel:* The determination of the states NREM sleep (top), REM sleep (middle), and wakefulness (lower panel) in the mouse is based on the distinctive patterns of the EEG and EMG. NREM sleep is characterized by the prevalence of high amplitude slow waves in the EEG concomitant with a low and invariable muscle tone in the EMG. The REM sleep EEG is dominated by highly regular theta oscillations with muscle atonia with occasional twitches (note the 10 Hz heartbeat artifact in the EMG). EEG activity during wakefulness is generally of low amplitude and mixed frequency together with a high and variable EMG. Both signals vary depending on the animal's waking behavior. Sleep-wake states were annotated for consecutive 4 s epochs. Twenty-second traces were obtained in an individual male mouse of mixed genetic background (C57BL/6J x 129P/Ola) within an undisturbed 12 h light period. The vertical scale bar delimits 400 μV and applies to both signals. *Right panel:* Spectral content of the signals depicted on the left was quantified with the fast Fourier routine (FFT). Spectra are averaged over the five spectra calculated within each of the 4 s epochs (sampling rate 200 Hz; 0.25 Hz frequency bins). Examples clearly illustrate the dominance of delta activity during NREM sleep, that theta oscillations are the main (only?) rhythmic component during REM sleep, and that in the waking EEG in this particular example no clear oscillation can be visually discerned. *Reprinted from Xie et al. (2005) with permission.*

GENETICS OF THE HUMAN EEG: TWIN, LINKAGE, AND CANDIDATE-GENE STUDIES

The earliest indications that genetic factors play a role in shaping sleep and the EEG come from studies comparing phenotypic variance between monozygotic (MZ) and dizygotic (DZ) twins. In these studies, a higher concordance in a phenotype between MZ twins as compared to DZ twins is taken as proof for a contribution of genetic factors because MZ twins are considered genetically identical, whereas DZ twins share, on average, only half of their segregating genes. Resemblance within twin pairs can also be due to shared environmental influences and an additional assumption is that this shared environment accounts equally for the similarity in MZ and DZ twins. With these assumptions twin studies provide information about the heritability of a trait, that is, the fraction of the total phenotypic variance that can be attributed to additive genetic factors (Boomsma, Busjahn, & Peltonen, 2002).

Sleep patterns and the duration of sleep within MZ twins have a higher concordance than within DZ twin pairs or unrelated subjects. For these sleep traits heritabilities between 40% and 60% can be estimated (Gedda & Brenci, 1979, 1983; Hori, 1986; Linkowski, 1999; Webb & Campbell, 1983). For aspects of the EEG activity within a behavioral state, twin studies revealed that additive genetic factors importantly outweigh environmental influences and heritabilities up to 90% have been reported (Christian et al., 1996; Juel-Nielsen & Harvald, 1958; Lykken, Tellegen, & Thorkelson, 1974; Stassen, Lykken, & Bomben, 1988; Stassen, Lykken, Propping, & Bomben, 1988; van Beijsterveldt & Boomsma, 1994; Vogel, 1970). Most EEG studies in twins concern the spectral composition of the EEG signal during wakefulness quantified with FFT. The contribution of the alpha rhythm (8–13 Hz) during waking has received much attention, but the activity in the delta, theta, and beta (15–35 Hz; the precise frequency range delimiting beta varies among studies) frequency ranges were also found to be highly heritable (>80%). Two more recent twin studies confirmed that the EEG spectral profile of NREM sleep, especially in frequencies below 16 Hz, also shows a heritability estimate of 96% (Ambrosius et al., 2008; De Gennaro et al., 2008), indicating that all of its phenotypic variance can be attributed to genetic factors. With these strikingly high heritabilities, EEG traits qualify as the most heritable traits in humans, matched only by heritabilities obtained for brain architecture such as has been demonstrated for the distribution of gray matter in the cerebral cortex (Thompson et al., 2001). It has been suggested that these two traits might well be interrelated in that common genetic factors underlie functional brain connectivity as well as rhythmic brain activity (Posthuma et al., 2005). Contributing factors could concern the inter-individual differences in the rate at which the brain matures as revealed by the EEG (Buchmann et al., 2011).

The contribution of alpha activity to the waking EEG is a clear example of an EEG trait that is both highly variable among individuals and highly stable over time within an individual. Moreover, twin and family studies have demonstrated that its variation is determined mainly by genetic factors. In fact, the differences in alpha activity found for an adult studied on two different occasions are similar to the differences (or rather the lack thereof) observed within a MZ twin pair (Lykken et al., 1974). One relatively common alpha activity variant, "low voltage alpha," seems to follow a simple Mendelian mode of inheritance with high penetrance (Anokhin et al., 1992; Vogel, Schalt, Kruger, Propping, & Lehnert, 1979). Genetic linkage studies could map the low voltage alpha trait to chromosome 20q in some families (Anokhin et al., 1992; Steinlein, Anokhin, Yping, Schalt, & Vogel, 1992; Steinlein, Fischer, Keil, Smigrodzki, & Vogel, 1992). This remains, surprisingly, the only genetic study in humans aimed at uncovering the genes contributing to an observed EEG trait. We will have to await the results of several currently ongoing genome-wide association studies (GWAS) to learn about the allelic variants that shape the EEG in the general population.

Meanwhile, a handful of candidate gene studies do convincingly illustrate that known polymorphisms can importantly impact the spectral composition of the EEG in humans (Landolt, 2011). For example, a single nucleotide polymorphism at position 22 (22G > A) in the gene encoding the enzyme adenosine deaminase (*Ada*) affects the contribution of delta activity to the EEG of both NREM and REM sleep with higher levels reached in heterozygous G/A allele carriers than in homozygous G/G carriers (Bachmann, Klaus, et al., 2011; Retey et al., 2005). The G/A genotype is associated with lower ADA activity (Riksen et al., 2008) and thus could result in altered extracellular adenosine levels (Hirschhorn, Yang, Israni, Huie, & Ownby, 1994). Also a 1083T > C polymorphism in the gene encoding the adenosine A2A receptor (*Adora2a*) affects the EEG (Bodenmann et al., 2011; Retey et al., 2007). In the theta/low-alpha frequency range (7–10 Hz) subjects with the T/T genotype displayed higher EEG power than C/C subjects during waking as well as during sleep. These results are of interest in the context of a role of the adenosine signaling pathway in the homeostatic regulation of sleep (Benington & Heller, 1995;

Landolt, 2008; Porkka-Heiskanen, Kalinchuk, Alanko, Urrila, & Stenberg, 2003).

Another example is the gene encoding the catecholamine-metabolizing enzyme catechol-O-methyltransferase (*Comt*) for which a common polymorphism exists altering an amino acid in its protein at codon 158 from valine to methionine (Val158Met). Val/Val individuals show higher COMT activity and lower dopaminergic signaling in prefrontal cortex than Met/Met individuals (Akil et al., 2003; Chen et al., 2004; Slifstein et al., 2008) and EEG activity in the alpha/sigma (11–13 Hz) frequency range is lower in Val/Val compared to Met/Met carriers (Bodenmann & Landolt, 2010; Bodenmann, Rusterholz, et al., 2009; Bodenmann, Xu, et al., 2009). A fourth example concerns a nonsynonymous Val66Met polymorphism in the gene encoding brain-derived neurotrophic factor (*Bdnf*) that was also found to affect the EEG (Bachmann, Klein, et al., 2012). Alpha activity in wakefulness doubled in Val/Val subjects compared to Val/Met carriers. In REM sleep, besides alpha, EEG activity in the theta and sigma frequency bands was also higher in Val/Val carriers. In NREM sleep EEG activity in the entire delta and theta range was enhanced in the Val/Val compared to Val/Met carriers while power in the alpha and sigma frequencies was reduced. The Val66Met polymorphism of *Bdnf* is associated with a pronounced neuroanatomical phenotype affecting many brain areas including grey matter volume in hippocampus and prefrontal cortical areas (Toro et al., 2009). This again illustrates the possibility that developmental differences in brain maturation underlie the trait-like individual EEG "fingerprints." Finally, homozygous carriers of five copies of a tandem-repeat in the circadian gene Period-3 ($Per3^{5/5}$) display higher EEG delta power during NREM sleep, higher EEG alpha power in REM sleep, and higher theta/alpha activity compared to homozygous carriers of only four repeats ($Per3^{4/4}$) (Dijk & Archer, 2010; Viola et al., 2007). The molecular functional relevance of this tandem-repeat polymorphism remains to be identified.

GENETICS OF THE EEG: MOUSE STUDIES

In genetics, the segregation of a trait of interest is followed in recombinant offspring and subsequent mapping aims at identifying the underling genes using polymorphic markers spanning the entire genome. At present, so-called single nucleotide polymorphism (SNP) arrays, which contain several hundred thousands of SNPs, are currently the norm to genotype individual mice or mouse lines and guide mapping of genes or genomic regions (i.e., quantitative trait loci or QTL) co-segregating with the traits of interest. In such genome-wide genetic screens no *a priori* assumptions are made on the gene systems involved allowing for the discovery of novel genes and gene pathways. QTL analysis is used to genetically dissect complex traits such as sleep and the EEG because it allows mapping of multiple naturally occurring allelic variants or gene mutations each contributing with small effect to the phenotypic variance (Abiola et al., 2003; Darvasi, 1998; E. S. Lander & Botstein, 1989). QTL analysis can be performed in a variety of segregating mouse populations including inter- and backcross and recombinant inbred (RI) strains of mice. Two or more inbred mouse strains differing for the trait of interest are crossed and their F1 offspring are then either intercrossed to generate F2 offspring or backcrossed to one of the progenitor strains to generate backcross populations. RI sets are generated by inbreeding F2 mice until homozygosity, thereby "fixing" a unique set of recombinations in several inbred lines. A promising set of RI lines, the "collaborative cross" (CC), is derived from eight inbred lines of mice in a balanced breeding design (Churchill et al., 2004). Inbreeding and genotyping of several hundred CC lines has been finalized and awaits phenotyping of EEG traits (Philip et al., 2011). With this number of lines, number of variable alleles, and available high-density SNP markers, mapping small QTLs (explaining 5% of the variance) at a 1-cM resolution is thought to be feasible (Valdar, Flint, & Mott, 2006). QTL analyses can also be performed in populations of outbred mice requiring the genotyping and phenotyping of each individual mouse (Yalcin et al., 2010). As such, using outbred mice is equivalent of a GWAS study in humans.

Besides QTL analysis, which aims at identifying naturally occurring allelic variants or gene mutations affecting a trait of interest, in mutagenesis the implication of specific genes in a phenotype is assessed by randomly inducing point-mutations using mutagens like N-ethyl-N-nitrosurea (ENU). With high-throughput screening of several hundreds of offspring for either dominant or recessive mutations, a major effect on a given trait can be identified. The feasibility of this approach in the mouse was demonstrated by the isolation of the circadian gene *Clock* (King et al., 1997; Vitaterna et al., 1994). Thus far, no attempts have been made to use this approach for EEG traits although, through phenotyping for abnormal locomotor activity and gait, mutagenesis screens identified a mutation in the gene encoding the neuronal voltage-gated sodium channel *Scn8a* to cause EEG spike-wave discharges (Papale et al., 2009).

Genetic studies of sleep in the mouse were pioneered by Valatx in the early 1970s (Valatx & Bugat, 1974; Valatx, Bugat, & Jouvet, 1972). Initial comparisons of inbred strains of mice revealed that many aspects of sleep greatly differed among strains and, as in humans, heritabilities for sleep durations and its distribution between 40% and 60% were obtained. The EEG spectral profiles also greatly differed among inbred strains. In a panel of six strains of mice we identified large differences in the relative contribution of EEG delta oscillations to the NREM sleep EEG and

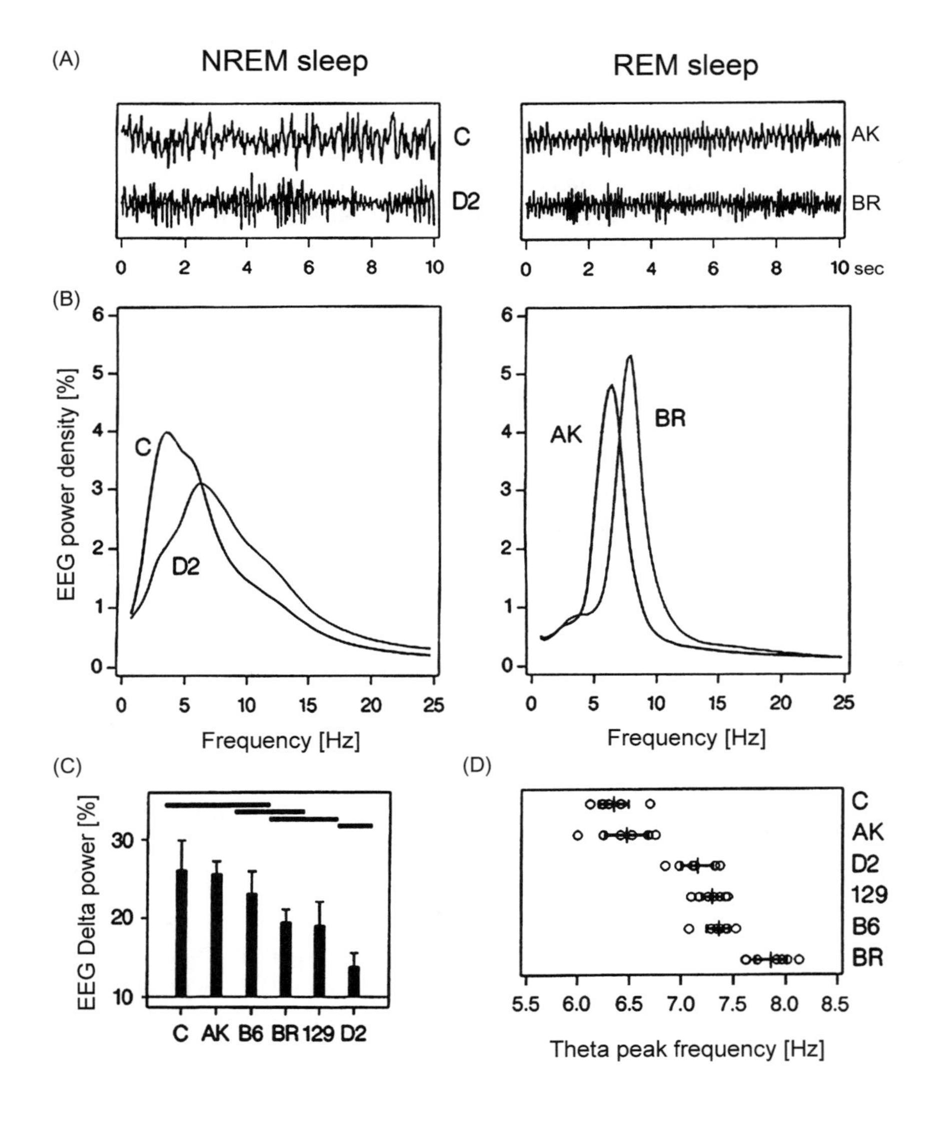

in the prevailing frequency of EEG theta oscillations during REM sleep (Franken, Malafosse, & Tafti, 1998) (Fig. 4.2). Like human twin pairs, mice of a particular inbred strain can be considered genetically identical clones that differ from other inbred strains. Unlike humans, mice are amenable to detailed genetic and subsequent functional analysis, and different approaches can, and have been, successfully employed to find genes implicated in sleep and the EEG.

Almost 20 years after Valatx's group collected sleep data in a panel of CXB RI mice, we used these data to report the first QTL sleep study (Tafti et al., 1997). CXB RI mice are derived from the inbred strains BALB/cBy (C) and C57BL/6By (B). Due to the small number of strains (i.e., 7 lines) the results had no power in yielding significant QTLs. Two years later, in a larger CXB RI set consisting of 13 lines, Toth and Williams identified several QTLs related to both NREM and REM sleep duration, none of them significant at a genome-wide level (Toth & Williams, 1999). Subsequent mapping of sleep-related QTLs in a set of BXD RI lines was more successful (Andretic, Franken, & Tafti, 2008). BXD RIs are derived from the inbred strains C57BL/6J (B) and DBA/2J (D). When

Figure 4.2 ***Spectral composition of the sleep EEG is determined by genetic factors.*** *Panels A:* Visual inspection of the EEG signals in a panel of six inbred strains of mice readily identified qualitative differences in the main contributing frequency components; the prevalence and amplitude of EEG slow waves during NREM sleep (left panel) is visibly reduced in DBA/2J (D2) mice as compared to Balb/cByJ (C) mice rendering the EEG "faster" in D2 mice. Also the frequency of the theta oscillations that characterize REM sleep (right panel) is visibly faster ("denser") in C57BR/cdJ (BR) mice as compared to AKR/J (AK) mice. *Panels B:* These qualitative EEG differences can be quantified with spectral analyses [FFT; average spectra for all 4 s epochs scored as NREM sleep (left) or REM sleep (right panel) in a 24 h baseline recording; n = 7/strain] and clearly show the reduction of the contribution of delta activity in D2 mice compared to C mice during NREM sleep and the right shift of the theta peak in BR mice compared to AK mice during REM sleep. *Panel C:* Quantification of the relative contribution of EEG delta power to the EEG spectrum during NREM sleep in all six strains. The horizontal bars on top group strains for which EEG delta power similarly contributed to the NREM sleep EEG. The B-D strain difference led to the mapping of a polymorphism in the promoter of the *Rarb* gene in a panel of BXD recombinant inbred mice (Maret et al., 2005). *Panel D:* Individual and mean (±2SEM) values of theta peak frequency for all six inbred strains. Three significantly distinct groups can be distinguished for which individual theta peak frequency values did not overlap: slow theta (AK and C; <6.8 Hz), fast theta (D2, C57BL/6J (B6), and 129P/Ola (129); 6.8–7.6 Hz), and "very" fast theta (BR; >7.6 Hz) mice. The B-C strain difference led to the mapping in inter- and backcross panels derived from B and C mice of a spontaneous deletion mutation in this C substrain (i.e., Balb/cByJ) in the gene *Acads* (Tafti et al., 2003). *Original data published in Franken et al. (1998).*

we initiated these experiments we had access to 25 BXD RI lines. In the meantime, this resource has been substantially augmented and a total of some 90 lines have become available making this a powerful resource for mapping QTLs with moderate to large effects (Peirce, Lu, Gu, Silver, & Williams, 2004). In the BXD RI set, we identified a genome-wide significant QTL on chromosome 14 for the contribution of EEG delta activity to the NREM sleep EEG. Subsequent haplotype mapping in several unrelated inbred strains and the candidate-gene approach identified retinoic-acid receptor beta (*Rarb*) as the underlying gene thereby implicating retinoic-acid signaling in modulating cortical synchrony during NREM sleep (Maret et al., 2005).

Another example of a successful QTL EEG study concerned the frequency of theta oscillations characteristic of REM sleep. Inbred strains vary greatly for this trait and "slow" and "fast" theta strains could be distinguished with 6.75 Hz as the frequency separating the two groups (Franken, et al., 1998; Tafti et al., 2003) (Fig. 4.2). The segregation of this trait was followed in a panel of inter- and backcrosses between BALB/cByJ, a "slow-theta" strain, and C57BL/6J, a "fast-theta" strain. A single gene was identified on chromosome 5 that was tightly linked to theta frequency. Subsequent fine mapping through selective phenotyping revealed short-chain acyl-coenzyme A dehydrogenase (*Acads*) as the responsible gene. BALB/cByJ mice acquired a deficiency in *Acads* that slows theta (Tafti et al., 2003). The isogenic substrain, BALB/cBy, which does not carry this spontaneous mutation, has "fast" theta oscillations, further confirming *Acads* as the gene causing this EEG trait.

These two examples demonstrate that QTL analysis, as a "genetics" approach, can be successful in identifying signaling pathways previously not thought to play a role in the generation of neuronal oscillations. Retinoic-acid receptors are important in brain development (Jacobs et al., 2006; Lane & Bailey, 2005) further supporting the idea that the EEG spectral profile could be used to index brain maturation. Alternatively, retinoic-acid receptors are implicated in several signaling pathways in the adult brain (Bremner & McCaffery, 2008; Krezel et al., 1998) and, like the *Comt* polymorphism mentioned above (Bodenmann, Rusterholz, et al., 2009), could have effects on EEG delta power through altered dopaminergic neurotransmission (e.g., Dimpfel, 2008; Kitaoka et al., 2007). The role of fatty-acid beta oxidation, in which *Acads* plays a key role, in determining the frequency of hippocampal theta oscillations is also unexpected because this pathway was not thought to be of importance in the adult brain.

Candidate-gene approaches also referred to as "reverse genetics" (i.e., from "genotype-to-phenotype"), are, strictly speaking, not genetic approaches in the sense that they do not allow for the identification of allelic variants or mutations underlying a trait. Since the introduction of the term reverse genetics, genetic approaches have sometimes been dubbed "forward genetics" (i.e., from "phenotype-to-genotype"). In the mouse, reverse genetics entail genetically altering a gene of interest and studying the effects of this intervention on a trait under investigation. Mice can be engineered to either lack a functional gene, overexpress a gene product, have an altered gene construct inserted, or express genes from other species. Extensive use of these techniques has been made in sleep research, which has been adequately reviewed elsewhere (Franken & Tafti, 2003; Wisor & Kilduff, 2005). Most concern genes involved in neurotransmitter signaling, endocrine and para-/autocrine signaling, and ion channels. Not surprisingly, disruption of ion channels can result in striking differences in the EEG spectral profiles as observed in mice lacking the small conductance calcium-activated potassium channel *Kcnn2* (Cueni et al., 2008). Although transgenic animals are mostly used to confirm the implication of existing pathways in sleep and the EEG complementing pharmacological and neuroanatomical approaches, sometimes unexpected results are obtained that lead to new insights. An example of this is the discovery that the transcription factors involved in circadian rhythm generation (or clock genes) also play a role in the homeostatic regulation of sleep (see below).

REGULATION OF SLEEP: TIME-OF-DAY VERSUS TIME-SPENT-AWAKE

Two main processes are generally considered when studying the regulation of sleep: a circadian process that sets internal time-of-day and assures proper entrainment of behavior and physiology to the daily light-dark cycle, and a homeostatic process that tracks and signals the propensity or need for sleep (Borbely, 1982; Daan et al., 1984). In mammals, a self-sustained circadian oscillation is generated in the suprachiasmatic nucleus (SCN) of the hypothalamus (Klein, Moore, & Reppert, 1991), which is considered the master circadian clock. The oscillating output of this clock gives time-context to most physiological processes and behaviors including sleep. Thus, the distribution of sleep over the 24-hour day is strongly determined by the circadian system. The homeostatic process tracks sleep need such that the need for sleep and the propensity to initiate sleep increases while awake and decreases as a function of time-spent-asleep. The two processes are thought to be generated independently but their interaction determines the timing, duration, and quality of both sleep and wakefulness (Dijk & Franken, 2005). So-called forced-desynchrony protocols have been especially instrumental in quantifying the interaction between the two processes (Dijk & Czeisler, 1994, 1995). These studies demonstrated that the circadian system is more than a device dictating when to sleep; it actively generates a sleep-wake propensity rhythm that is timed to oppose homeostatic changes in sleep drive, enabling us to stay awake and alert throughout the day despite an accumulating need for sleep, and asleep during the night despite a waning of sleep need (Dijk & Franken, 2005).

Of the various aspects of sleep that are homeostatically regulated, EEG delta power during NREM sleep most reliably varies as a function of time awake and asleep (Dijk, Beersma, & Daan, 1987; Franken et al., 2001; Franken, Tobler, & Borbély, 1991; Tobler & Borbely, 1986; Werth, Dijk, Achermann, & Borbely, 1996). In mouse, rat, and human, changes in EEG delta power are so predictable that its time course can be mathematically calculated in detail solely based on the sleep-wake distribution both under baseline conditions and after sleep deprivation (Achermann & Borbely, 2003; Franken et al., 2001; Franken, Tobler, et al., 1991; Huber, Deboer, & Tobler, 2000a) (Fig. 4.3), further supporting the view that changes in EEG delta power mainly reflect a sleep-wake dependent, homeostatic process. Studies in animals rendered arrhythmic by lesioning the SCN or through a shift of the light-dark cycle show that the sleep-deprivation induced increase in EEG delta power is unaffected and thus does not depend on a functioning circadian system (Easton, Meerlo, Bergmann, & Turek, 2004; Larkin, Franken, & Heller, 2002; Trachsel, Edgar, Seidel, Heller, & Dement, 1992). Results from the above-mentioned forced desynchrony experiments in humans confirmed that the circadian contribution to the time course of EEG delta power during NREM sleep is very small albeit not zero (Dijk & Czeisler, 1995). These and other observations let to the established notion that sleep homeostasis and circadian rhythm generation are separate processes.

It should be noted, however, that because changes in EEG delta power are so predictable and salient, often the homeostatic regulation of sleep is equated with the sleep-wake dependent changes in delta power thereby overlooking the homeostatic regulation of other aspects of sleep such as the duration of sleep, especially that of REM sleep. This is also reflected in the prevailing hypotheses concerning sleep function that exclusively

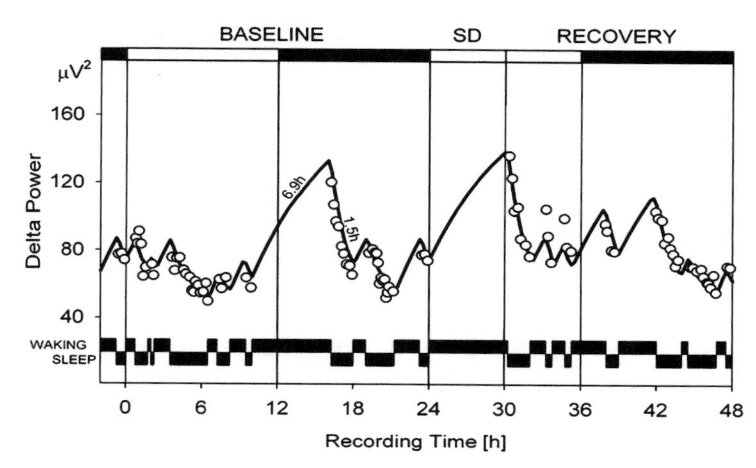

Figure 4.3 ***Changes in EEG delta power during NREM sleep can be accounted for by the sleep-wake distribution and index homeostatic sleep need.*** EEG delta power during NREM sleep depends on the prior sleep-wake distribution in that levels are high after periods of extended wakefulness; the longer the period of waking, the higher subsequent EEG delta power is. During NREM sleep, EEG delta power decreases exponentially. This is illustrated here for one individual male Balb/cByJ mouse. Open symbols depict absolute values of EEG delta power reached in individual NREM sleep episodes (>5 min) during a 24 h baseline recording followed by a 6 h sleep deprivation (SD) starting at light onset, and 18 h of recovery sleep. Both spontaneous (e.g., between 10–16 h) and enforced (24–30 h) periods of wakefulness of similar duration are followed by similar levels of EEG delta power. Note the rapid decline (recovery?) during periods of sleep. These dynamics suggest that EEG delta power reflects a yet to be identified homeostatically regulated recovery process, Process S (Daan et al., 1984). The dynamics of this process can be simulated iteratively by assuming that the need for NREM sleep increases during wakefulness and REM sleep (both states devoid of EEG delta oscillations) and decreases during NREM sleep according to exponential (saturating) functions. The simulated levels of Process S (solid line) predict the observed EEG delta power values in great detail and >80% of the variance in EEG delta power can thus be accounted for by the sleep-wake distribution. Optimal fit to empirical data in this mouse was obtained with 6.9 h and 1.5 h for the time constants of the buildup and decrease, respectively. The time constants quantifying the increase of Process S varied greatly among inbred strains (Franken et al., 2001). The black horizontal bars on top indicate the 12 h dark periods. *Graph reprinted from Franken et al. (2001) with permission.*

concern the dynamics of EEG delta power (Tononi & Cirelli, 2006). A clear dissociation of the homeostatic regulation of sleep amount and that of EEG delta power is illustrated with the following sleep-deprivation experiment carried out in the rat (Franken, Dijk, Tobler, & Borbély, 1991). Twenty-four hours without sleep resulted in the expected and immediate increase in EEG delta power that quickly subsided over the first 4 hours of the recovery period. The duration of NREM sleep was also increased but this increase lasted for most of the 48 hours for which recovery was monitored. As a result of the increase in NREM sleep time, EEG delta power fell to values below those reached under baseline conditions (i.e., "negative rebound"). Although both the immediate positive and subsequent negative rebound in EEG delta power in this study could be readily explained based on the altered sleep-wake distribution (Franken, Tobler, et al., 1991), the following question presents itself: If EEG delta power indeed reflects homeostatic sleep need then why should animals continue to sleep more compared to baseline when sleep need is below baseline? Equally overlooked are the low predictive value of EEG delta power in modeling the effects of longer-term sleep restrictions on cognitive performance and attention (Van Dongen, Maislin, Mullington, & Dinges, 2003).

While variations in EEG delta power during NREM sleep measured over the day are mostly sleep-wake driven other frequency components of NREM sleep EEG are also influenced by circadian factors. Notably EEG activity in spindle or sigma frequency range reveals, besides a marked sleep-wake dependent variation, an equally marked circadian dependent variation (Dijk & Czeisler, 1995). For the waking EEG, modulation of alpha, theta, and beta activity according to circadian time were reported (Cajochen, Wyatt, Czeisler, & Dijk, 2002). Also in the rat a circadian modulation of EEG activity in specific frequency bands can be observed in all three sleep-wake states (Yasenkov & Deboer, 2011).

GENETICS OF THE HOMEOSTATIC REGULATION OF SLEEP

Comparisons of the sleep-wake dependent dynamics in EEG delta power among six inbred strains of mice revealed that the rate at which homeostatic sleep need increases during wakefulness varied greatly according to genetic background (Franken et al., 2001) (Fig. 4.4). In the BXD panel of RI lines introduced above we followed the segregation of the increase in delta power both after sleep deprivation and after sleep onset under baseline conditions. Both traits yielded overlapping QTLs on chromosome 13 that reached genome-wide significance. We named this QTL *Dps1* (for delta-power-sleep-1) (Fig. 4.4). The *Dps1* QTL is specific for the homeostatic regulation of EEG delta power in that the sleep-deprivation induced rebounds in NREM sleep and in REM sleep duration did not map to this region (our unpublished data). Moreover, the *Dps1* QTL differed from

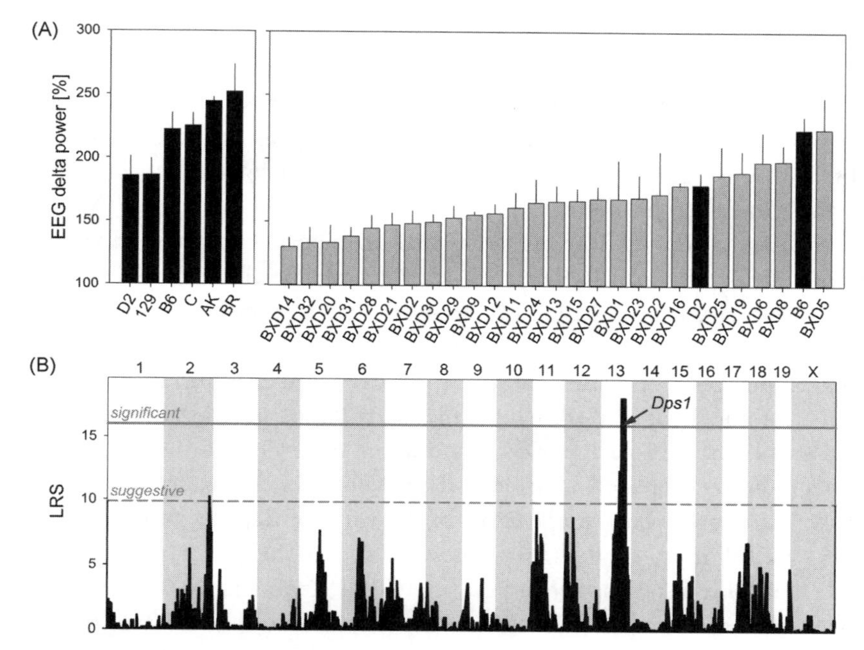

Figure 4.4 ***The homeostatic regulation of NREM sleep EEG delta power is under genetic control.*** *Panels A:* Inbred strains greatly differ for the level of EEG delta power reached after a 6h sleep deprivation (SD). An initial screen of 6 strains (left panel; n=7/strain, 10–12 weeks old males) showed that in DBA/2J (D2) and 129P/Ola (129) mice lower EEG delta power values were reached in the first 30min of recovery sleep after SD as compared to AKR/J (AK) and C57BR/cdJ (BR) mice while in C57BL/6J (B6) and Balb/cByJ (C) intermediary levels were reached. In 25 recombinant inbred (RI) strains of mice derived from the B and D parental strains (i.e., BXD1-31; n=4–7/strain; right panel) an even larger variability in this increase in EEG delta power after SD was observed surpassing the parental differences (black bars), especially towards lower range values. Values were expressed as % of individual values reached in the last 4h of the preceding baseline light period (=100%), that is, the time the lowest levels are reached in the baseline. *Panel B:* QTL mapping of the segregation of the EEG delta power level reached after SD in the BXD RI panel (parentals included) yielded two loci; one "suggestive" on chromosome 2 and one "significant" on chromosome 13 (the *Dps1* QTL, see text). Significant and suggestive thresholds (horizontal grey lines) refer to likelihood ratio statistic (LRS) values corresponding to genome-wide probability levels of 5% and 63%, respectively, of falsely rejecting the null hypothesis of no linkage (E. Lander & Kruglyak, 1995). LRS quantifies the association between trait differences and differences in a particular DNA sequence and is calculated as the ratio of the probability of an association versus no association (e.g., an LRS of 3 indicates a 1000 times more likelihood of an association). The *Dps1* QTL explained 49% of the variance in the SD trait suggesting the presence of a "major" gene governing the increase of the homeostatic need for sleep during wakefulness. *Data in panels A taken from Franken, Malafosse, & Tafti (1999) and Franken et al. (2001), respectively. Association data depicted in panel B were generated with the use of WebQTL (www.genenetwork.org; Record ID: 10143).*

the chromosome 14 QTL obtained in the same mice for the contribution of EEG delta power to the NREM sleep EEG (see above under *Rarb*). This indicates that different aspects of the same variable can be governed by different genes. This also indicates that differences in EEG delta power observed among individuals do not necessarily equate to functional differences in the sleep homeostat.

In an extensive set of micro-array analysis experiments we then searched for genes for which mRNA levels in the brain paralleled the sleep-wake dependent changes in EEG delta power; that is, their expression should decrease (or increase; we did not impose an *a priori* direction-of-change as long as changes were consistent with the sleep-wake distribution) over the course of the light period when mice predominantly sleep, and increase during the active or dark period. In addition, these changes in gene expression should be increased with sleep deprivation dependent on the duration of the deprivation but independent of the time-of-day the deprivation was performed. Finally, changes in expression should be genotype dependent such that inbred strains that responded with a larger increase in EEG delta power after sleep deprivation should also have a larger increase in mRNA. The transcript *Homer1a* matched all these requirements (Maret et al., 2007) (Fig. 4.5). Of great interest is that *Homer1a* is also the best candidate gene in the *Dps1* region (Mackiewicz, Paigen, Naidoo, & Pack, 2008; Maret et al., 2007). Formal genetic proof confirming the involvement of this isoform of *Homer1* in the *Dps1* QTL could come from mice lacking *Homer1a*, specifically. Interest in *Homer1a* comes from its role in homeostatic synaptic scaling (Hu et al., 2010) and neuroprotection (Szumlinski, Kalivas, & Worley, 2006), both suggested as possible functions of sleep (Mongrain et al., 2010; Tononi & Cirelli, 2006).

This example demonstrates that only the combination of several approaches can yield new and meaningful insights. The combination of genome, transcriptome, and phenotypic data at multiple levels of organization to understand complex traits has been referred to as "systems genetics." With respect to sleep such approaches have only now begun to be implemented (Millstein et al., 2011; Winrow et al., 2009).

A NON-CIRCADIAN ROLE FOR CLOCK GENES IN SLEEP HOMEOSTASIS

Although sleep homeostasis and circadian rhythm generation are considered separate processes (see above), several examples indicate the presence

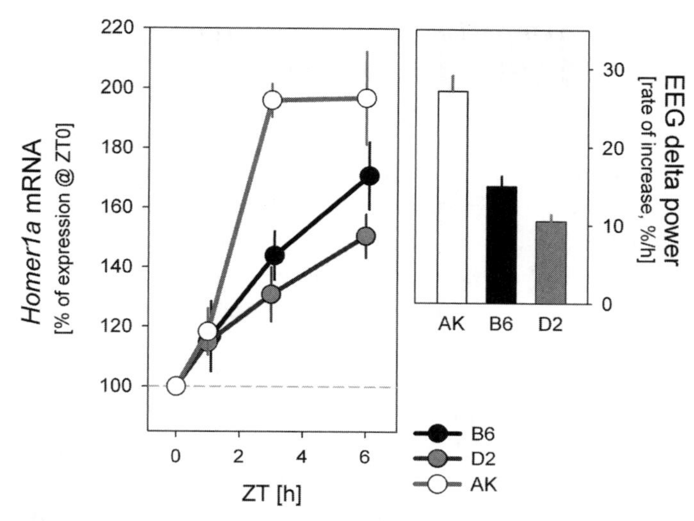

Figure 4.5 ***The genotype specific rate of increase in Homer1a expression matched the dynamics of EEG delta power.*** *Left panel:* The time course of *Homer1a* mRNA levels in the forebrain of male mice of three inbred strains after 1, 3, and 6 h of sleep deprivation starting at light onset (0 h; n = 4/strain/time-point). *Homer1a* expression in AKR/J (AK) mice accumulated at a faster rate than in C57BL/6J (B6) and DBA2/J (D2) mice. In AK mice after 3 h of sleep deprivation maximal levels of mRNA were already reached while in the remaining two strains this increase was more gradual reaching lower levels. *Right panel:* These strain differences in the increase rate of *Homer1a* match the strain specific increase rate in EEG delta power during wakefulness (n = 7/strain). The sleep-wake dependent changes in EEG delta power during NREM sleep are thought to reflect changes in sleep need, thereby implying that *Homer1a*, which is localized within the *Dps1* QTL (Fig. 4.4), is, at least, a reliable correlate of homeostatic sleep need. *Data taken from Franken et al. (2001) and Maret et al. (2007). Figure reprinted from Andretic et al. (2008), with permission.*

of at least some "cross talk" between the two; for example, sleep deprivation is able to phase shift the circadian clock in hamsters (Antle & Mistlberger, 2000) and firing rates of SCN neurons decrease during NREM sleep and are negatively correlated with the levels of EEG delta power attained during this state (Deboer, Detari, & Meijer, 2007; Deboer, Vansteensel, Detari, & Meijer, 2003). Our finding that circadian clock genes play a role in sleep homeostasis further obscures this distinction and suggest that the same molecular circuitry used to set internal time-of-day might be equally utilized to track and anticipate sleep need (Franken & Dijk, 2009).

Mutagenesis screens in *Drosophila* and the mouse identified the genes *Period* and *Clock*, respectively (King et al., 1997; Konopka & Benzer, 1971; Reddy et al., 1984; Vitaterna et al., 1994). Subsequent studies revealed that these transcriptional regulators engage in molecular negative feedback loops that are thought to underlie circadian rhythm generation at the cellular level (Lowrey & Takahashi, 2004). The core of this self-sustained molecular oscillation in mammals consists of the positive elements *Clock*, *Npas2*, and *Bmal1* and of the negative elements *Period1* and *-2* (*Per1, Per2*) and *Cryptochrome1* and *-2* (*Cry1, Cry2*). BMAL1 can form transcriptionally active heterodimers by partnering with either CLOCK or NPAS2. Both dimers initiate the transcription of the *Per* and *Cry* genes. After translation and reentry into the cell nucleus, PER and CRY protein complexes suppress CLOCK:BMAL1- and NPAS2:BMAL1-mediated transcription thus providing the negative feedback enabling the initiation of a new cycle of transcription-translation. The involvement of these clock genes in circadian rhythm generation has been demonstrated by constructing mice carrying targeted disruptions for one or a combination of these genes. Thus mice lacking *Bmal1*, *Clock* and *Npas2*, *Cry1* and *Cry1*, or *Per1* and *Per2* all lack circadian organization of overt behavioral rhythms when kept under constant dark conditions (Lowrey & Takahashi, 2004).

In an attempt to further establish the independence of circadian and sleep homeostatic processes we examined sleep in *Cry1,2* knockout mice (*Cry1,2*$^{-/-}$) (Wisor et al., 2002). As expected, these mice are behaviorally arrhythmic under constant conditions (van der Horst et al., 1999; Vitaterna et al., 1999). Besides this anticipated circadian phenotype these mice also had an unexpected and distinct homeostatic phenotype with more time spent in NREM sleep, more consolidated sleep, and higher EEG delta power during baseline while the rebound in EEG delta power after sleep deprivation was attenuated (Wisor et al., 2002). Aberrant sleep homeostatic phenotypes have been observed for other mouse lines carrying targeted disruptions for one or two clock genes. Mice homozygous for the *Bmal1* deletion showed increases in total sleep time, sleep fragmentation, and EEG delta power under baseline conditions, and an attenuated compensatory response to sleep deprivation (Laposky et al., 2005). For *Clock* mutant mice decreases in NREM sleep time and consolidation were reported under baseline conditions (Naylor et al., 2000). Mice lacking *Npas2* slept less during baseline and the compensatory response in NREM sleep time and EEG delta power (0.75–2.0 Hz) after an 8 hour sleep deprivation was reduced (Franken et al., 2006) (Fig. 4.6). Sleep regulation was also assessed in mice lacking *Per1* and *Per2* although results were inconclusive; single *Per1* and *Per2* knock-out mice showed a smaller increase in EEG delta power after sleep deprivation (Kopp, Albrecht, Zheng, & Tobler, 2002), while in *Per1,2* double mutant mice EEG delta power seemed enhanced to a greater extent after sleep deprivation compared to wild-type mice (Shiromani et al., 2004). Evidence of clock genes being implicated in the homeostatic regulation of sleep was also found in *Drosophila* (Hendricks et al., 2003; Hendricks et al., 2001; Shaw, Tononi, Greenspan, & Robinson, 2002) and in humans (Viola et al., 2007) pointing to a possible evolutionary conserved pathway.

Another type of observation in support of a role for clock genes in sleep homeostasis is that in the forebrain expression of both *Per1* and *Per2* is increased after enforced waking (Franken et al., 2006; Franken, Thomason, Heller, & O'Hara, 2007; Maret et al., 2007; Mongrain et al., 2010; Wisor et al., 2002). *Per1* and *Per2* expression increased linearly as a function of the duration of the time mice were kept awake (Franken et al., 2007) and one study reported a positive correlation between the mRNA changes in *Per1* and *Per2* and EEG delta power (Wisor et al., 2008). High forebrain levels of *Per* expression seem also to be associated with increased sleep need under baseline conditions and under conditions where the sleep-wake distribution was altered (Abe, Honma, Namihira, Masubuchi, & Honma, 2001; Abe et al., 2001; Dudley et al., 2003; Masubuchi et al., 2000; Mrosovsky, Edelstein, Hastings, & Maywood, 2001; Reick, Garcia, Dudley, & McKnight, 2001; Wakamatsu et al., 2001).

In situ hybridization studies revealed that *Per1* and *Per2* expression after sleep deprivation was affected the most in brain areas where the delta as well as sigma/spindle oscillations characteristic of the NREM sleep EEG are generated, that is, the cortex and thalamus (Franken, et al., 2007; Steriade, 2003; Wisor et al., 2008). NPAS2, which seems important for coupling *Per2* expression to the sleep-wake distribution (Franken et al., 2006; Reick et al., 2001), is abundantly expressed in these areas (Zhou et al., 1997) (Fig. 4.7). The EEG during NREM sleep in *Npas2*$^{-/-}$ mice displayed an overall reduction in sigma power (Fig. 4.6). Also EEG activity in delta frequency range was affected with a shift from activity in the slow delta frequencies (1.0–2.25 Hz) to faster delta frequencies (2.25–4.0 Hz), which was especially pronounced during NREM sleep immediately following a prolonged period including an 8 hour sleep deprivation (Franken et al., 2006) (Fig. 4.6). Other studies also reported that slow and fast delta oscillations were differentially modulated by prolonged waking (Deboer, Fontana, &

Tobler, 2002; Huber, Deboer, & Tobler, 2000b). One type of oscillation that contributes to the activity in the delta frequency range originates from thalamocortical neurons (Amzica & Steriade, 1998). When the membrane potential of these neurons reaches levels of hyperpolarization characteristic of deep NREM sleep (stages 3 and 4 in humans, i.e., slow-wave sleep), their frequency becomes faster and their contribution to delta activity at the level of the EEG greater (Amzica & Steriade, 1998; Dossi, Nunez, & Steriade, 1992). This could have contributed to the transient shift to faster delta frequencies immediately after long periods of wakefulness, because

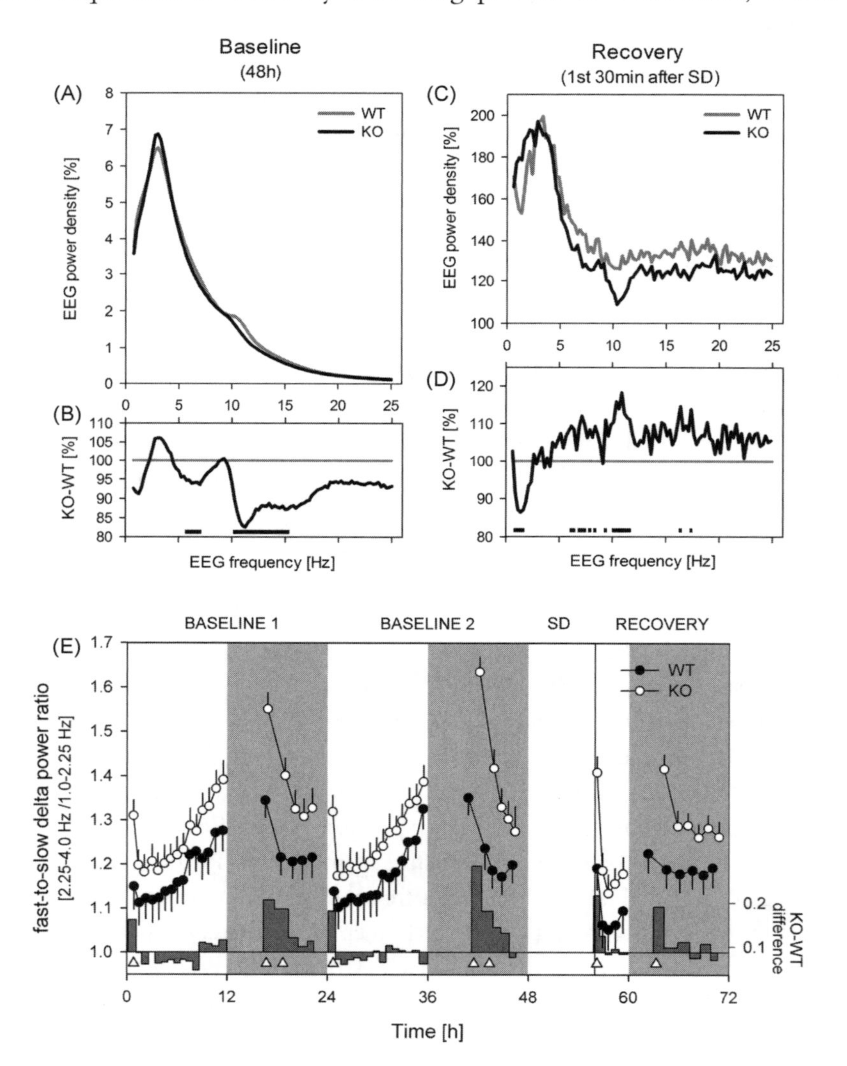

NREM sleep is then deepest, and hyperpolarization greatest (Dossi et al., 1992). Following this conjecture, the higher fast-to-slow delta power ratio in *Npas2*$^{-/-}$ mice suggests that membrane potential of thalamocortical neurons during NREM sleep, on average, is more hyperpolarized, which is consistent with the reduction in sleep spindles that predominantly occur at intermediate levels of membrane hyperpolarization (Steriade, McCormick, & Sejnowski, 1993). These observations suggest an unexpected role for a circadian transcription factor in the generation of EEG rhythms of thalamocortical origin. The mechanisms through which this NPAS2 affects thalamocortical and corticocortical activity deserve further investigation. These analyses also underscore that activity in the delta frequencies does not uniformly respond to prior wake duration. In addition, NPAS2 might play a role in postnatal thalamocortical developmental; *Npas2* is expressed first after postnatal week 1 (Zhou et al., 1997), immediately preceding the time at which, in rats, the first slow waves and spindles appear (Davis, Frank, & Heller, 1999).

Figure 4.6 ***The circadian clock gene NPAS2 modulates thalamocortical oscillations.*** *Panel A*: Mean spectral EEG profiles calculated over two baseline days (48 hours) in *Npas2*$^{-/-}$ (KO; n = 19) and wild-type (WT; n = 19) control mice averaged for all 4-second epochs scored as NREM sleep. EEG spectra were normalized to total EEG power. *Panel B*: Spectral differences as percent change for KO (black line) versus WT (grey line) mice. The largest genotype effect was observed in the spindle frequency range (11–15 Hz; significant differences indicated by the horizontal black bar). The concomitant decrease in EEG power in slow delta frequencies (1.0–2.5 Hz) and the increase in fast delta frequencies (2.5–4.0 Hz) resulted in a significant shift in power within the delta band towards faster frequencies (see panel E). *Panel C*: Spectral changes in the NREM sleep EEG in the first 30 minutes of recovery sleep after an 8-hour sleep deprivation (SD). Significant increases in EEG power were observed in both genotypes over a wide frequency range especially prominent in the delta frequencies. Values expressed as % of corresponding baseline values (=100%). *Panel D:* Genotype differences for the increase in EEG power during NREM sleep after SD. KO mice displayed a smaller increase in the slow delta frequencies specifically, while the fast delta activity did not differ thereby even further augmenting the fast-to-slow delta power ratio (see panel E). The increase in theta and low sigma activity was larger in KO mice. *Panel E:* Genotype- and time-dependent changes in the fast-to-slow delta power ratio during NREM sleep. KO values were, in general, higher than WT and ratios were high immediately after a period of sustained wakefulness (i.e., first values of the light period, first value of the sleep episode during the dark (i.e., the "nap"), and first value after SD). At these sleep-onset times the ratio was further increased in KO mice (dark grey areas) and genotypic differences became significant (triangles). Light-grey areas denote 12-hour dark periods. *Data in panels A–D were taken from Franken et al. (2006). Panel E reprinted from Supplementary data (Franken et al., 2006), with permission.*

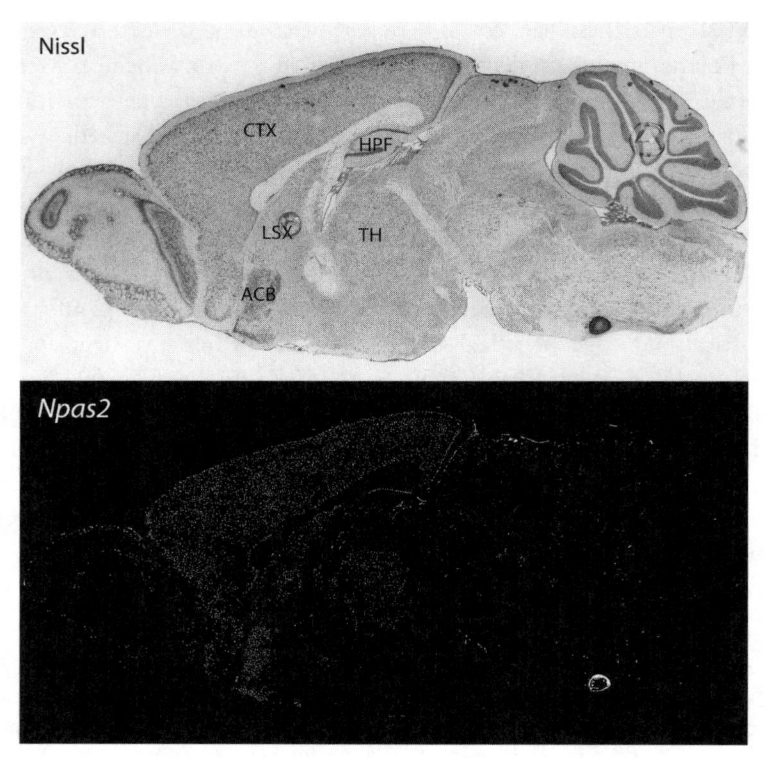

Figure 4.7 Npas2 ***in the mouse brain is most abundantly expressed in the cerebral cortex.*** *Upper panel:* Nissl stain for a sagittal section of the brain of a male C57Bl/6J mouse aged 8 weeks. *Lower panel: In situ* hybridization for *Npas2* in the same section. Expression was limited to the forebrain and was most abundant in the neocortex, nucleus accumbens, and thalamus confirming earlier work (Garcia et al., 2000; Zhou et al., 1997). Abbreviations: HPF = Hippocampal formation, LSX = Lateral septal complex, ACB = Nucleus accumbens, CTX = cerebral cortex (neocortex), TH = thalamus. *Images (ID: 70593327) were downloaded from the Allan brain atlas (www.brain-map.org) with permission.*

Twin studies in humans and comparisons of inbred strains of mice, candidate-gene studies in humans for natural occurring polymorphisms and in mice carrying genetically engineered allele constructs, and QTL and linkage studies all demonstrate that the spectral composition of the EEG both during sleep and wakefulness greatly depend on genetic factors. These factors are likely to underlie the surprisingly stable EEG spectra within an individual over time (EEG "fingerprint"). The genetically determined differences in rhythmic brain activity could be related to the genetic programs in place for "wiring" the brain, that is, guiding functional neuronal connections during development and maturation (Buchmann et al., 2011; Mitchell, 2007). Genetically determined EEG differences need not be functional in the context of sleep-wake regulation in that, for example, differences in EEG delta power are not necessarily related to altered homeostatic regulation of sleep as was illustrated in a panel of RI mouse where one QTL affected the level of EEG delta power while another modulated the dynamics of the sleep-wake dependent changes in EEG delta power (Franken et al., 2001; Maret et al., 2005). Results of forward, reverse, and molecular genetic approaches applied to sleep homeostasis have contributed to hypotheses on sleep function as illustrated for *Homer1a* (Maret et al., 2007) and for clock genes (Franken & Dijk, 2009). This integrated or systems genetics approach will be instrumental in further unraveling the genes and gene pathways shaping the EEG in the mouse, human, and, possibly, *Drosophila* (Marley & Baines, 2011; Nitz, van Swinderen, Tononi, & Greenspan, 2002).

REFERENCES

Abe, H., Honma, S., Namihira, M., Masubuchi, S., & Honma, K. (2001). Behavioural rhythm splitting in the CS mouse is related to clock gene expression outside the suprachiasmatic nucleus. *The European Journal of Neuroscience*, *14*(7), 1121–1128.

Abe, H., Honma, S., Namihira, M., Masubuchi, S., Ikeda, M., Ebihara, S., et al. (2001). Clock gene expressions in the suprachiasmatic nucleus and other areas of the brain during rhythm splitting in CS mice. *Brain Research. Molecular Brain Research*, *87*(1), 92–99.

Abiola, O., Angel, J. M., Avner, P., Bachmanov, A. A., Belknap, J. K., Bennett, B., et al. (2003). The nature and identification of quantitative trait loci: A community's view. *Nature Reviews. Genetics*, *4*(11), 911–916. doi:10.1038/nrg1206.

Achermann, P., & Borbely, A. A. (2003). Mathematical models of sleep regulation. *Frontiers in Bioscience*, *8*, s683–s693.

Akil, M., Kolachana, B. S., Rothmond, D. A., Hyde, T. M., Weinberger, D. R., & Kleinman, J. E. (2003). Catechol-O-methyltransferase genotype and dopamine regulation in the human brain. *The Journal of Neuroscience : The Official Journal of the Society for Neuroscience*, *23*(6), 2008–2013.

Ambrosius, U., Lietzenmaier, S., Wehrle, R., Wichniak, A., Kalus, S., Winkelmann, J., et al. (2008). Heritability of sleep electroencephalogram. *Biological Psychiatry*, *64*(4), 344–348. doi:10.1016/j.biopsych.2008.03.002.

Amzica, F., & Steriade, M. (1998). Electrophysiological correlates of sleep delta waves. *Electroencephalography and Clinical Neurophysiology*, *107*(2), 69–83.

Andretic, R., Franken, P., & Tafti, M. (2008). Genetics of sleep. *Annual Review of Genetics*, *42*, 361–388. doi:10.1146/annurev.genet.42.110807.091541.

Anokhin, A., Steinlein, O., Fischer, C., Mao, Y., Vogt, P., Schalt, E., et al. (1992). A genetic study of the human low-voltage electroencephalogram. *Human Genetics*, *90*(1–2), 99–112.

Antle, M. C., & Mistlberger, R. E. (2000). Circadian clock resetting by sleep deprivation without exercise in the Syrian hamster. *The Journal of Neuroscience*, *20*(24), 9326–9332.

Bachmann, V., Klaus, F., Bodenmann, S., Schafer, N., Brugger, P., Huber, S., et al. (2011). Functional ADA polymorphism increases sleep depth and reduces vigilant attention in Humans. *Cerebral Cortex*. doi:10.1093/cercor/bhr173.

Bachmann, V., Klein, C., Bodenmann, S., Schäfer, N., Berger, W., Brugger, P., et al. (2012). The BDNF Val66Met polymorphism modulates sleep intensity: EEG frequency- and state-specificity. *Sleep*, *35*(3), 335–344.

Benington, J. H., & Heller, H. C. (1995). Restoration of brain energy metabolism as the function of sleep. *Progress in Neurobiology*, *45*(4), 347–360.

Bodenmann, S., Hohoff, C., Freitag, C., Deckert, J., Retey, J. V., Bachmann, V., et al. (2011). Polymorphisms of ADORA2A modulate psychomotor vigilance and the effects of caffeine on neurobehavioral performance and sleep EEG after sleep deprivation. *British Journal of Pharmacology*. doi:10.1111/j.1476-5381.2011.01689.x.

Bodenmann, S., & Landolt, H. P. (2010). Effects of modafinil on the sleep EEG depend on Val158Met genotype of COMT. *Sleep*, *33*(8), 1027–1035.

Bodenmann, S., Rusterholz, T., Durr, R., Stoll, C., Bachmann, V., Geissler, E., et al. (2009). The functional Val158Met polymorphism of COMT predicts interindividual differences in brain alpha oscillations in young men. *The Journal of Neuroscience*, *29*(35), 10855–10862. doi:10.1523/JNEUROSCI.1427-09.2009.

Bodenmann, S., Xu, S., Luhmann, U. F., Arand, M., Berger, W., Jung, H. H., et al. (2009). Pharmacogenetics of modafinil after sleep loss: Catechol-O-methyltransferase genotype modulates waking functions but not recovery sleep. *Clinical Pharmacology and Therapeutics*, *85*(3), 296–304. doi:10.1038/clpt.2008.222.

Boomsma, D., Busjahn, A., & Peltonen, L. (2002). Classical twin studies and beyond. *Nature Reviews. Genetics*, *3*(11), 872–882. doi:10.1038/nrg932.

Borbely, A. A. (1982). A two process model of sleep regulation. *Human Neurobiology*, *1*(3), 195–204.

Borbely, A. A., Baumann, F., Brandeis, D., Strauch, I., & Lehmann, D. (1981). Sleep deprivation: Effect on sleep stages and EEG power density in man. *Electroencephalography and Clinical Neurophysiology*, *51*(5), 483–495.

Bremner, J. D., & McCaffery, P. (2008). The neurobiology of retinoic acid in affective disorders. *Progress in Neuro-Psychopharmacology & Biological Psychiatry*, *32*(2), 315–331. doi:10.1016/j.pnpbp.2007.07.001.

Buchmann, A., Ringli, M., Kurth, S., Schaerer, M., Geiger, A., Jenni, O. G., et al. (2011). EEG sleep slow-wave activity as a mirror of cortical maturation. *Cerebral Cortex*, *21*(3), 607–615. doi:10.1093/cercor/bhq129.

Buckelmuller, J., Landolt, H. P., Stassen, H. H., & Achermann, P. (2006). Trait-like individual differences in the human sleep electroencephalogram. *Neuroscience*, *138*(1), 351–356. doi:10.1016/j.neuroscience.2005.11.005.

Cajochen, C., Wyatt, J. K., Czeisler, C. A., & Dijk, D. J. (2002). Separation of circadian and wake duration-dependent modulation of EEG activation during wakefulness. *Neuroscience*, *114*(4), 1047–1060.

Chen, J., Lipska, B. K., Halim, N., Ma, Q. D., Matsumoto, M., Melhem, S., et al. (2004). Functional analysis of genetic variation in catechol-O-methyltransferase (COMT): Effects on mRNA, protein, and enzyme activity in postmortem human brain. *American Journal of Human Genetics*, *75*(5), 807–821. doi:10.1086/425589.

Christian, J. C., Morzorati, S., Norton, J. A., Jr., Williams, C. J., O'Connor, S., & Li, T. K. (1996). Genetic analysis of the resting electroencephalographic power spectrum in human twins. *Psychophysiology*, *33*(5), 584–591.

Chrobak, J. J., Lorincz, A., & Buzsaki, G. (2000). Physiological patterns in the hippocampo-entorhinal cortex system. *Hippocampus*, *10*(4), 457–465. (10.1002/1098-1063(2000)10:4<457::AID-HIPO12>3.0.CO;2-Z)

Churchill, G. A., Airey, D. C., Allayee, H., Angel, J. M., Attie, A. D., Beatty, J., et al. (2004). The collaborative cross, a community resource for the genetic analysis of complex traits. *Nature Genetics*, *36*(11), 1133–1137.

Cueni, L., Canepari, M., Lujan, R., Emmenegger, Y., Watanabe, M., Bond, C. T., et al. (2008). T-type Ca2+ channels, SK2 channels and SERCAs gate sleep-related oscillations in thalamic dendrites. *Nature Neuroscience*, *11*(6), 683–692. (doi: 10.1038/nn.2124)

Daan, S., Beersma, D. G., & Borbely, A. A. (1984). Timing of human sleep: Recovery process gated by a circadian pacemaker. *The American Journal of Physiology*, *246*(2 Pt 2), R161–183.

Darvasi, A. (1998). Experimental strategies for the genetic dissection of complex traits in animal models. *Nature Genetics*, *18*(1), 19–24. doi:10.1038/ng0198-19.

Davis, F. C., Frank, M. G., & Heller, H. C. (1999). Ontogeny of sleep and circadian rhythms. In F. W. Turek, & P. Zee (Eds.), *Regulation of sleep and circadian rhythms* (pp. 19–79). New York; Basel: Marcel Dekker, Inc..

De Gennaro, L., Ferrara, M., Vecchio, F., Curcio, G., & Bertini, M. (2005). An electroencephalographic fingerprint of human sleep. *NeuroImage*, *26*(1), 114–122. doi:10.1016/j.neuroimage.2005.01.020.

De Gennaro, L., Marzano, C., Fratello, F., Moroni, F., Pellicciari, M. C., Ferlazzo, F., et al. (2008). The electroencephalographic fingerprint of sleep is genetically determined: A twin study. *Annals of Neurology*, *64*(4), 455–460. doi:10.1002/ana.21434.

Deboer, T., Detari, L., & Meijer, J. H. (2007). Long term effects of sleep deprivation on the mammalian circadian pacemaker. *Sleep*, *30*(3), 257–262.

Deboer, T., Fontana, A., & Tobler, I. (2002). Tumor necrosis factor (TNF) ligand and TNF receptor deficiency affects sleep and the sleep EEG. *Journal of Neurophysiology*, *88*(2), 839–846.

Deboer, T., Vansteensel, M. J., Detari, L., & Meijer, J. H. (2003). Sleep states alter activity of suprachiasmatic nucleus neurons. *Nature Neuroscience*, *6*(10), 1086–1090.

Dijk, D. J., & Archer, S. N. (2010). PERIOD3, circadian phenotypes, and sleep homeostasis. *Sleep Medicine Reviews*, *14*(3), 151–160. doi:10.1016/j.smrv.2009.07.002.

Dijk, D. J., Beersma, D. G., & Daan, S. (1987). EEG power density during nap sleep: Reflection of an hourglass measuring the duration of prior wakefulness. *Journal of Biological Rhythms*, *2*(3), 207–219.

Dijk, D. J., & Czeisler, C. A. (1994). Paradoxical timing of the circadian rhythm of sleep propensity serves to consolidate sleep and wakefulness in humans. *Neuroscience Letters*, *166*(1), 63–68.

Dijk, D. J., & Czeisler, C. A. (1995). Contribution of the circadian pacemaker and the sleep homeostat to sleep propensity, sleep structure, electroencephalographic slow waves, and sleep spindle activity in humans. *The Journal of Neuroscience*, *15*(5 Pt 1), 3526–3538.

Dijk, D. J., & Franken, P. (2005). Interaction of sleep homeostasis and circadian rhythmicity: Dependent or independent systems?. In M. H. Meir, H. Kryger, T. Roth, & W. Dement (Eds.), *Principles and practice of sleep medecine* (4th ed.). Saunders/Elsevier Philadelphia, PA, pp. 418–434.

Dimpfel, W. (2008). Pharmacological modulation of dopaminergic brain activity and its reflection in spectral frequencies of the rat electropharmacogram. *Neuropsychobiology*, *58*(3–4), 178–186. doi:10.1159/000191124.

Dossi, R. C., Nunez, A., & Steriade, M. (1992). Electrophysiology of a slow (0.5–4 Hz) intrinsic oscillation of cat thalamocortical neurones in vivo. *The Journal of Physiology*, *447*, 215–234.

Dudley, C. A., Erbel-Sieler, C., Estill, S. J., Reick, M., Franken, P., Pitts, S., et al. (2003). Altered patterns of sleep and behavioral adaptability in NPAS2-deficient mice. *Science*, *301*(5631), 379–383.

Easton, A., Meerlo, P., Bergmann, B., & Turek, F. W. (2004). The suprachiasmatic nucleus regulates sleep timing and amount in mice. *Sleep*, *27*(7), 1307–1318.

Finelli, L. A., Achermann, P., & Borbely, A. A. (2001). Individual 'fingerprints' in human sleep EEG topography. *Neuropsychopharmacology: Official Publication of the American College of Neuropsychopharmacology*, *25*(5 Suppl), S57–62. doi:10.1016/S0893-133X(01)00320-7.

Franken, P., Chollet, D., & Tafti, M. (2001). The homeostatic regulation of sleep need is under genetic control. *The Journal of Neuroscience*, *21*(8), 2610–2621.

Franken, P., & Dijk, D. J. (2009). Circadian clock genes and sleep homeostasis. *The European Journal of Neuroscience*, *29*(9), 1820–1829. doi:10.1111/j.1460-9568.2009.06723.x.

Franken, P., Dijk, D. J., Tobler, I., & Borbély, A. A. (1991). Sleep deprivation in rats: Effects on EEG power spectra, vigilance states, and cortical temperature. *American Journal of Physiology*, *261*(1 Pt 2), R198–208.

Franken, P., Dudley, C. A., Estill, S. J., Barakat, M., Thomason, R., O'Hara, B. F., et al. (2006). NPAS2 as a transcriptional regulator of non-rapid eye movement sleep: Genotype and sex interactions. *Proceedings of the National Academy of Sciences of the United States of America*, *103*(18), 7118–7123.

Franken, P., Malafosse, A., & Tafti, M. (1998). Genetic variation in EEG activity during sleep in inbred mice. *American Journal of Physiology*, *275*(4 Pt 2), R1127–1137.

Franken, P., Malafosse, A., & Tafti, M. (1999). Genetic determinants of sleep regulation in inbred mice. *Sleep*, *22*(2), 155–169.

Franken, P., & Tafti, M. (2003). Genetics of sleep and sleep disorders. *Frontiers in Bioscience*, *8*, e381–397.

Franken, P., Thomason, R., Heller, H. C., & O'Hara, B. F. (2007). A non-circadian role for clock-genes in sleep homeostasis: A strain comparison. *BMC Neuroscience*, *8*, 87.

Franken, P., Tobler, I., & Borbély, A. A. (1991). Sleep homeostasis in the rat: Simulation of the time course of EEG slow-wave activity. *Neuroscience Letters*, *130*(2), 141–144. (published erratum appears in Neurosci Lett 1991 Nov 11;132(2):279)

Garcia, J. A., Zhang, D., Estill, S. J., Michnoff, C., Rutter, J., Reick, M., et al. (2000). Impaired cued and contextual memory in NPAS2-deficient mice. *Science*, *288*(5474), 2226–2230.

Gedda, L., & Brenci, G. (1979). Sleep and dream characteristics in twins. *Acta Geneticae Medicae et Gemellologiae*, *28*(3), 237–239.

Gedda, L., & Brenci, G. (1983). Twins living apart test: Progress report. *Acta Geneticae Medicae et Gemellologiae*, *32*(1), 17–22.

Hasan, S., Pradervand, S., Ahnaou, A., Drinkenburg, W., Tafti, M., & Franken, P. (2009). How to keep the brain awake? The complex molecular pharmacogenetics of wake promotion. *Neuropsychopharmacology*, *34*(7), 1625–1640. doi:10.1038/npp.2009.3.

Hendricks, J. C., Lu, S., Kume, K., Yin, J. C., Yang, Z., & Sehgal, A. (2003). Gender dimorphism in the role of cycle (BMAL1) in rest, rest regulation, and longevity in Drosophila melanogaster. *Journal of Biological Rhythms*, *18*(1), 12–25.

Hendricks, J. C., Williams, J. A., Panckeri, K., Kirk, D., Tello, M., Yin, J. C., et al. (2001). A non-circadian role for cAMP signaling and CREB activity in Drosophila rest homeostasis. *Nature Neuroscience*, *4*(11), 1108–1115.

Hirschhorn, R., Yang, D. R., Israni, A., Huie, M. L., & Ownby, D. R. (1994). Somatic mosaicism for a newly identified splice-site mutation in a patient with adenosine deaminase-deficient immunodeficiency and spontaneous clinical recovery. *American Journal of Human Genetics*, *55*(1), 59–68.

Hori, A. (1986). Sleep characteristics in twins. *The Japanese Journal of Psychiatry and Neurology*, *40*(1), 35–46.

Hu, J. H., Park, J. M., Park, S., Xiao, B., Dehoff, M. H., Kim, S., et al. (2010). Homeostatic scaling requires group I mGluR activation mediated by Homer1a. *Neuron*, *68*(6), 1128–1142. doi:10.1016/j.neuron.2010.11.008.

Huber, R., Deboer, T., & Tobler, I. (2000). Effects of sleep deprivation on sleep and sleep EEG in three mouse strains: Empirical data and simulations. *Brain Research*, *857*(1–2), 8–19.

Huber, R., Deboer, T., & Tobler, I. (2000). Topography of EEG dynamics after sleep deprivation in mice. *Journal of Neurophysiology*, *84*(4), 1888–1893.

Jacobs, S., Lie, D. C., DeCicco, K. L., Shi, Y., DeLuca, L. M., Gage, F. H., et al. (2006). Retinoic acid is required early during adult neurogenesis in the dentate gyrus. *Proceedings of the National Academy of Sciences of the United States of America*, *103*(10), 3902–3907. doi:10.1073/pnas.0511294103.

Juel-Nielsen, N., & Harvald, B. (1958). The electroencephalogram in uniovular twins brought up apart. *Acta Genetica et Statistica Medica*, *8*(1), 57–64.

King, D. P., Zhao, Y., Sangoram, A. M., Wilsbacher, L. D., Tanaka, M., Antoch, M. P., et al. (1997). Positional cloning of the mouse circadian clock gene. *Cell*, *89*(4), 641–653.

Kitaoka, K., Hattori, A., Chikahisa, S., Miyamoto, K., Nakaya, Y., & Sei, H. (2007). Vitamin A deficiency induces a decrease in EEG delta power during sleep in mice. *Brain Research*, *1150*, 121–130. doi:10.1016/j.brainres.2007.02.077.

Klein, D. C., Moore, R. Y., & Reppert, S. M. (1991). *Suprachiasmatic nucleus. The mind's clock*. New York: Oxford University Press.

Konopka, R. J., & Benzer, S. (1971). Clock mutants of Drosophila melanogaster. *Proceedings of the National Academy of Sciences of the United States of America*, *68*(9), 2112–2116.

Kopp, C., Albrecht, U., Zheng, B., & Tobler, I. (2002). Homeostatic sleep regulation is preserved in mPer1 and mPer2 mutant mice. *The European Journal of Neuroscience*, *16*(6), 1099–1106.

Krezel, W., Ghyselinck, N., Samad, T. A., Dupe, V., Kastner, P., Borrelli, E., et al. (1998). Impaired locomotion and dopamine signaling in retinoid receptor mutant mice. *Science*, *279*(5352), 863–867.

Krueger, J. M., Rector, D. M., Roy, S., Van Dongen, H. P., Belenky, G., & Panksepp, J. (2008). Sleep as a fundamental property of neuronal assemblies. *Nature Reviews Neuroscience*, *9*(12), 910–919.

Lander, E., & Kruglyak, L. (1995). Genetic dissection of complex traits: Guidelines for interpreting and reporting linkage results. *Nature Genetics*, *11*(3), 241–247. doi:10.1038/ng1195-241.

Lander, E. S., & Botstein, D. (1989). Mapping mendelian factors underlying quantitative traits using RFLP linkage maps. *Genetics*, *121*(1), 185–199.

Landolt, H. P. (2008). Sleep homeostasis: A role for adenosine in humans? *Biochemical Pharmacology*, *75*(11), 2070–2079. doi:10.1016/j.bcp.2008.02.024.

Landolt, H. P. (2011). Genetic determination of sleep EEG profiles in healthy humans. *Progress in Brain Research*, *193*, 51–61. doi:10.1016/B978-0-444-53839-0.00004-1.

Lane, M. A., & Bailey, S. J. (2005). Role of retinoid signalling in the adult brain. *Progress in Neurobiology*, *75*(4), 275–293. doi:10.1016/j.pneurobio.2005.03.002.

Laposky, A., Easton, A., Dugovic, C., Walisser, J., Bradfield, C., & Turek, F. (2005). Deletion of the mammalian circadian clock gene BMAL1/Mop3 alters baseline sleep architecture and the response to sleep deprivation. *Sleep*, *28*(4), 395–409.

Larkin, J. E., Franken, P., & Heller, H. C. (2002). Loss of circadian organization of sleep and wakefulness during hibernation. *American Journal of Physiology. Regulatory, Integrative and Comparative Physiology*, *282*(4), R1086–1095.

Linkowski, P. (1999). EEG sleep patterns in twins. *Journal of Sleep Research*, *8*(*Suppl 1*), 11–13.

Lowrey, P. L., & Takahashi, J. S. (2004). Mammalian circadian biology: Elucidating genome-wide levels of temporal organization. *Annual Review of Genomics and Human Genetics*, *5*, 407–441.

Lykken, D. T., Tellegen, A., & Thorkelson, K. (1974). Genetic determination of EEG frequency spectra. *Biological Psychology*, *1*(4), 245–259.

Mackiewicz, M., Paigen, B., Naidoo, N., & Pack, A. I. (2008). Analysis of the QTL for sleep homeostasis in mice: Homer1a is a likely candidate. *Physiological Genomics*, *33*(1), 91–99. doi:10.1152/physiolgenomics.00189.2007.

Mang, G., & Franken, P. (2012). Sleep and EEG phenotyping in mice. *Current Protocols in Neuroscience 2*, 54–74. doi:10.1002/9780470942390.mo110126.

Maret, S., Dorsaz, S., Gurcel, L., Pradervand, S., Petit, B., Pfister, C., et al. (2007). Homer1a is a core brain molecular correlate of sleep loss. *Proceedings of the National Academy of Sciences of the United States of America, 104*(50), 20090–20095.

Maret, S., Franken, P., Dauvilliers, Y., Ghyselinck, N. B., Chambon, P., & Tafti, M. (2005). Retinoic acid signaling affects cortical synchrony during sleep. *Science, 310*(5745), 111–113.

Marley, R., & Baines, R. A. (2011). Increased persistent Na+ current contributes to seizure in the slamdance bang-sensitive drosophila mutant. *Journal of Neurophysiology, 106*(1), 18–29. doi:10.1152/jn.00808.2010.

Masubuchi, S., Honma, S., Abe, H., Ishizaki, K., Namihira, M., Ikeda, M., et al. (2000). Clock genes outside the suprachiasmatic nucleus involved in manifestation of locomotor activity rhythm in rats. *The European Journal of Neuroscience, 12*(12), 4206–4214.

McCormick, D. A., & Bal, T. (1997). Sleep and arousal: Thalamocortical mechanisms. *Annual Review of Neuroscience, 20*, 185–215. doi:10.1146/annurev.neuro.20.1.185.

Millstein, J., Winrow, C. J., Kasarskis, A., Owens, J. R., Zhou, L., Summa, K. C., et al. (2011). Identification of causal genes, networks, and transcriptional regulators of REM sleep and wake. *Sleep, 34*(11), 1469–1477. doi:10.5665/sleep.1378.

Mitchell, K. J. (2007). The genetics of brain wiring: From molecule to mind. *PLoS Biology, 5*(4), e113. doi:10.1371/journal.pbio.0050113.

Mongrain, V., Hernandez, S. A., Pradervand, S., Dorsaz, S., Curie, T., Hagiwara, G., et al. (2010). Separating the contribution of glucocorticoids and wakefulness to the molecular and electrophysiological correlates of sleep homeostasis. *Sleep, 33*(9), 1147–1157.

Mrosovsky, N., Edelstein, K., Hastings, M. H., & Maywood, E. S. (2001). Cycle of period gene expression in a diurnal mammal (Spermophilus tridecemlineatus): Implications for nonphotic phase shifting. *Journal of Biological Rhythms, 16*(5), 471–478.

Naylor, E., Bergmann, B. M., Krauski, K., Zee, P. C., Takahashi, J. S., Vitaterna, M. H., et al. (2000). The circadian clock mutation alters sleep homeostasis in the mouse. *The Journal of Neuroscience, 20*(21), 8138–8143.

Nitz, D. A., van Swinderen, B., Tononi, G., & Greenspan, R. J. (2002). Electrophysiological correlates of rest and activity in drosophila melanogaster. *Current Biology: CB, 12*(22), 1934–1940.

Papale, L. A., Beyer, B., Jones, J. M., Sharkey, L. M., Tufik, S., Epstein, M., et al. (2009). Heterozygous mutations of the voltage-gated sodium channel SCN8A are associated with spike-wave discharges and absence epilepsy in mice. *Human Molecular Genetics, 18*(9), 1633–1641. doi:10.1093/hmg/ddp081.

Peirce, J. L., Lu, L., Gu, J., Silver, L. M., & Williams, R. W. (2004). A new set of BXD recombinant inbred lines from advanced intercross populations in mice. *BMC Genetics, 5*, 7.

Philip, V. M., Sokoloff, G., Ackert-Bicknell, C. L., Striz, M., Branstetter, L., Beckmann, M. A., et al. (2011). Genetic analysis in the collaborative cross breeding population. *Genome Research, 21*(8), 1223–1238. doi:10.1101/gr.113886.110.

Porkka-Heiskanen, T., Kalinchuk, A., Alanko, L., Urrila, A., & Stenberg, D. (2003). Adenosine, energy metabolism, and sleep. *Scientific World Journal, 3*, 790–798.

Posthuma, D., de Geus, E. J., Mulder, E. J., Smit, D. J., Boomsma, D. I., & Stam, C. J. (2005). Genetic components of functional connectivity in the brain: The heritability of synchronization likelihood. *Human Brain Mapping, 26*(3), 191–198. doi:10.1002/hbm.20156.

Reddy, P., Zehring, W. A., Wheeler, D. A., Pirrotta, V., Hadfield, C., Hall, J. C., et al. (1984). Molecular analysis of the period locus in Drosophila melanogaster and identification of a transcript involved in biological rhythms. *Cell, 38*(3), 701–710.

Reick, M., Garcia, J. A., Dudley, C., & McKnight, S. L. (2001). NPAS2: An analog of clock operative in the mammalian forebrain. *Science*, *293*(5529), 506–509.

Retey, J.V., Adam, M., Honegger, E., Khatami, R., Luhmann, U. F., Jung, H. H., et al. (2005). A functional genetic variation of adenosine deaminase affects the duration and intensity of deep sleep in humans. *Proceedings of the National Academy of Sciences of the United States of America*, *102*(43), 15676–15681. doi:10.1073/pnas.0505414102.

Retey, J. V., Adam, M., Khatami, R., Luhmann, U. F., Jung, H. H., Berger, W., et al. (2007). A genetic variation in the adenosine A2A receptor gene (ADORA2A) contributes to individual sensitivity to caffeine effects on sleep. *Clinical Pharmacology and Therapeutics*, *81*(5), 692–698. doi:10.1038/sj.clpt.6100102.

Riksen, N. P., Franke, B., van den Broek, P., Naber, M., Smits, P., & Rongen, G. A. (2008). The 22G>A polymorphism in the adenosine deaminase gene impairs catalytic function but does not affect reactive hyperaemia in humans in vivo. *Pharmacogenetics and Genomics*, *18*(10), 843–846. doi:10.1097/FPC.0b013e328305e630.

Sehgal, A., & Mignot, E. (2011). Genetics of sleep and sleep disorders. *Cell*, *146*(2), 194–207. doi:10.1016/j.cell.2011.07.004.

Shaw, P. J., Tononi, G., Greenspan, R. J., & Robinson, D. F. (2002). Stress response genes protect against lethal effects of sleep deprivation in Drosophila. *Nature*, *417*(6886), 287–291.

Shiromani, P. J., Xu, M., Winston, E. M., Shiromani, S. N., Gerashchenko, D., & Weaver, D. R. (2004). Sleep rhythmicity and homeostasis in mice with targeted disruption of mPeriod genes. *American Journal of Physiology. Regulatory, Integrative and Comparative Physiology*, *287*(1), R47–57.

Slifstein, M., Kolachana, B., Simpson, E. H., Tabares, P., Cheng, B., Duvall, M., et al. (2008). COMT genotype predicts cortical-limbic D1 receptor availability measured with [11C] NNC112 and PET. *Molecular Psychiatry*, *13*(8), 821–827. doi:10.1038/mp.2008.19.

Stassen, H. H., Lykken, D. T., & Bomben, G. (1988). The within-pair EEG similarity of twins reared apart. *European Archives of Psychiatry and Neurological Sciences*, *237*(4), 244–252.

Stassen, H. H., Lykken, D. T., Propping, P., & Bomben, G. (1988). Genetic determination of the human EEG. Survey of recent results on twins reared together and apart. *Human Genetics*, *80*(2), 165–176.

Steinlein, O., Anokhin, A., Yping, M., Schalt, E., & Vogel, F. (1992). Localization of a gene for the human low-voltage EEG on 20q and genetic heterogeneity. *Genomics*, *12*(1), 69–73.

Steinlein, O., Fischer, C., Keil, R., Smigrodzki, R., & Vogel, F. (1992). D20S19, linked to low voltage EEG, benign neonatal convulsions, and fanconi anaemia, maps to a region of enhanced recombination and is localized between CpG islands. *Human Molecular Genetics*, *1*(5), 325–329.

Steriade, M. (2003). The corticothalamic system in sleep. *Frontiers in Bioscience*, *8*, D878–899.

Steriade, M., McCormick, D. A., & Sejnowski, T. J. (1993). Thalamocortical oscillations in the sleeping and aroused brain. *Science*, *262*(5134), 679–685.

Szumlinski, K. K., Kalivas, P. W., & Worley, P. F. (2006). Homer proteins: Implications for neuropsychiatric disorders. *Current Opinion in Neurobiology*, *16*(3), 251–257. doi:10.1016/j.conb.2006.05.002.

Tafti, M., Franken, P., Kitahama, K., Malafosse, A., Jouvet, M., & Valatx, J. L. (1997). Localization of candidate genomic regions influencing paradoxical sleep in mice. *Neuroreport*, *8*(17), 3755–3758.

Tafti, M., Petit, B., Chollet, D., Neidhart, E., de Bilbao, F., Kiss, J. Z., et al. (2003). Deficiency in short-chain fatty acid beta-oxidation affects theta oscillations during sleep. *Nature Genetics*, *34*(3), 320–325.

Thompson, P. M., Cannon, T. D., Narr, K. L., van Erp, T., Poutanen, V. P., Huttunen, M., et al. (2001). Genetic influences on brain structure. *Nature Neuroscience*, *4*(12), 1253–1258. doi:10.1038/nn758.

Tobler, I., & Borbely, A. A. (1986). Sleep EEG in the rat as a function of prior waking. *Electroencephalography and Clinical Neurophysiology, 64*(1), 74–76.

Tobler, I., Kopp, C., Deboer, T., & Rudolph, U. (2001). Diazepam-induced changes in sleep: Role of the alpha 1 GABA(A) receptor subtype. *Proceedings of the National Academy of Sciences of the United States of America, 98*(11), 6464–6469. doi:10.1073/pnas.111055398.

Tononi, G., & Cirelli, C. (2006). Sleep function and synaptic homeostasis. *Sleep Medicine Reviews, 10*(1), 49–62.

Toro, R., Chupin, M., Garnero, L., Leonard, G., Perron, M., Pike, B., et al. (2009). Brain volumes and Val66Met polymorphism of the BDNF gene: Local or global effects? *Brain Structure & Function, 213*(6), 501–509. doi:10.1007/s00429-009-0203-y.

Toth, L. A., & Williams, R. W. (1999). A quantitative genetic analysis of slow-wave sleep and rapid-eye movement sleep in CXB recombinant inbred mice. *Behavior Genetics, 29*(5), 329–337.

Trachsel, L., Edgar, D. M., Seidel, W. F., Heller, H. C., & Dement, W. C. (1992). Sleep homeostasis in suprachiasmatic nuclei-lesioned rats: Effects of sleep deprivation and triazolam administration. *Brain Research, 589*(2), 253–261.

Valatx, J. L., & Bugat, R. (1974). [Genetic factors as determinants of the waking-sleep cycle in the mouse (author's transl)]. *Brain Research, 69*(2), 315–330.

Valatx, J. L., Bugat, R., & Jouvet, M. (1972). Genetic studies of sleep in mice. *Nature, 238*(5361), 226–227.

Valdar, W., Flint, J., & Mott, R. (2006). Simulating the collaborative cross: Power of quantitative trait loci detection and mapping resolution in large sets of recombinant inbred strains of mice. *Genetics, 172*(3), 1783–1797. doi:10.1534/genetics.104.039313.

van Beijsterveldt, C. E., & Boomsma, D. I. (1994). Genetics of the human electroencephalogram (EEG) and event-related brain potentials (ERPs): A review. *Human Genetics, 94*(4), 319–330.

van der Horst, G. T., Muijtjens, M., Kobayashi, K., Takano, R., Kanno, S., Takao, M., et al. (1999). Mammalian Cry1 and Cry2 are essential for maintenance of circadian rhythms. *Nature, 398*(6728), 627–630.

Van Dongen, H. P., Maislin, G., Mullington, J. M., & Dinges, D. F. (2003). The cumulative cost of additional wakefulness: Dose-response effects on neurobehavioral functions and sleep physiology from chronic sleep restriction and total sleep deprivation. *Sleep, 26*(2), 117–126.

Vertes, R. P., & Kocsis, B. (1997). Brainstem-diencephalo-septohippocampal systems controlling the theta rhythm of the hippocampus. *Neuroscience, 81*(4), 893–926.

Viola, A. U., Archer, S. N., James, L. M., Groeger, J. A., Lo, J. C. Y., Skene, D. J., et al. (2007). PER3 polymorphism predicts sleep structure and waking performance. *Current Biology, 17* doi:10.1016/j.cub.2007.01.073.

Vitaterna, M. H., King, D. P., Chang, A. M., Kornhauser, J. M., Lowrey, P. L., McDonald, J. D., et al. (1994). Mutagenesis and mapping of a mouse gene, Clock, essential for circadian behavior. *Science, 264*(5159), 719–725.

Vitaterna, M. H., Selby, C. P., Todo, T., Niwa, H., Thompson, C., Fruechte, E. M., et al. (1999). Differential regulation of mammalian period genes and circadian rhythmicity by cryptochromes 1 and 2. *Proceedings of the National Academy of Sciences of the United States of America, 96*(21), 12114–12119.

Vogel, F. (1970). The genetic basis of the normal human electroencephalogram (EEG). *Humangenetik, 10*(2), 91–114.

Vogel, F., Schalt, E., Kruger, J., Propping, P., & Lehnert, K. F. (1979). The electroencephalogram (EEG) as a research tool in human behavior genetics: Psychological examinations in healthy males with various inherited EEG variants. I. Rationale of the study. Material. Methods. Heritability of test parameters. *Human Genetics, 47*(1), 1–45.

Wakamatsu, H., Yoshinobu, Y., Aida, R., Moriya, T., Akiyama, M., & Shibata, S. (2001). Restricted-feeding-induced anticipatory activity rhythm is associated with a phase-shift of the expression of mPer1 and mPer2 mRNA in the cerebral cortex and hippocampus but not in the suprachiasmatic nucleus of mice. *The European Journal of Neuroscience*, *13*(6), 1190–1196.

Webb, W. B., & Campbell, S. S. (1983). Relationships in sleep characteristics of identical and fraternal twins. *Archives of General Psychiatry*, *40*(10), 1093–1095.

Werth, E., Dijk, D. J., Achermann, P., & Borbely, A. A. (1996). Dynamics of the sleep EEG after an early evening nap: Experimental data and simulations. *American Journal of Physiology. Regulatory, Integrative and Comparative Physiology*, *271*(3), R501–510.

Winrow, C. J., Williams, D. L., Kasarskis, A., Millstein, J., Laposky, A. D., Yang, H. S., et al. (2009). Uncovering the genetic landscape for multiple sleep-wake traits. *PloS ONE*, *4*(4), e5161. doi:10.1371/journal.pone.0005161.

Winsky-Sommerer, R. (2009). Role of GABAA receptors in the physiology and pharmacology of sleep. *The European Journal of Neuroscience*, *29*(9), 1779–1794. doi:10.1111/j.1460-9568.2009.06716.x.

Wisor, J. P., & Kilduff, T. S. (2005). Molecular genetic advances in sleep research and their relevance to sleep medicine. *Sleep*, *28*(3), 357–367.

Wisor, J. P., O'Hara, B. F., Terao, A., Selby, C. P., Kilduff, T. S., Sancar, A., et al. (2002). A role for cryptochromes in sleep regulation. *BMC Neuroscience*, *3*, 20.

Wisor, J. P., Pasumarthi, R. K., Gerashchenko, D., Thompson, C. L., Pathak, S., Sancar, A., et al. (2008). Sleep deprivation effects on circadian clock gene expression in the cerebral cortex parallel electroencephalographic differences among mouse strains. *The Journal of Neuroscience*, *28*(28), 7193–7201.

Xie, X., Dumas, T., Tang, L., Brennan, T., Reeder, T., Thomas, W., et al. (2005). Lack of the alanine-serine-cysteine transporter 1 causes tremors, seizures, and early postnatal death in mice. *Brain Research*, *1052*(2), 212–221. doi:10.1016/j.brainres.2005.06.039.

Yalcin, B., Nicod, J., Bhomra, A., Davidson, S., Cleak, J., Farinelli, L., et al. (2010). Commercially available outbred mice for genome-wide association studies. *PLoS Genetics*, *6*(9) doi:10.1371/journal.pgen.1001085.

Yasenkov, R., & Deboer, T. (2011). Interrelations and circadian changes of electroencephalogram frequencies under baseline conditions and constant sleep pressure in the rat. *Neuroscience*, *180*, 212–221. doi:10.1016/j.neuroscience.2011.01.063.

Zhou, Y. D., Barnard, M., Tian, H., Li, X., Ring, H. Z., Francke, U., et al. (1997). Molecular characterization of two mammalian bHLH-PAS domain proteins selectively expressed in the central nervous system. *Proceedings of the National Academy of Sciences of the United States of America*, *94*(2), 713–718.

CHAPTER 5

Evoked Electrophysiological and Vascular Responses across Sleep

David M. Rector
Department of Veterinary and Comparative Anatomy, Pharmacology and Physiology, Washington State University

INTRODUCTION

Sleep is traditionally defined by a collection of metrics assessed in concert to determine an organism's state. These include posture, EEG waveforms, cardio-respiratory rate, eye movement, and muscle tone, to name a few. However, sleep state characteristics are often difficult to interpret during abnormal sleep and under conditions of sleep pathology because those processes that normally produce the standard metrics are disturbed in various ways. For example, during total sleep deprivation, delta wave intrusions are present during periods that otherwise appear as waking, and may define a putative "microsleep" state (e.g., Grenèche et al., 2008). Microsleep may represent brief sleep periods and encompass processes similar to sleep. Leg movements during restless-leg syndrome occur during sleep, in spite of the fact that muscles are supposed to be relaxed. Mahowald and Schenck (2005) propose that individuals with parasomnias, such as sleepwalking, are simultaneously awake and asleep. Such conditions make sleep difficult to study because the normal collection of metrics is not present.

Traditional sleep characteristics may also represent only those mechanisms that have developed under niche appropriate conditions. Specifically, normal conditions dictate that an organism maximize the amount of sleep it acquires during times when other functions such as foraging and reproduction may not be optimal. Since sleep is a highly vulnerable state for most animals, it must occur at specific times when food is not available and be undisturbed by predators. Humans, however, have changed these rules with the need to be productive 24 hours a day, 7 days a week, and 356 days a year (Balkin et al., 2004). The new rules have produced conditions where we fail to get enough sleep, resulting in a number of pathologies such as increased sleep-loss related accidents, insomnia, daytime sleepiness, obstructive sleep apnea, diabetes, heart failure, depression, and

Sleep and Brain Activity
DOI: http://dx.doi.org/10.1016/B978-0-12-384995-3.00005-8

many other illnesses that may have their origins in, or be exacerbated by sleep loss (Grandner et al., 2010).

Additionally, sleep is not necessarily a whole brain phenomenon (Krueger and Obal, 1993; Kavanau, 1996; Pigarev et al., 1997; Krueger et al., 2008). An increasing number of studies show aspects similar to sleep and wake can occur simultaneously or at different levels in different brain regions. First demonstrated in humans by stimulating one hand more than another (Kattler et al., 1994), this is particularly evident in rodents that use their whiskers in the dark, and visual system during the light; they exhibit greater amounts of slow wave activity in the respective brain regions that are used more (Yasuda et al., 2005). Many other examples have followed in both humans and animals (e.g., Huber et al., 2006). In principle, if sleep is restorative, then any region that is used more should exhibit characteristics of sleep more often than other regions. With this perspective in mind, traditional sleep markers may be inadequate to define sleep, especially under abnormal conditions. Thus, the development of new sleep markers is needed. Ideally, these markers should be more closely related to those processes involved in controlling sleep.

There are at least two main issues that must be addressed in the next phases of sleep research. First, as we begin to understand the cellular and physiological consequences of sleep, we must identify better markers of sleep state and need. Second, given that humans will increasingly be driven to deprive themselves of sleep, we may identify additional long and short term consequences to sleep loss that have not yet been appreciated. More specifically, people who are driven to increase productivity are frequently frustrated by sleepiness, and use many physical and pharmacological mechanisms to stay awake (Wesensten et al., 2004). While they may experience short term gains in productivity (Van Dongen et al., 2003), detrimental long term consequences could build up over time.

Our studies have started to address these issues from several different perspectives. Evoked electrical and hemodynamic markers are predictive of sleep states, and in some ways may be better than traditional measures since they can be localized to specific brain regions experiencing differential use, and on average, correlate with whole animal state. We have also investigated the relationship between vascular compliance and sleep, leading to a sleep control theory that may help to integrate many long standing ideas of energy restoration and recent discoveries involving ATP, cytokines, and sleep, with the potential for severe long term consequences if energy restoration is not permitted during forced extended waking.

ELECTRICAL MARKERS OF SLEEP AND SLEEPINESS

Traditional electrical sleep state markers have focused on passive EEG measurements, which categorize the amount of power in different frequency bands. Since the EEG is principally derived from synchronous tissue activity, much research has investigated the source of oscillations at the various frequencies. However, due to the electrical diffusion in skin and scalp, EEG signals tend to be an average over large regions. The introduction of high density EEG (e.g., Massimini et al., 2009) and mathematical source localization algorithms to reduce the bulk averaging of the electrical signals has shown that such EEG characteristics are indeed not observed globally across the brain, but rather have greater power in those areas that have been used more during waking.

However, the passive measurement of tissue activity synchronization may not be enough to detect subtle changes in tissue state. Especially since similar synchronized activity can be observed under a mixture of different states. Observations beginning many years ago probed tissue state by stimulating it with specific input, then related differential electrical evoked responses across different sleep states (for review see Colrain and Campbell, 2007). Indeed, the electrical excitability of the cortex is inherently dependent on many factors including sleep state as well as arousal level and attention, which all may be inherently related to those processes that regulate sleep.

A Focus on Cortical Columns

One major hurdle when investigating regional sleep is the difficulty in defining the least significant component that can sleep. Since evidence is rapidly building that something less than the whole brain can be asleep at any given moment in time, how far do we have to probe to find the smallest component that sleeps? If sleep is defined as simple rest/activity cycles, then single cells may sleep. However, since cells within cortical columns are more highly interconnected than between columns (Shaw et al., 1982), the cortical column may represent a functional processing unit in the cortex (Koch, 2004). Additionally, vessels within microcapillary beds appear to be regulated at the cortical column level (Fehm et al., 2006). Thus, for a number of practical reasons, including ease of cortical surface access, our investigations to date have focused on markers that originate from surface cortical columns.

Methods of Recording Electrical Markers

Previous electrical measures to assess sleep typically use passive measures. By recording and calculating the frequency content of EEG, EOG, EKG, and

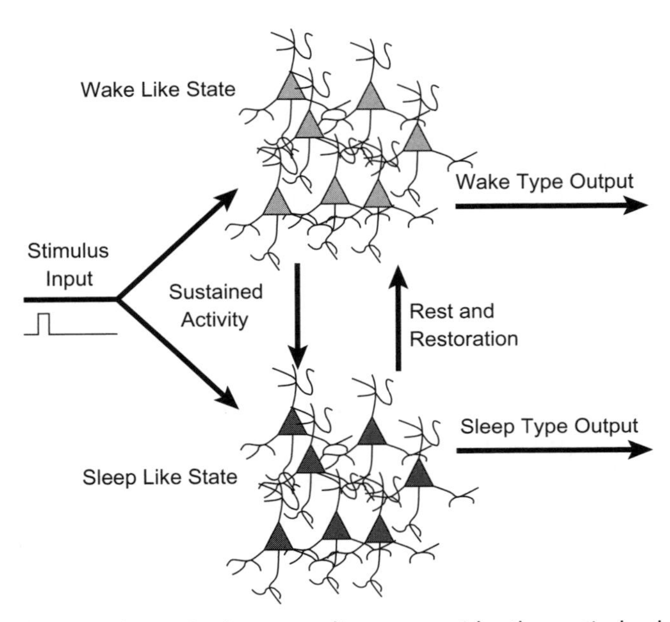

Figure 5.1 As a starting point in our studies, we consider the cortical column to be an operational unit of tightly interconnected cells within the brain for functional purposes (see text). Additionally, microcapillary beds are arranged such that key regulatory points serve individual columns and are sensitive to activity within the cortical column. When a cortical column receives a stimulus input, it produces an output response. The electrical consequences of the response can be detected as a surface evoked response potential because most of the cells act together creating a large electrical field, recorded with a cortical surface electrode or EEG electrode. Microvessel dilation is also triggered by enhanced electrical activity, generating an evoked hemodynamic response. Given identical input characteristics, the output response will be different depending on whether the column is in its wake-like or sleep-like state. The differential responses provide a marker for state that can be probed with the input stimulus, as long as the stimulus does not change the animal's state.

EMG, sleep can generally be inferred from stereotypical patterns. As mentioned earlier, these patterns break down under abnormal sleep or in the context of sleep pathologies, thus new measures are needed. We have focused on using active measures to probe sleep states that presume that the input/output relationships of the cortex will be different during different states (Fig. 5.1).

To record electrical evoked responses from cortical columns, we have used several approaches. First, the differences across cortical columns can be mapped using high density electrode arrays placed on the cortical surface to record the electrocorticogram (ECoG). By stimulating specific sensory modalities, then identifying the particular electrode that produces the

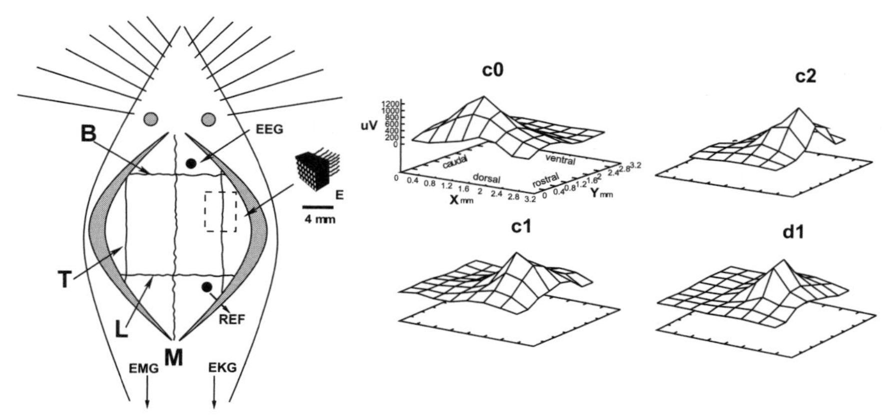

Figure 5.2 We placed cortical surface electrode arrays over the somatosensory cortex of the rat and recorded the two-dimensional pattern of electrical activity after stimulation. The 64-channel electrode array was used to map field potentials on the somatosensory cortical surface while the rat's whiskers were stimulated. Screw electrodes were used for recording frontal and parietal EEG. Stainless steel wire was threaded across the rib cage to record EKG and inserted into the neck muscles to record EMG. Several anchor screws were placed around the skull for additional headstage support. For some rats, two pairs of stainless steel screw electrodes were placed 4 mm apart over the somatosensory cortex on the temporal ridge (T), 1 mm caudal to bregma (B). The 64 channel electrode array was constructed from black Delrin and 0.2 mm stainless steel wires spaced 0.4 mm apart. A squared opening in the skull over the somatosensory cortex, outlined by the dotted line, allowed placement of the array on the cortical surface (E). One screw electrode was placed rostral to bregma near the midline (M) for frontal lobe EEG and one was placed near lambda (L) for ground reference (REF). The location of each whisker barrel was mapped by twitching whiskers in sequence while generating 2D surface maps of electrical potentials through the electrode array. To illustrate, twitching whiskers c0, c1, c2, and d1 each produced unique maps (right panels) with the initial peak amplitude corresponding to the location of the whisker barrel associated with that whisker. Subsequent analysis used the electrode within the array that corresponded to the location closest to the cortical column for the stimulated whisker.

earliest and largest response, a surface electrode array can be used to associate particular electrodes with a given cortical column (Fig. 5.2). Second, the mapping procedure is not always required since specific stimuli will necessarily activate a particular cortical column, and the earliest components of an evoked response recorded with an EEG electrode will correspond to a specific location (Penfield, 1958). Either way, electrical evoked responses triggered by stimulus events can be associated with activation of specific cortical regions, and used to assess the state of that region. We have used both somatosensory stimulation in rats through whisker stimulation and auditory stimulation through speaker clicks to elicit evoked responses.

Since low level stimuli do not elicit arousal (Phillips et al., 2011a), they can be used to probe cortical state during sleep.

State Dependent Electrical Evoked Responses

The amplitude and latency of the electrical evoked response from any sensory modality has been clearly linked to different conscious states beginning with Weitzman and Kremen (1965). Specifically, when using low level stimuli, evoked responses are significantly larger on average during sleep than during wake and REM sleep (Rector et al., 2005). Since the same stimulus (input) produces a different evoked response (output) dependent on sleep state, we postulate that evoked response characteristics can be used as a marker for state, at least within individual cortical columns. Furthermore, if we imagine with Sherrington that each cortical column can be represented by a light (Sherrington, 1951), "on" when active and "off" when quiescent, then the brain could be seen as a constellation of twinkling lights, usually "on" when awake, and usually "off" when asleep, but with the possibility to flip on or off depending on the need for attention or sleep, respectively (Roy et al., 2008). Many studies on attention and arousal control circuits provide a basis for turning the cortical columns "on" (Saper et al., 2001), and most likely they are configured to keep most of the brain awake during critical periods when the organism requires maximum alertness. Central control circuits may exist specifically to maintain an animal's vigilance during those periods when it is most efficient to do so. However, when pushed into abnormal conditions (e.g., trauma or extended waking), those processes that drive the columns to turn "off" and enter their sleep-like state may compete with arousal mechanisms, increasing the likelihood that a given column may enter the sleep-like state while the whole animal is awake. In the following sections, we will postulate that a lack of resources could potentially be at the root of this drive. Further investigations into the mechanisms that produce the state dependent input/output relationships will provide tremendous insight into those processes that regulate sleep.

Slow Waves, Cortical Column Cell Membrane Potential, and Evoked Responses

Initial studies into the mechanisms that underlie the state dependent evoked response characteristics suggest that baseline membrane potential is a principal component. We hypothesize that cells are depolarized during the active, wake-like state, and hyperpolarized during the quiescent,

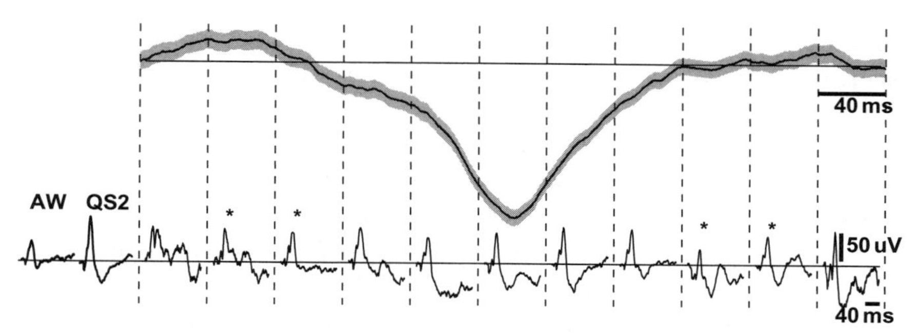

Figure 5.3 During quiet sleep periods with high amplitude slow waves, we filtered the EEG between 0 and 4 Hz, then identified the time point for each nadir in the slow waves. By registering the nadir of each slow wave in time, we created an average slow wave trace (top trace). The gray region around the average slow wave represents the standard error of the mean for each sample point. For each stimulus, the evoked response was averaged based on its timing relative to the nearest nadir in 40 ms bins. Evoked responses that occurred 80 ms before or after the nadir were significantly larger (*, $p < 0.1$) than responses that occurred 120 to 160 ms before or after the nadir. Evoked response traces below the average slow wave represent average data from one animal with the average active wake (AW) and quiet sleep (QS2) ERP traces plotted on the left for comparison to the typical state-related responses.

sleep-like state. Evoked responses have lower latency and are lower amplitude during the wake-like depolarized state because synaptic activation can occur more rapidly when the membrane potential is depolarized, and the change in membrane potential is smaller, resulting in a smaller averaged evoked response. Evoked responses have longer latency and are larger in amplitude in the hyperpolarized, sleep-like state because it takes more time to activate the synapses from a lower membrane potential. When activated, the total change in membrane potential is larger, starting from a lower initial value, resulting in a larger evoked response.

Many experiments on anesthetized animals (Steriade et al., 1993; Timofeev et al., 1996) and later in waking and sleeping animals (Steriade et al., 2001), support the hypothesis stated above. Delta slow waves, which are characteristic of sleep, result from the rhythmic alteration of membrane potential between the depolarized and hyperpolarized states. Work by Massimini et al. (2003) and ourselves (Rector et al., 2009a) show that evoked responses generated during the up (depolarized) state of slow wave sleep are remarkably similar to those generated during waking, while evoked potentials generated during the down (hyperpolarized) state are similar to those during sleep (Fig. 5.3). Further analysis on individual evoked responses demonstrates that roughly half the evoked responses during slow wave sleep are much larger than the average responses observed during sleep, while the other half are small and wake-like, as would be expected from a roughly 50% duty cycle within the slow wave rhythm (Rector et al., 2009a). When averaged across the entire record, the evoked responses during sleep appear larger than during wake because the sleep-like responses are more than twice the amplitude of the wake-like responses.

Several key implications of the state dependent evoked responses are important to consider. First, the results imply that a relative measure of average baseline membrane potential of cells within a cortical column can be inferred from the size and latency of the evoked response. This is a remarkable possibility considering that the measurement can be noninvasive from the scalp surface. Second, stimuli should be low amplitude, and should not arouse the animal statistically more often than without the stimuli (Phillips et al., 2011a). However, if animals do wake up during our stimulus protocol, they tend to wake up from stimuli presented during the up states of slow wave sleep, corresponding to the low amplitude evoked response, but not when presented during the down state (Phillips et al., 2011a). This indicates that additional sensory integration and processing is possible during sleep, but only when the cells are in the depolarized (wake-like or up) state.

Lack of Cortical Gain Control during Sleep

Our more recent experiments delve even deeper into the state dependent input/output relationships and show additional important features. Stimuli presented during wake and REM sleep produce modulated evoked responses as would be expected by typical stimulus/response curves. However, during sleep, the evoked responses are constant amplitude, regardless of stimulus intensity (Fig. 5.4). These results suggest that when the cells within the column are in their hyperpolarized or down state, they are not able to adequately reflect the stimulus intensity. Since the evoked response amplitude is constant, regardless of stimulus intensity, it appears that the down/hyperpolarized state of the cortical column does not provide input gain control (Phillips et al., 2011b).

Other Considerations

There is much work remaining to support the notion that evoked response characteristics represent "wake-like" and "sleep-like" states, especially during slow wave sleep. Up and down states during slow wave sleep might be physiologically different from the fluctuations between silent and desynchronized states characteristic of waking. Indeed the cholinergic

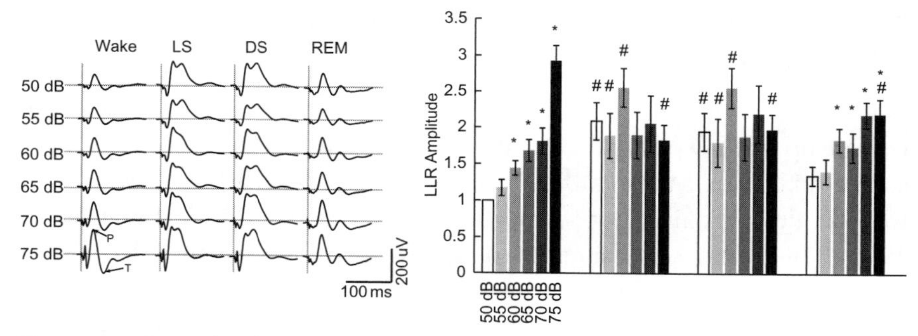

Figure 5.4 Example electrical evoked response traces beginning 150 ms after the stimulus from one animal show the state and stimulus intensity changes in the cortical response to auditory stimulation. The amplitudes increased consistently with increased stimulation intensity during Wake and rapid eye movement sleep (REM), but were at a constant level during light sleep (LS) and deep sleep (DS). We also observed a significantly longer trough latency during LS and DS due to the presence of a second waveform that extended the duration of the response. The average response amplitudes for all animals across state and stimulus intensity were divided by the average 50 dB wake value in order to normalize values across animal and plotted in bar graph form on the right. During Wake and rapid eye movement sleep (REM), increased stimulation intensity elicited larger amplitude components. No significant changes in amplitude were seen during light sleep (LS) or deep sleep (DS). This result caused the low level stimuli (50 to 60 dB) to appear larger during LS and DS than Wake, and high level stimuli (70 to 75 dB) to appear smaller. During rapid eye movement sleep (REM), the response at the highest stimulus intensity was significantly lower when compared to the Wake value. Significant differences across stimulation intensity compared to the 50 dB level within a state are marked with a (*), $p < 0.05$ Mann-Whitney U test. Significant differences across state compared to Wake at the same intensity level are marked with a (#).

inputs are different, and connectivity between distant columns is also reduced, perhaps related to inhibition of arousal systems (Szymusiak and McGinty, 2008). Finally, the thalamocortical circuit is most certainly synchronizing if not orchestrating these fluctuations within the intact preparation. While the up state during slow wave sleep is not necessarily identical to the waking state, there is much striking evidence to support the notion that the up state during slow wave sleep and the waking state share many physiological and functional similarities. Fundamentally, this relationship will help us eventually understand performance deficits during sleep deprivation, and other issues found in many sleep pathologies. Since hyperpolarized cells exhibit lower metabolism and energy demands, it is possible that this state may be related to the restorative aspects of sleep (Benington and Heller, 1995). In the next section, we investigate a vascular sleep state marker with direct ties to the membrane potential changes.

EVOKED VASCULAR AND HEMODYNAMIC RELATIONSHIPS TO SLEEP

With cholinergic activating/arousal systems driving cells into their depolarized/up state, we do not yet have a mechanism for driving cells into a sleep-like state. Many studies have shown that overall brain metabolism is lower during quiet sleep compared to waking and REM (Braun et al., 1997). This makes sense since cortical activity is lower during sleep. Thus, if sleep is a restorative process, then energy stores need to be replenished at a rate faster than they are used during sleep. Since metabolic requirements of the depolarized/up state are higher due to a large amount of spontaneous activity (Steriade et al., 2001; Vyazovskiy et al., 2009), it is possible that a reduction in metabolite reserves, including oxygen and sugar, makes it difficult for the cells to maintain a depolarized membrane potential, and recovery occurs during the hyperpolarized/down state. Additionally, blood perfusion limits may reduce the amount of heat that can be removed from active tissue, further compromising cellular function (Lydig, 1987; McGinty and Szymusiak, 1990). To test this hypothesis, we used pulse oximetry to record changes in oxy- and deoxyhemoglobin concentration during evoked responses across sleep and waking.

Evoked Hemodynamic Responses

Locally regulated neurovascular coupling increases blood volume and flow to active brain tissue for just-in-time metabolite delivery and waste removal (Roy and Sherrington, 1890; Filosa and Blanco, 2007; Buxton et al., 2004). Conditions of impaired perfusion and overdriven cells underlie significant injury described for pathological conditions including stroke, epilepsy, head trauma, and obstructive sleep apnea (Macey et al., 2008). Yet everyday activities, such as extended time spent on a task and mild sleep deprivation, cause cognitive and performance deficits. Fundamental limitations in vascular compliance may extend beyond pathological conditions and provide a unified model in which cells are routinely exposed to limited perfusion, with consequences to performance, and must hyperpolarize for recovery (Fig. 5.5).

While compliance is a factor in recent Windkessel models of vascular dynamics (Huppert et al., 2007), saturation of blood delivery is only considered under extreme conditions such as epilepsy. However, that notion developed from animals under anesthesia or sedation (Devor et al., 2003), conditions characterized by vascular dilation and significantly lower metabolism.

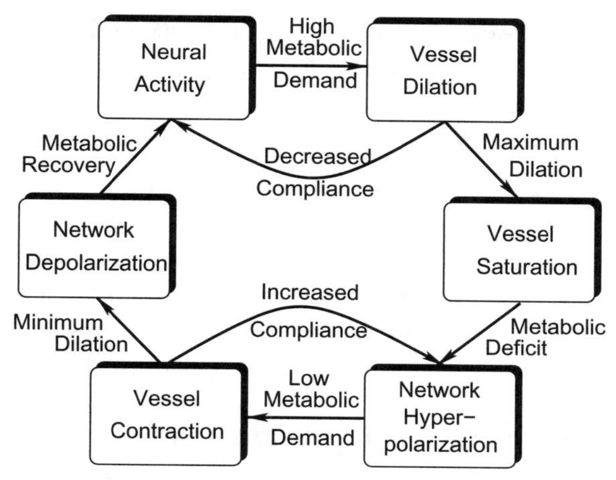

Figure 5.5 Our model of vascular regulation across wake and sleep-like states suggests physical limits in the ability of blood to deliver metabolites to tissue may underlie state transitions. Neural activity demands metabolites, initiating vasodilation through many well-known neural and chemical mechanisms. High metabolic demand associated with sustained activity will gradually decrease vessel compliance, limiting the ability to deliver oxygen and glucose. When vessels reach saturation, the potential for metabolic deficit may trigger a protective mechanism such that the network hyperpolarizes (down, sleep-like state) with lower metabolic demand, and restoration of vessel compliance. After sufficient recovery, the tissue can depolarize, entering its up (wake-like) state. This general model goes beyond pathological conditions and predicts everyday performance limits.

During normal waking, vascular smooth muscle compliance and expansion under prolonged neural activity may be limited by tissue compression (Schei et al., 2009; Rector et al., 2009b; Behzadi and Liu, 2005). Thus, physical vascular limits to deliver blood and remove waste products within local activated brain regions may ultimately restrict the normal period over which brain cells can operate and will impose functional consequences in that region. This hypothesis challenges conventional notions about the limits of vascular compliance and provides a comprehensive, unified physiological explanation for cognitive and performance deficits resulting from extended use of neuronal pathways, time-on-task effects, and sleep deprivation. In theory, if the resources required for optimal and sustained performance were understood, we might be able to devise methods to counteract the consequences of neural activity and sleep loss, and increase sustained activity periods. However, neural tissue is confined within the skull, and ultimately limited by hydrodynamic forces, vascular expansion, nutrient delivery, and heat removal. If these limits are exceeded over the long term, cellular trauma could result.

Methods to Record Evoked Vascular Responses

Most hemodynamic studies during sleep used global cerebral blood flow measures (e.g., Braun et al., 1997), or other slow measures of localized blood volume changes such as fMRI BOLD or PET imaging methods. We implanted LEDs and photodiodes into animals to obtain continuous spectroscopic measurements of oxy- and deoxyhemoglobin concentrations, similar to pulse oximetry. Using this technique, evoked changes in hemoglobin concentration can be assessed continuously at greater than 1000 samples per second, providing high resolution measurements of oxy- and deoxyhemoglobin concentration during sleep/wake cycles and concomitant with evoked response.

Evoked Hemodynamic Changes across State

Our initial recordings revealed that the evoked hemodynamic response was much larger during sleep than during wake (Fig. 5.6). Initially we assumed this was due to the larger evoked electrical response or due to increased synchronized activity, however, the increase was many times larger than the corresponding evoked electrical signal. This result, combined with the significantly larger evoked hemodynamic response observed during anesthesia-related suppressed states, led us to the hypothesis that when tissue is less active, blood vessels are in a more relaxed and compliant state, enabling an increased evoked blood response to stimulation.

Evoked Hemodynamic Responses are Muted after Sleep Deprivation

To further test vascular compliance limits, we measured the evoked hemodynamic response after increasing the amount of sleep deprivation. With increasing deprivation, evoked hemodynamic responses decrease significantly, suggesting that the vessels are stretched to their limit during the deprivation, and that it takes some time to recover (Fig. 5.7). Thus, silent or hyperpolarized periods may be required to restore vessel compliance. However, to understand the mechanisms involved in this response, many more experiments are required to image the dynamics of vascular expansion.

Cytokine and ATP Involvement in Vascular Control

Cytokine molecules such as adenosine, NO, TNF, and IL1 are responsible for changes in blood flow and are also involved in cortical-column state changes, thus linking metabolism, blood flow, and state (Krueger et al., 2008). Cytokines are produced during normal activity as well as during

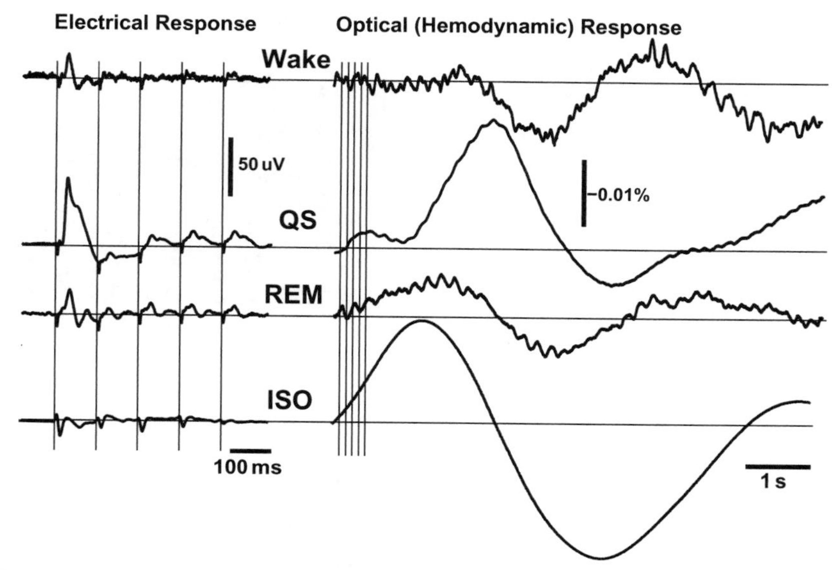

Figure 5.6 The traces in the left panel illustrate typical electrical evoked responses during auditory stimulation during Wake, quiet sleep (QS), rapid eye movement sleep (REM), and during isoflurane anesthesia. Five consecutive stimuli are presented with 100 ms interstimulus interval (ISI), represented by the vertical lines. Traces in the right panel show the corresponding evoked hemodynamic response as recorded by changes in 660 nm reflect light. The size of both the electrical and optical responses increases during sleep and then decreases back to waking levels during REM sleep. We initially thought that the increased hemodynamic response during sleep was caused by the increased evoked electrical response, however, under isoflurane anesthesia, the electrical response is smaller than during quiet sleep, yet the optical response is much larger than sleep. These results suggest that when baseline neural activity high (wake and REM sleep), the concurrent hemodynamic response is muted due to vascular stretching. We hypothesize that this muted hemodynamic response is due to the possibility that blood vessels are less compliant.

cellular stress, including metabolic deficits. Thus sleep promoting cytokines may be responsible for modulating blood supply to the tissue based on demand by expanding vessels and increasing flow. As the currency for cellular energy, ATP represents the ultimate measure of energy expenditure. When tissue expends energy during the depolarized state, cytokines initiate increased blood volume through vessel expansion until the vessels can no longer supply sufficient metabolites. Increased cytokine production could then trigger cells within the cortical column to enter the hyperpolarized state, both as a neuroprotective mechanism, and to give the cells a chance to restore their energy stores and allow time for vessels to relax.

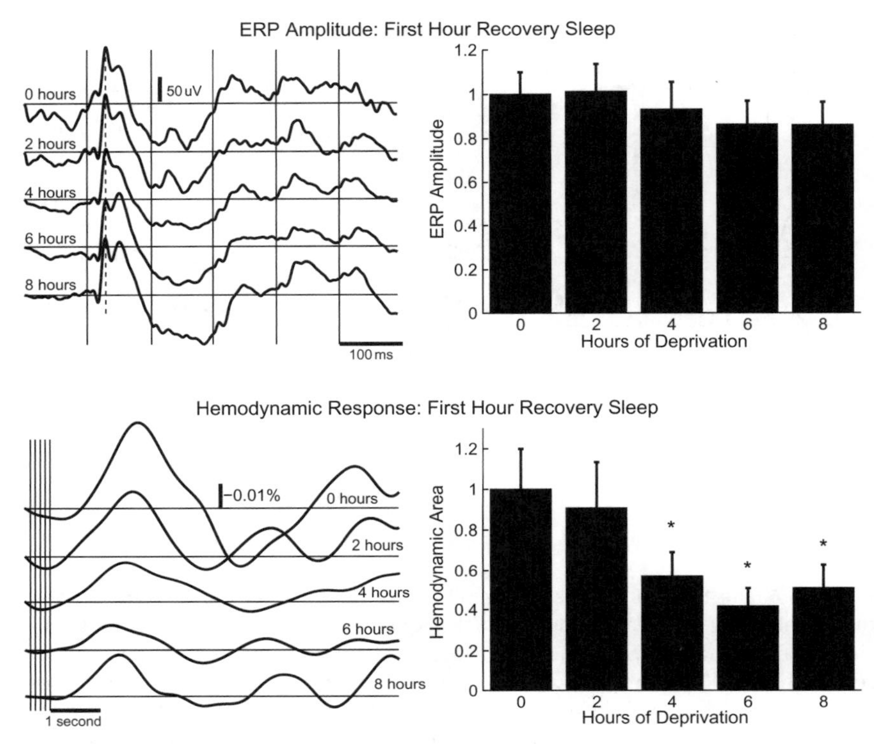

Figure 5.7 To further test possible saturation of the evoked vascular response due to reduced compliance, we subjected animals to sleep deprivation and recorded the evoked electrical and optical responses during subsequent recovery sleep. During the first hour of recovery sleep after sleep deprivation, the electrical evoked response is characteristically large, and does not depend significantly on the amount of prior sleep deprivation. The traces in the upper left panel plot the average electrical evoked response during the first hour of recovery sleep after increasing periods of sleep deprivation from a typical recording. The upper right bar graph depicts the average electrical response following different amounts of sleep deprivation from 12 animals. However, the evoked optical/hemodynamic response is muted with increasing amounts of sleep deprivation (lower panels). These data fit our hypothesis since blood vessels should be more dilated with increasing time awake, and thus the evoked hemodynamic response should be smaller as blood vessels dilate to their maximum volume, resulting in a decreased ability to deliver blood to activated tissue.

CONCLUSIONS

Increasing evidence points to limits in energy stores that relate to sleep need and restoration of these stores during sleep. It is possible that the hyperpolarized/down state is required as a neuroprotective mechanism to prevent metabolite deficit and enable heat removal. Development of agents to temporarily maintain attention, cellular activity states, and performance may help to keep people awake longer, but the long-term consequences of maintaining alert states may have adverse affects that go far beyond currently studied processes. In particular, if the limits of vascular compliance are constantly reached, metabolic deficit may limit the ability of the tissue to maintain activity levels and process incoming information. More dramatically, chronically reaching these limits over the long term may result in cellular trauma, oxidative stress, and cell death. If maintained over many years, such trauma could lead to decreased brain mass in critical areas that were overused by the person, and subsequent degradation of capacity (Macey et al., 2008).

Arguably, many more experiments are required to test these ideas; however, much evidence has already accumulated in support of the mechanisms proposed here. A focused effort to study vascular compliance across normal and sleep deprived conditions will reveal some important results. The experiments conducted to date recorded blood volume and component changes in response to stimulus evoked changes. However, these are indirect measures of blood vessel compliance. By imaging microvessels directly with implantable microscopy technology (Rector & Harper, 1991), future studies should be able to assess the contributions of blood vessel stretching more directly.

ACKNOWLEDGMENTS

This research was supported by the W.M. Keck Foundation, NIH MH60263, and NIH MH71830.

REFERENCES

Balkin, T. J., Bliese, P. D., Belenky, G., Sing, H., Thorne, D. R., Thomas, M., et al. (2004). Comparative utility of instruments for monitoring sleepiness-related performance decrements in the operational environment. *Journal of Sleep Research*, *13*(3), 219–227.

Behzadi, Y., & Liu, T. T. (2005). An arteriolar compliance model of the cerebral blood flow response to neural stimulus. *Neuroimage*, *25*(4), 1100–1111.

Benington, J. H., & Heller, H. C. (1995). Restoration of brain energy metabolism as the function of sleep. *Progress in Neurobiology*, *45*(4), 347–360.

Braun, A. R., Balkin, T. J., Wesenten, N. J., Carson, R. E., Varga, M., Baldwin, P., et al. (1997). Regional cerebral blood flow throughout the sleep-wake cycle. An $H_2(15)O$ pet study. *Brain*, *120*(Pt 7), 1173–1197.

Buxton, R. B., Uludağ, K., Dubowitz, D. J., & Liu, T. T. (2004). Modeling the hemodynamic response to brain activation. *Neuroimage*, *23*(Suppl. 1), S220–233.

Colrain, I. M., & Campbell, K. B. (2007). The use of evoked potentials in sleep research. *Sleep Medicine Reviews*, *11*(4), 277–293.

Devor, A., Dunn, A. K., Andermann, M. L., Ulbert, I., Boas, D. A., & Dale, A. M. (2003). Coupling of total hemoglobin concentration, oxygenation, and neural activity in rat somatosensory cortex. *Neuron, 39*(2), 353–359.

Fehm, H. L., Kern, W., & Peters, A. (2006). The selfish brain: Competition for energy resources. *Progress in Brain Research, 153*, 129–140.

Filosa, J. A., & Blanco, V. M. (2007). Neurovascular coupling in the mammalian brain. *Experimental Physiology, 92*(4), 641–646.

Grandner, M. A., Patel, N. P., Gehrman, P. R., Perlis, M. L., & Pack, A. I. (2010). Problems associated with short sleep: Bridging the gap between laboratory and epidemiological studies. *Sleep Medicine Reviews, 14*(4), 239–247.

Grenèche, J., Krieger, J., Erhardt, C., Bonnefond, A., Eschenlauer, A., Muzet, A., et al. (2008). EEG spectral power and sleepiness during 24 h of sustained wakefulness in patients with obstructive sleep apnea syndrome. *Clinical Neurophysiology, 119*(2), 418–428.

Huber, R., Ghilardi, M. F., Massimini, M., Ferrarelli, F., Riedner, B. A., Peterson, M. J., et al. (2006). Arm immobilization causes cortical plastic changes and locally decreases sleep slow wave activity. *Nature Neuroscience, 9*, 1169–1176.

Huppert, T. J., Allen, M. S., Benav, H., Jones, P. B., & Boas, D. A. (2007). A multicompartment vascular model for inferring baseline and functional changes in cerebral oxygen metabolism and arterial dilation. *Journal of Cerebral Blood Flow and Metabolism, 27*(6), 1262–1279.

Kattler, H., Dijk, D. J., & Borbely, A. A. (1994). Effect of unilateral somatosensory stimulation prior to sleep on the sleep EEG in humans. *Journal of Sleep Research, 3*, 1599–1604.

Kavanau, J. L. (1996). Memory sleep and dynamic stabilization of neural circuitry: Evolutionary perspectives. *Neuroscience and Biobehavioral Reviews, 20*, 289–311.

Koch, C. (2004). *The Quest for Consciousness*. Englewood, Colorado: Roberts and Company.

Krueger, J. M., & Obal, F., Jr. (1993). A neuronal group theory of sleep function. *Journal of Sleep Research, 2*, 63–69.

Krueger, J. M., Rector, D. M., Roy, S., Van Dongen, H. P., Belenky, G., & Panksepp, J. (2008). Sleep as a fundamental property of neuronal assemblies. *Nature Reviews Neuroscience, 9*(12), 910–919.

Lydic, R. (1987). State-dependent aspects of regulatory physiology. *The FASEB Journal, 1*(1), 6–15.

Macey, P. M., Kumar, R., Woo, M. A., Valladares, E. M., Yan-Go, F. L., & Harper, R. M. (2008). Brain structural changes in obstructive sleep apnea. *Sleep, 31*(7), 967–977.

Mahowald, M. W., & Schenck, C. H. (2005). Insights from studying human sleep disorders. *Nature, 437*, 1279–1285.

Massimini, M., Rosanova, M., & Mariotti, M. (2003). EEG Slow (~1 Hz) waves are associated with nonstationarity of Thalamo-Cortical sensory processing in the sleeping human. *Journal of Neurophysiology, 89*, 1205–1213.

Massimini, M., Tononi, G., & Huber, R. (2009). Slow waves, synaptic plasticity and information processing: Insights from transcranial magnetic stimulation and high-density EEG experiments. *The European Journal of Neuroscience, 29*(9), 1761–1770.

McGinty, D., & Szymusiak, R. (1990). Keeping cool: A hypothesis about the mechanisms and functions of slow-wave sleep. *Trends in Neurosciences, 13*(12), 480–487.

Penfield, W. (1958). Functional localization in temporal and deep sylvian areas. *Research Publications – Association for Research in Nervous and Mental Disease, 36*, 210–226.

Phillips, D. J., Schei, J. L., Meighan, P. C., & Rector, D. M. (2011). Cortical evoked responses associated with arousal from sleep. *Sleep, 34*(1), 65–72.

Phillips, D. J., Schei, J. L., Meighan, P. C., & Rector, D. M. (2011). State-dependent changes in cortical gain control as measured by auditory evoked responses to varying intensity stimuli. *Sleep, 34*(11), 1527–1537.

Pigarev, I. N., Nothdurf, H. C., & Kastner, S. (1997). Evidence for asynchronous development of sleep in cortical areas. *Neuroreport, 8*, 2557–2560.

Rector, D., & Harper, R (1991). Imaging of hippocampal neural activity in freely behaving animals. *Behavioural Brain Research, 42*(2), 143–149.
Rector, D. M., Topchiy, I. A., Carter, K. M., & Rojas, M. J. (2005). Local functional state differences between rat cortical columns. *Brain Research, 1047*, 45–55.
Rector, D. M., Schei, J. L., & Rojas, M. J. (2009a). Mechanisms underlying state dependent surface-evoked response patterns. *Neuroscience, 159*, 115–126.
Rector, D. M., Schei, J. L., Van Dongen, H. P., Belenky, G., & Krueger, J. M. (2009b). Physiological markers of local sleep. *The European Journal of Neuroscience, 29*(9), 1771–1778.
Roy, S., Krueger, J. M., Rector, D. M., & Wan, Y. (2008). Network models for activity-dependent sleep regulation. *Journal of Theoretical Biology, 253*, 462–468.
Roy, C. S., & Sherrington, C. (1890). On the regulation of the blood supply of the brain. *The Journal of Physiology, 11*, 85–108.
Saper, C. B., Chou, T. C., & Scammel, T. E. (2001). The sleep switch: Hypothalamic control of sleep and wakefulness. *TINS, 24*(12), 726–731.
Szymusiak, R., & McGinty, D. (2008). Hypothalamic regulation of sleep and arousal. *Annals of the New York Academy of Sciences, 1129*, 275–286.
Shaw, G. L., Harth, E., & Scheibel, A. B. (1982). Cooperativity in brain function: Assemblies of approximately 30 neurons. *Experimental Neurology*, 77(2), 324–358.
Schei, J. L., Foust, A. J., Rojas, M. J., Navas, J. A., & Rector, D. M. (2009). State-dependent auditory evoked hemodynamic responses recorded optically with indwelling photodiodes. *Applied Optics, 48*(10), D121–D129.
Steriade, M., McCormick, D. A., & Sejnowski, T. J. (1993). Thalamocortical oscillations in the sleeping and aroused brain. *Science, 262*, 679–685.
Steriade, M., Timofeev, I., & Grenier, F. (2001). Natural waking and sleep states: A view from inside neocortical neurons. *Journal of Neurophysiology, 85*, 1969–1985.
Sherrington, C. S. (1951). *Man on his nature* (2nd ed.). Cambridge, MA: Cambridge University Press.
Timofeev, I., Contreras, D., & Steriade, M. (1996). Synaptic responsiveness of cortical and thalamic neurones during various phases of slow sleep oscillation in cat. *The Journal of Physiology, 494*(1), 265–278.
Van Dongen, H. P., Maislin, G., Mullington, J. M., & Dinges, D. F. (2003). The cumulative cost of additional wakefulness: Dose-response effects on neurobehavioral functions and sleep physiology from chronic sleep restriction and total sleep deprivation. *Sleep, 26*(2), 117–126.
Vyazovskiy, V. V., Olcese, U., Lazimy, Y. M., Faraguna, U., Esser, S. K., Williams, J. C., et al. (2009). Cortical firing and sleep homeostasis. *Neuron, 63*, 865–878.
Weitzman, E. D., & Kremen, H. (1965). Auditory evoked responses during different stages of sleep in man. *Electroencephalography and Clinical Neurophysiology, 18*, 65–70.
Wesensten, N. J., Belenky, G., Thorne, D. R., Kautz, M. A., & Balkin, T. J. (2004). Modafinil vs. caffeine: Effects on fatigue during sleep deprivation. *Aviation, Space, and Environmental Medicine, 75*(6), 520–525.
Yasuda, T., Yasuda, K., Brown, R. A., & Krueger, J. M. (2005). State-dependent effects of light-dark cycle on somatosensory and visual cortex EEG in rats. *The American Journal of Physiology, 289*, R1083–R1089.

Sleep and Learning in Birds: Rats! There's More to Sleep

Daniel Margoliash[1,2] and Timothy P. Brawn[1]
[1]Department of Psychology, University of Chicago, Chicago, IL, USA, [2]Department of Organismal Biology and Anatomy, University of Chicago, Chicago, IL, USA

Why should a neuroscientist study sleep in birds? Considering mammals, there is a broad literature describing neuronal and genetic mechanisms of sleep regulation, evidence in rodents and humans supporting a role for sleep in synaptic homeostasis, a well-developed rodent model of spatial processing involving hippocampal sleep reactivations, a visual system model for developmental effects of sleep in cats, and extensive behavioral, imaging, and polysomnographic evidence of sleep consolidation in humans. Why study birds, let alone flies or worms: What is missing?

SLEEP RESEARCH FROM THE ETHOLOGICAL PERSPECTIVE

There are many reasons beyond certain manifest technical advantages to study sleep in birds, but here we identify four conceptual issues. Because sleep has a complex physiology that broadly influences behavior, understanding how the physiology of sleep influences a specific behavior requires an appropriate animal model of that behavior. With regard to sleep and learning, the focus of mammalian animal research has been more on neurophysiological mechanisms of plasticity, and less on the behavioral consequences of sleep processes. Conversely, though our understanding of the neurophysiology of bird sleep is still in its infancy, some important results have already been achieved by connecting strongly with ethologically grounded behaviors. For example, recent behavioral and neurophysiological studies have connected developmental sensorimotor vocal learning to sleep mechanisms, with surprising conclusions that reshape our thinking about offline components of vocal learning, and skill learning in general. The well-established similarities between aspects of song learning and language acquisition make direct predictions on human behavior, which have gained some support in recent studies. This conclusion emphasizes a broader one, that since sleep is manifest broadly in the animal kingdom, it can profitably

Sleep and Brain Activity
DOI: http://dx.doi.org/10.1016/B978-0-12-384995-3.00006-X

be studied in relation to specific behavioral adaptations in a broad range of animal species. Conversely, if the goal is to inform human behavior, comparative studies bear the burden to assess similarity both in behavioral and physiological traits in relation to humans, ultimately evaluated in terms of evolutionary processes. All comparative work bears such a burden, including work in mammals, but whereas there are mammalian systems for studying sleep consolidation effects or neuronal plasticity as is observed in humans, the two approaches have yet to be combined in a single system.

A second conceptual issue concerns the relation between the memory systems thought to be engaged in human studies of sleep consolidation compared to animal studies. The most consistent evidence of sleep consolidation in humans has been for nondeclarative memory tasks. Though the effects of sleep are increasingly being studied for other types of memory, there is a disconnect between studies of human memory and rodent studies that are largely focused on patterns of hippocampal place cell activity. Moreover, the pattern of performance changes observed across waking and sleep in adult humans has only been compellingly replicated recently, in an adult songbird (European starling) model. If the functional distinction of memory systems is important in assessing learning and memory behaviors, and if identifying similar patterns of performance in humans and animals facilitates understanding the mechanisms of human behavior, this currently recommends the starling model. We develop this concept and line of inquiry in a separate section that follows.

A third issue is more comparative and phylogenetic. There has been a longstanding misconception regarding the organization of bird pallium and its relation to (mammalian) neocortex. The avian cortex is not a layered structure but is organized into regions or fields, often separated by fiber tracts (Fig. 6.1, *left panel*). Early anatomists mistakenly assumed that the entire avian pallium was a hypertrophied striatum (Jarvis et al., 2005). A recent broad reevaluation of these relations considering anatomical, physiological, behavioral, and molecular data has identified homologies between bird cortex and neocortex, facilitating the placement of bird studies in a meaningful mammalian context (Reiner et al., 2004). For example, numerous hodological characteristics and histological markers identify a limited, circumscribed region of the avian pallium as the proper avian striatum (Fig. 6.1, *right panel*). In other regions of the avian cortex a series of canonical sensory pathways mimic the canonical neocortical circuit. Each avian pathway includes a granule cell layer receiving ascending input from a different dorsal thalamic nucleus, projections from that layer (possibly through intermediate local

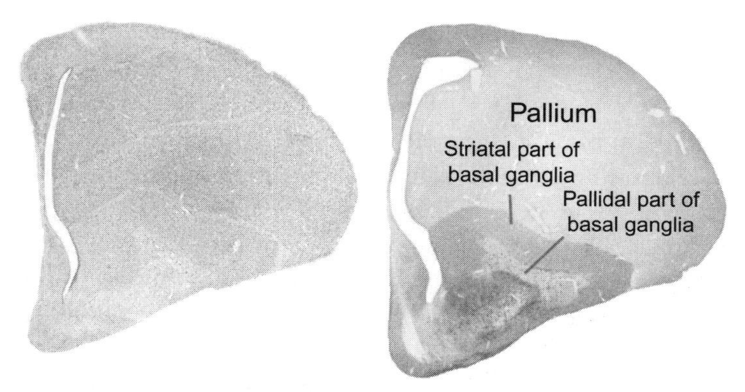

Figure 6.1 ***The avian telencephalon is dominated by a cortex (pallium), not striatum.*** *Left panel:* A sagittal section through pigeon telencephalon, showing regions separated by fiber tracts. Not having neocortical laminar organization, avian telencephalon was incorrectly thought to represent a massively enlarged striatum, not cortex. *Right panel:* Pigeon telencephalon labeled with choline acetyltransferase, one of a very large panel of chemical and molecular markers indicating that avian striatum is restricted to the ventral region of the telencephalon. *Up*, dorsal; *left*, medial. *Adapted from Reiner et al. (2004).*

connections) to secondary neurons (akin to neocortical layer 2/3 neurons), which in turn project to neurons that project out of the cortex (akin to neocortical layer 5 neurons). Neocortical layer-specific molecular markers are selectively expressed in corresponding structures of the avian cortex (Dugas-Ford, 2009). The avian forebrain also has a full complement of ascending modulatory systems arising from brainstem, midbrain, and basal forebrain, as is seen in mammalian cortex (Ball & Balthazart, 2010). Remarkably, recent data have demonstrated radial, columnar-like organization in the avian auditory cortex (Wang, Brzozowska-Prechtl, & Karten, 2010).

Collectively, these observations compellingly support the hypothesis that cortical cells and circuits in birds and mammals share a common schema and evolutionary history (Karten, 1997). The central conclusion of this profound hypothesis is that the basic pattern of forebrain connectivity shared between birds and mammals arises from cells and circuits sharing a common ancestor. This helps to explain why all vertebrate systems share common principles of functional organization (Ulinski, 1984). This conceptualization of the avian cortex helps to interpret physiological/functional data collected in avian auditory cortex that was difficult to reconcile with the old hierarchical organization scheme. It additionally challenges uniqueness claims for neocortex and brings equivalent claims regarding mammalian sleep under scrutiny.

Of relevance to the value of avian sleep studies are recent observations showing that sleep in songbirds shares many features with mammalian sleep (see below). Not all aspects of human sleep physiology have been observed in birds (nor are all aspects of mammalian nonhuman (e.g., rodent) sleep prominent in humans). Whether the similarity between songbird and mammalian sleep arises from deep homology shared between birds and mammals remains an open question given the current understanding of sleep in basal birds, basal mammals, and in reptiles. Nevertheless the homology between avian and mammalian cortical structures coupled with a similar sleep structure make songbirds an attractive model system for sleep research. Furthermore, the exceptionally rich song production system of songbirds along with their well-developed auditory system allows the work to be placed in a comparative and ethological context supported by rigorous experimental observation.

A final conceptual issue is that these converging lines of evidence suggest deep homologies linking birds and mammals. Presumably these arise from shared molecular mechanisms for pattern formation along the neuroaxis, mechanisms that are broadly expressed in invertebrates as well as all vertebrates, which are somehow differentially expressed in higher vertebrates. Understanding how the expression of these mechanisms leads to the formation of an elaborated pallium and sleep structure is a fundamental question for sleep research (Rattenborg, 2006). Restricting the study of sleep to mammals limits the phylogenetic scope of the work, yet the deepest mysteries of sleep—why do most if not all animals sleep (Cirelli & Tononi, 2008; but see Siegel, 2008), and what is the relation between an elaborated pallium and an elaborated sleep structure—are best addressed with a broad evolutionary perspective.

THE ORGANIZATION OF SLEEP IN BIRDS

Sleep in birds, as in mammals, is associated with species-specific behavioral postures and may be triggered by environmental releasers, features of innate behavior well known in ethology. Beyond such overt behaviors, birds also exhibit complex EEG patterns that are associated with different stages of sleep. Birds produce slow wave activity associated with slow wave sleep (SWS) as well as rapid eye movements (REM) with occasional correlated head movements associated with REM sleep. The weight of evidence is that both SWS and REM sleep are absent in reptiles and fish, suggesting that birds are the only nonmammalian species expressing both SWS and REM. Sleep homeostasis, which is commonly observed in mammals, is also observed in birds (Jones, Vyazovskiy, Cirelli, Tononi, & Benca, 2008). Likewise, a tendency for REM sleep to increase throughout the night has been observed in a number of bird species (Low, Shank, Sejnowski, & Margoliash, 2008; Szymczak, 1987; Tobler & Borbely, 1988).

The sleep structure of numerous avian species has been characterized with EEG recordings (see Campbell & Tobler, 1984; Rattenborg et al., 2002). Sleep in most avian species has been characterized by non-REM (NREM) activity, with REM sleep observed in brief and infrequent periods. The songbirds (passerine, or perching birds) appear to be outliers to this overall pattern. The several songbird species studied showed more REM than observed in nonpasserine species, systematic variation of SWS and REM sleep, and other features that are distinct from the typical avian pattern (e.g., Szymczak et al., 1996; Jones et al., 2008). In a striking example, an ultradian pattern was observed in adult male zebra finches that was characterized by SWS density falling from 50% at the onset of sleep to 25% by the end of the night. Concurrently, the number and duration of REM bouts systematically increased so that REM accounted for 30% of sleep and individual REM bouts averaged 15 seconds by the end of the night. This pattern is more complex than that seen for nonpasserine species and is more similar to mammalian sleep than has previously been observed for birds.

It is valuable to consider these observations through the lens of phylogeny. The pattern of sleep-related cortical activity common among mammals is not observed in basal mammals, which nevertheless exhibit a similar pattern of sleep-related brainstem activity. Thus, one hypothesis is that increased encephalization within the mammalian lineage was associated with the emergence of sleep-related cortical activity. A similar process may have occurred in the avian lineage. Approximately 5,000 of the more than 9,000 avian species are passerines. Of these, almost 4,000 are oscine passerines, so-called true songbirds. There are additionally some 350 species of parrots (psittacines). Parrots and songbirds have been proposed to be sister groups in a recent molecular phylogenetic study of birds. It remains to be seen if the more complex cortical patterns of sleep recently observed in a few songbird species obtain more generally in oscines, in passerines, or even in psittacines. If complex cortical patterns of sleep are associated with increased encephalization, then we would expect this to emerge in the lineage including songbirds and parrots, which include the most cognitively advanced avian species. Thus the principle distinction between avian and mammalian sleep might simply be related to the timing of increased encephalization within each lineage, which occurred earlier in mammals than in birds. This also hints at a conserved feature of sleep (perhaps brainstem mechanisms) that might be shared with reptiles, albeit a study in turtles failed to identify brainstem activity associated with REM sleep (Eiland, Lyamin, & Siegel, 2001). Given the emerging understanding of avian and mammalian forebrain homology, this motivates continued study of reptilian sleep.

TOWARDS AN ANIMAL MODEL: SLEEP-DEPENDENT MEMORY CONSOLIDATION IN HUMANS

Here, we take up the challenge for comparative research investigating the role of sleep in memory consolidation. First we describe a substantial literature that helps to define sleep-dependent memory consolidation in humans. These observations, which have helped shape our efforts to develop a corresponding animal model system, are described in the following section.

Memory consolidation describes a process in which a newly acquired memory is transformed from a labile state, where it may be susceptible to interference or decay, to a more stable and strengthened form. Sleep is widely believed to play a fundamental role in the consolidation of memories (Diekelmann & Born, 2010; Walker & Stickgold, 2006), a position that is broadly supported by behavioral studies of human memory. The standard approach for uncovering behavioral evidence of sleep consolidation has been to train participants on a memory task in the morning or evening. The participants are then retested after a 12 hr retention interval that consists entirely of wakefulness or that includes a normal night of sleep in order to determine whether performance after sleeping retention is better than after waking retention. Using this approach, sleep consolidation has been reported for a variety of memory tasks, including associative memory (Ellenbogen, Hulbert, Jiang, & Stickgold, 2009; Ellenbogen, Hulbert, Stickgold, Dinges, & Thompson-Schill, 2006), emotional memory (Hu, Stylos-Allan, & Walker, 2006; Payne, Stickgold, Swanberg, & Kensinger, 2008), prospective memory (Scullin & McDaniel, 2010), and episodic memory (Racsmany, Conway, & Demeter, 2010).

Though studies relying on 12 hr retention intervals can indicate the differential effects of waking and sleeping retention on the consolidation of a newly formed memory, interpretation of the results is confounded by circadian factors because training and testing for the two conditions occur at different times of day. Consequently, performance changes could reflect circadian influences on memory rather than an underlying consolidation process if the ability to acquire or perform a task is different in the morning and evening. Accordingly, behavioral studies of sleep consolidation often include conditions with 24 hr retention intervals in addition to the 12 hr retention periods. In these conditions, participants are trained and tested in the morning or evening and retested 24 hrs later, ensuring that performance cannot be attributed to the circadian time. Human studies have provided extensive behavioral evidence of sleep-dependent consolidation for a broad range of memory tasks with this more complete experimental design. For example, sleep has been shown to benefit motor-sequence learning (Brawn, Fenn, Nusbaum, & Margoliash, 2010; Korman et al., 2007; Walker, Brakefield, Morgan, Hobson, & Stickgold, 2002), sensorimotor learning (Brawn, Fenn, Nusbaum, & Margoliash, 2008; Robertson, Pascual-Leone, & Press, 2004), visual texture discrimination learning (Gais, Plihal, Wagner, & Born, 2000), the perceptual learning of synthetic speech (Fenn, Nusbaum, & Margoliash, 2003), spatial navigation memory (Ferrara et al., 2008), spatial associative learning (Talamini, Nieuwenhuis, Takashima, & Jensen, 2008), and relational memory (Ellenbogen, Hu, Payne, Titone, & Walker, 2007).

Variants of this approach have further confirmed a role for sleep in memory consolidation. In the nap paradigm, participants receive task training in the morning and are retested later in the day after a retention period that either includes or does not include a nap. Naps have been shown to benefit memory in a manner similar to a full night of sleep (e.g., Korman et al., 2007; Mednick, Nakayama, & Stickgold, 2003; Nishida & Walker, 2007; Tucker et al., 2006). These studies represent a powerful approach for investigating sleep consolidation. On the one hand, nap studies avoid circadian confounds because the time-of-training and time-of-testing are identical for the nap and no-nap conditions. On the other hand, nap studies provide evidence against a purely time-dependent mechanism of consolidation. According to the time-dependent hypothesis, memory consolidation merely requires the passage of an appropriate amount of time. Any memory benefits that appear after a night of sleep would result from a time-dependent mechanism that may act more efficiently across many hours of sleep due to a lack of interfering experiences. However, nap studies establish that memory benefits can arise after a short retention period as long as it includes sleep, indicating that the occurrence of sleep rather than the passage of time is the critical factor.

Finally, sleep deprivation studies have provided additional support for a sleep-dependent rather than time-dependent consolidation mechanism. In the sleep deprivation paradigm, participants are trained on a task and then retested at least two days later. Participants in a deprivation condition are not allowed to sleep on the first night after training but are allowed a full night of recovery sleep on the second night before being retested the following day. This recovery night of sleep ensures that any performance impairments during the retest cannot be attributed to the many confounds that would result from being tested while in a sleep-deprived state. If a strictly time-dependent process governed consolidation, memory benefits should appear even after sleep deprivation because the passage of time, not sleep, would be responsible for the consolidation. Yet, deprivation studies have demonstrated that the expected memory benefits do not appear when participants are deprived of sleep on the first night after training (e.g., Fischer, Nitschke, Melchert, Erdmann, & Born, 2005; Gais, Lucas, & Born, 2006; Stickgold, James, & Hobson, 2000), suggesting a necessary role for sleep.

Overall, there is compelling human behavioral evidence that sleep is important for the consolidation of newly acquired memories. While the exact effects of waking and sleeping retention on task performance depend on the type of memory probed and the experimental procedures used, many human studies highlight a pattern of consolidation defined by performance deterioration across waking retention periods prior to sleep followed by performance enhancement and stabilization after sleep.

THE HIPPOCAMPAL SYSTEM AND SLEEP-DEPENDENT MEMORY CONSOLIDATION: LIMITED EVIDENCE OF BEHAVIORAL CONSEQUENCES

Human studies have presented convincing behavioral evidence of sleep-dependent consolidation but are ultimately limited at uncovering the underlying neural mechanisms, which will require appropriate animal models. To date, most animal studies of sleep consolidation have explored the "standard model" of memory consolidation. According to the standard model, the hippocampal formation receives input from neocortical regions involved in the initial encoding of an experience, which binds the information into a coherent memory trace. During sleep, coordinated reactivation of the memory trace is thought to transfer the memory from a hippocampal to a neocortical representation (Frankland & Bontempi, 2005; Rattenborg, Martinez-Gonzalez, Roth, & Pravosudov, 2011). This model is supported by studies showing that the firing patterns expressed in hippocampal neurons of rats moving along a track are reactivated during subsequent sleep (e.g., Kudrimoti, Barnes, & McNaughton,

1999; Lee & Wilson, 2002; Wilson & McNaughton, 1994) and coincide with reactivations in the visual cortex (Ji & Wilson, 2007). Furthermore, sleep reactivation of waking neural activity has been identified in the medial prefrontal cortex after rats have learned a rule (Peyrache, Khamassi, Benchenane, Wiener, & Battaglia, 2009) or been trained to run to a sequence of locations (Euston, Tatsuno, & McNaughton, 2007). These hippocampal and cortical sleep reactivations have been interpreted as evidence of a hippocampus-to-cortex memory transfer that underlies memory consolidation and the associated memory benefits of sleep (Marshall & Born, 2007; O'Neill, Pleydell-Bouverie, Dupret, & Csicsvari, 2010; Rasch & Born, 2007). There are elegant studies relating hippocampal physiological activity to brain rhythms, mechanisms of plasticity, and behavioral state (see Poe, 2010).

Memory consolidation, however, cannot be inferred from neural events alone (Hennevin, Huetz, & Edeline, 2007). Reactivation of waking neural activity during subsequent sleep is not by itself evidence of a memory consolidation process. Learning may alter brain activity during subsequent sleep, but neural plasticity is only tenuously linked to the consolidation of a newly formed memory trace in studies that do not verify that sleep reactivations confer any memory benefit on the animal.

Recent studies have begun to address to this concern by incorporating behavioral measures of spatial tasks into studies of sleep reactivation. For example, in one study, an increase in hippocampal sharp wave ripples, which are associated with hippocampal reactivations, was correlated with performance on a place-reward association task (Ramadan, Eschenko, & Sara, 2009). In two other studies, electrical stimulation was used to disrupt sharp wave ripples during post-training sleep. These stimulations produced performance impairments compared to control conditions on a place-reward association task (Girardeau, Benchenane, Wiener, Buzsaki, & Zugaro, 2009) and a spatial navigation task (Ego-Stengel & Wilson, 2010). Given that the performance benefits only became apparent after several days of training and testing, however, this makes it difficult to determine the respective influences of sleep, waking time, and repeated training/testing. While these studies have forged a stronger link between sleep reactivation and memory than prior work, the relatively weak behavioral effects in these animal models are in stark contrast to human studies that reveal clear memory benefits after a single period of sleep.

We emphasize that our point is not that studies need to achieve some arbitrary threshold to be "acknowledged" as being "relevant" to learning. Our point is that without a strong and direct linkage to a learning phenomenon, it is difficult to relate changes in behavior with changes in neural activity; hence the mechanisms of learning remain unresolved. We amplify on these points by beginning to develop an example in the following section.

EUROPEAN STARLINGS AND SLEEP-DEPENDENT MEMORY CONSOLIDATION: BEHAVIOR

Though human and animal studies of sleep-dependent consolidation have been mutually beneficial, a fundamental discrepancy between the two lines of research has remained. Human studies have provided compelling behavioral evidence of sleep-dependent consolidation while being limited at uncovering the underlying mechanisms that account for the sleep-dependent performance benefits. Animal studies have provided neural data that could be the basis of memory consolidation but with limited evidence that adult animals receive any memory benefit from the proposed mechanisms. It remains unresolved whether adult animals express sleep-dependent memory benefits similar to those observed in humans.

To address this question, we developed a paradigm using European starlings (*Sturnus vulgaris*) that was modeled after the standard behavioral approach of human sleep-memory studies. Starlings are a species of songbird with complex vocalizations that consist of long sequences of temporally discrete motifs, with each motif itself ($\approx$1 s duration) being a complex sequence of syllables (Adret-Hausberger & Jenkins, 1988; Eens, 1997). Wild starlings maintain large repertoires of mostly unique motifs (Chaiken, Bohner, & Marler, 1993), and can learn to identify individuals by associating the production of certain motifs with specific individuals (Gentner & Hulse, 2000; Gentner, Hulse, Bentley, & Ball, 2000). Starlings are also advantageous in that they can be trained using operant techniques to classify auditory stimuli through differential reinforcement of responses to different stimuli. In the Go/No-Go paradigm, starlings are rewarded with food access when they respond to the "Go" stimulus but are given a lights-out punishment when they respond to the "No-Go" stimulus. Over the course of training, starlings learn to respond to the Go stimulus and withhold response from the No-Go stimulus, thus demonstrating the ability to learn and maintain auditory classifications. Thus, this approach takes advantage of natural song recognition behavior along with the ease in which starlings can perform operant learning tasks in the laboratory (Gentner, 2004). This results in an attractive model system for studying auditory perceptual learning and associative memory.

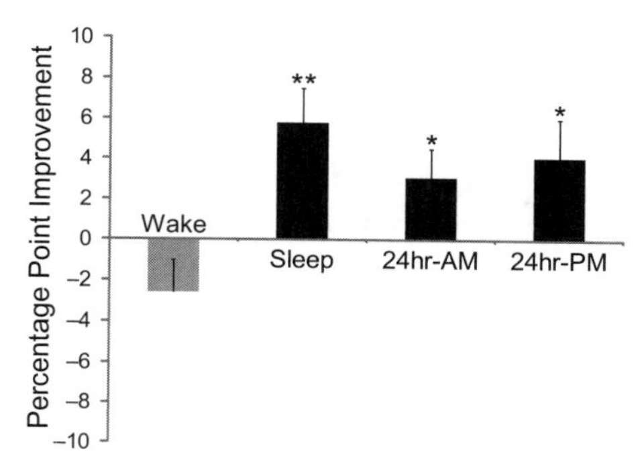

Figure 6.2 ***Auditory classification performance improvement in starlings.*** Starlings were trained to classify two 5 s segments of novel starling song and retested after a retention period that consisted of wakefulness (gray bar) or that included a night of sleep (black bars). Performance improvement scores were calculated as the difference between the post-retention and post-training test scores. Data are means ± SEM (*$p < 0.05$; **$p < 0.01$). *Data are from Brawn, Nusbaum et al. (2010).*

To determine whether starlings express behavioral evidence of sleep-dependent consolidation, 24 starlings each completed six experimental conditions in which they learned to classify pairs of 5 s segments of novel starling song. In each condition, starlings were given a 2 hr training session during which they could complete up to 200 trials followed by a 50 trial post-training test. The starlings were then given a 50 trial post-retention test after an interval that consisted of a full day awake or that included a night of sleep. In the "wake" condition, in which starlings were trained in the morning and retested the same evening, classification performance decreased nonsignificantly from the post-training test to the post-retention test. By comparison, the "sleep" condition, in which starlings were trained in the evening and retested the following morning, exhibited a significant performance improvement. Likewise, the "24-hr AM" and "24-hr PM" conditions, both of which included a night of sleep during the retention interval, expressed significant performance improvements. These results were then replicated in two conditions that were retested after both waking and sleeping retention. In the "AM-PM-AM" condition, classification performance declined nonsignificantly across wakefulness and then improved significantly after sleep. In the "PM-AM-PM" condition, performance improved significantly after sleep followed by a nonsignificant change across the day. The results demonstrate that sleep produces a pattern of memory benefits in starlings that is similar to that observed in humans (Fig. 6.2) (Brawn, Nusbaum, & Margoliash, 2010).

We note that to establish the sleep-dependent memory consolidation behavior in starlings (and in humans), it has been necessary to simplify the learning task so that there are significant performance increases at the end of training that forms a baseline for assessing any subsequent changes in performance. How these sleep-dependent learning processes interact with more cognitively complex learning processes that presumably recruit greater forebrain participation remains an open question.

One critical difference between the starling results and human studies is that human task performance often deteriorates significantly before sleep (Brawn et al., 2008; Brawn, Fenn, et al., 2010; Ellenbogen et al., 2009; Ellenbogen et al., 2006; Fenn et al., 2003; Payne et al., 2008), whereas the performance decline across waking retention failed to reach significance in the starlings. This difference could be attributed to interference from the richer waking experience in humans because daytime behavior in human studies is rarely controlled. In contrast, each starling only encountered a very familiar baseline stimulus set when it was not engaged in training or testing sessions, thus reducing potential interference. Because real-world learning experiences often involve the acquisition of similar skills or information that could interfere with each other, investigating memory processing under conditions of interference may prove to be a more informative approach for understanding how memories are consolidated.

To test this hypothesis, we extended the auditory classification paradigm to explore the interaction between interference and consolidation across waking and sleeping retention by training starlings on two similar, therefore putatively interfering, classification tasks (T. Brawn, unpublished data). Starlings each completed seven experimental conditions that followed an A-B-A (interference) or A-A (control) design. The interference training (additional training on new song stimuli) occurred immediately after completing task A ("Early Interference") or 4 hours later ("Late Interference"). As in the prior experiments, starlings were maintained on a simple baseline task to preclude other sources of interference. Thus we had strong experimental control of the amount and timing of interfering exposure. We observed that interference training caused significant declines in performance when tested on the first evening. This was observed for performance on both tasks, indicating that learning task B resulted in retroactive effects (learning B affects performance on A) and proactive effects (learning A affects performance on B). Despite these interference-induced impairments, performance on both tasks was significantly improved when the starlings were retested on the following day.

These studies establish a new paradigm to investigate memory consolidation. They demonstrate that acquiring new material in the presence of other, interfering material actually magnifies the memory benefit of sleep. This is consistent with the effects of associative interference on declarative memory in humans (Ellenbogen et al., 2009; Ellenbogen et al., 2006). These observations demonstrate that sleep consolidation separately enhances memory of interfering experiences, facilitating opportunistic daytime learning. Our ability to separately manipulate the two memories, potential differences in the time course for manifestation of proactive and retroactive interference, and the distinction between the labile memory representation on the first day and the consolidated memory after a period of sleep, should facilitate our search for neuronal correlates of these behaviors.

FUNCTIONAL ORGANIZATION OF THE STARLING AUDITORY SYSTEM

The auditory classification paradigm described above has provided compelling evidence that sleep produces behavioral memory benefits in starlings that are consistent with the patterns of performance changes across waking and sleep in humans. Though it is not yet known how sleep acts to benefit memory in starlings, the extensive knowledge of the songbird auditory system helps to guide such studies.

Hierarchical organization. We focus on the forebrain, which is the central locus of learning complex and species-specific vocalizations (Simpson & Vicario, 1990), and has descending influence on midbrain and brainstem responses (Dick, Lee, Nusbaum, & Price, 2011; Strait, Chan, Ashley, & Kraus, 2012; Suga & Ma, 2003). In the main avian ascending auditory pathways, nucleus ovoidalis (Ov) of the thalamus receives input from nucleus mesencephalicus lateralis pars dorsalis (MLd) of the midbrain (Karten, 1967) (see Fig. 6.3 for a schematic of the auditory system connections). Neurons in different subdivisions of Ov project to different subdivisions of field L (Karten, 1968), a pallial structure that has traditionally been considered analogous to the mammalian primary auditory cortex. Of the five subdivisions of field L (Fortune & Margoliash, 1992), L2a and L2b receive the majority of thalamic input from the core region of Ov, with a shell region of Ov projecting to subdivisions L1 and L3. Thus Ov core and shell represent different functional pathways (Durand, Tepper, & Cheng, 1992; Wild, Karten, & Frost, 1993). The field L subdivisions are highly interconnected and project to separate higher order structures in

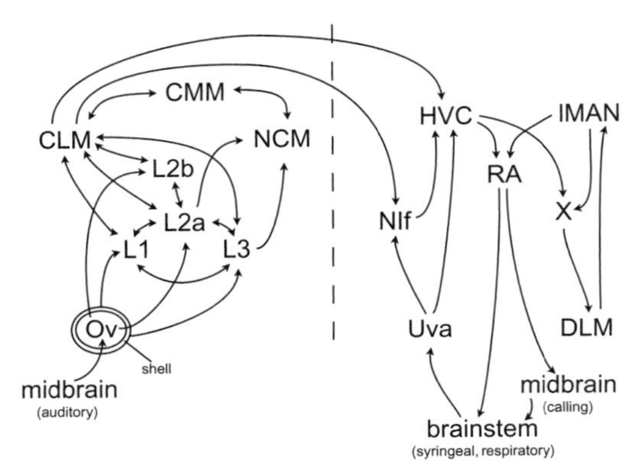

Figure 6.3 ***Connections of the auditory system and song system in a passerine bird.*** *Left side*: Auditory connections showing input from the thalamus (Ov) to subdivisions of Field L in the telencephalon and beyond to secondary auditory areas. There are multiple ascending pathways, roughly separated into one leading to NCM and anther leading to CM (CMM and CLM). *Right side*: Song system connections including a motor pathway (HVC and RA), a basal ganglia pathway (X, DLM, and lMAN), and ascending and descending connections from the motor pathway. Abbreviations: CLM, caudal lateral mesopallium; CMM, caudal medial mesopallium; DLM, nucleus dorsolateralis anterior, pars medialis of the thalamus; HVC, acronym is the proper name; lMAN, lateral magnocellular nucleus of anterior nidopallium; L1/2a/2b/3, subdivisions of field L; NCM, caudal medial nidopallium; NIf, nucleus interface of the nidopallium; Ov, nucleus ovoidalis; RA, robust nucleus of the arcopallium; Uva, nucleus uvaeformis of the thalamus; X, area X.

distinct pathways (Vates, Broome, Mello, & Nottebohm, 1996; Wild et al., 1993) (see Fig. 6.3).

A traditional view of the avian secondary auditory areas emphasizes a hierarchical organization, but this is under active revision. There are unidirectional field L projections to the caudomedial nidopallium (NCM) and reciprocal connections between field L and the caudolateral mesopallium (CLM). In the traditional view, these lead to the caudomedial mesopallium (CMM) at the top of the hierarchy, which has reciprocal connections with NCM and CLM but no direct connections with field L (Vates et al., 1996). Recent anatomical studies in auditory regions of chick brain slices, however, have suggested a revision of this view. Radial patterns of organization were identified with connections between CM (CMM and CLM) and field L that are reminiscent of neocortical columnar organization and connectivity (Wang et al., 2010). This anatomical work places field L3 as an output structure, whose neurons have properties similar to those of neocortical layer 5 neurons.

Changes in neuronal response properties along the auditory pathway have also been interpreted to reflect the traditional hierarchical scheme. An increase in nonlinearity and selectivity for complex acoustic features along the ascending auditory pathway has been noted comparing neuronal responses in the various subdivisions of field L and in CLM and CMM (Muller & Leppelsack, 1985; Sen, Theunissen, & Doupe, 2001; Theunissen et al., 2004). The most compelling physiological distinctions observed, however, have been restricted to comparisons of neurons in L2a/b with neurons in L1, L3, and secondary areas. In contrast, a recent survey of six auditory regions in the starling (L1, L3, CMM, CLM, and NCM) identified L3 as having more selective neurons than any other region in addition to a broader distribution of neurons tolerant to within-class distinction than any other region (C. D. Meliza, unpublished results). Selectivity for a class and tolerance for differences within a class are both features of object recognition. This result provides a physiological context for the recent anatomical findings.

Neurophysiology of song perceptual learning. The responses in the higher order auditory areas NCM, CMM, and CLM show differential effects of learning (Jeanne, Thompson, Sharpee, & Gentner, 2011; Thompson & Gentner, 2010). These are likely to be sites of physiological mechanisms that drive sleep-dependent memory consolidation. Recordings from CMM in operantly trained starlings seem particularly promising. Neurons in CMM rapidly learn to distinguish between different classes of song stimuli in an operant conditioning paradigm similar to that described above for sleep-dependent memory consolidation. Training starlings to classify starling songs using a Go/No-Go (GNG) or Two-Alternative Choice (2AC) task enhances the neural representation of the trained songs within CMM such that CMM neurons exhibit a significant preference, as quantified by a response strength index, for the trained songs compared to novel songs, as well as a preference for positively reinforced songs (Go stimuli) over negatively reinforced songs (No-Go stimuli) (Gentner & Margoliash, 2003). This indicates that CMM neurons are strongly influenced by the behavioral saliency of the song in addition to its spectrotemporal characteristics. Though some neurons respond to a large proportion of motifs in the training songs, CMM neurons are generally selective for a limited number of trained motifs, suggesting that a population of CMM neurons tuned to behaviorally salient motifs mediates song recognition and classification. Significantly, the enhanced response to training stimuli was broadly represented across the population of CMM neurons.

Our initial study of CMM had examined responses at a single point in time, immediately after birds had achieved asymptotic behavior on the GNG or 2AC training (Gentner & Margoliash, 2003). More recently we have investigated the time course of the classification learning (D. Zaraza, unpublished results). In this new study, starlings were extensively trained on a GNG task to classify a pair of starling songs, remaining on task for at least 10 days of asymptotic performance. Thereafter, awake-restrained starlings participated in multiple single-unit recording sessions, each conducted at the conclusion of a day's operant training session. Unexpectedly, the population of neurons did not show the strong, rate-driven differential responses between the overtrained and unfamiliar songs, which contrasts with the results described above. There was a residual trace of the training in the neuronal response, however, detected in the statistically significant preference of a subpopulation of neurons for familiar motifs over unfamiliar motifs.

In the design of the new study (D. Zaraza, unpublished results), starlings were then transferred to new training stimuli. Birds rapidly learned the new stimuli (in a single session) because they were already familiar with the GNG task. This is very similar to the rapid learning starlings exhibited in the sleep-memory experiments described above. It contrasts with our original study where starlings learned a single task that was more complicated (more songs to discriminate), and the initial task of learning the apparatus contributed to the time course of learning (Gentner & Margoliash, 2003).

We observed that after transferring to a second stimulus, CMM neurons showed the broad population-wide response to the new stimulus, as observed in the prior experiment (Gentner & Margoliash, 2003). Interestingly, the response to the old stimulus that birds were no longer training on also suddenly induced the broad population-wide response. Both facets of the population-wide response (to new and old stimuli) declined over multiple days of recordings, with the response to the new stimulus maintained only in a smaller population of neurons selective for familiar motifs. We speculate that if acquiring multiple memories on the same day induces a population-wide response for each of those memories, the interaction between those responses might be related to the behavioral interference between the memories that we have observed.

A hypothesis on sleep perceptual learning. The correlation between changes in classification performance and changes in CMM activity as starlings rapidly transitioned from chance to asymptotic performance highlights the connection between neural activity in CMM and classification memory. There appear to be two distinct processes associated with perceptual learning. Both selective responses to familiar motifs in limited numbers of neurons and population-wide increases in average responses to songs emerge after learning, but apparently only the population-wide response declines slowly over days.

These preliminary results have many features of memory consolidation that are consistent with a role of sleep. We hypothesize that the population-wide response (first process) helps "tune" or strengthen the motif selective response (second process). The first process is elicited suddenly after the onset of new training, induces responses broadly across the CMM population, induces the response for stimuli that had previously been learned, and decays over days. This suggests that the information is present in a broad network and can be recruited by autoassociative mechanisms. If so, this opens the possibility that the information might also be recruited in sleep replay, which requires correlated activity over populations of neurons. We speculate that an interaction between the two processes occurs during sleep, resulting in changes in the distribution of responses within CMM and across the auditory hierarchy. An important first step in testing these hypotheses will be to assess if changes in CMM neuronal responses over days are time-dependent or sleep-dependent.

Our hypothesis is based on preliminary results from neuronal recordings combined with established results from behavioral experiments. It is the structure of the behavioral results and the confidence in those results that guides our neurophysiological analyses. This helps to emphasize the principal point we are presenting in this chapter, the importance of behavioral studies for a mechanistic understanding of sleep memory consolidation.

AN INTRODUCTION TO BIRDSONG PRODUCTION LEARNING AND THE ROLE OF SLEEP

The majority of sleep-dependent memory consolidation studies in humans have been conducted in adults. Yet the most intensive periods of learning are experienced during development. Sleep is linked to developmental learning through many processes and finds support in the requirement of infants for extensive sleep (Tarullo, Balsam, & Fifer, 2011), and additional confirmation arising from perceptual learning experiments in infants (Gomez, Bootzin, & Nadel, 2006; Hupbach, Gomez, Bootzin, & Nadel, 2009).

Animal studies in birds provide further evidence linking sleep and memory consolidation. Here, we focus on adult song maintenance and developmental learning of bird song. Song learning involves perceptual and procedural (i.e., nondeclarative) learning, but this does not exclude the possibility that declarative memory also contributes to song learning. For example, adult birds use song for individual recognition (Falls, 1982; Stoddard, 1996), and this behavior could conceivably rely in part on declarative memories.

Developmental song learning is extensively studied and thoroughly reviewed (Konishi, 1978; Kroodsma & Miller, 1982; Margoliash, 2002; Marler, 1997), so we provide only a brief description, emphasizing aspects related to sleep-dependent processes. In the classical description, a juvenile songbird still in the altricial phase forms a sensory memory of the songs produced by its father or other adult conspecifics. This represents the sensory phase of song learning, and the memory of song is often conceived of as an acquired sensory "template" used to guide subsequent vocal learning. The sensorimotor phase of song learning begins with subsong, akin to human babbling, which is dominated by relatively long and variable sequences of relatively unstructured sounds. In this phase, the bird produces low amplitude vocalizations with increased spectrotemporal complexity and greater variability in duration and spectral content than innate calls. Subsong transitions to more structured singing where individual syllables and sequences of syllables can be recognized, albeit with variation in both. This plastic singing continues to be modified until a bird crystalizes a song pattern, typically at the time of sexual maturity. In zebra finches, the subject of many developmental studies, the crystallized song starts with introductory notes which are followed by a fixed sequence of syllables called a motif. Unlike the starling song described above, a zebra finch learns only one motif, and a song bout consists of singing that motif one or more times.

This impoverished description of song learning gains vitality when elaborated by the biological richness afforded in the natural history of 4,000 species of oscine passerines. A wealth of studies in numerous species, however, suggests a constant feature. Apparently, every one of these species requires auditory feedback for song learning (Kroodsma, 1982; Kroodsma & Baylis, 1982). In many species, including zebra finches, adult individuals express a reduced requirement for auditory feedback to maintain their songs. In contrast, suboscine species likely show variability in this pattern, with some species not requiring auditory feedback for developmental song learning (Kroodsma & Konishi, 1991) while others likely do (Saranathan, Hamilton, Powell, Kroodsma, & Prum, 2007).

THE BIRD SONG SYSTEM

All oscine passerines express specialized forebrain nuclei associated with song learning and production. One forebrain "motor" pathway, which consists of projections from the association/secondary motor cortical nucleus HVC to the motor output nucleus RA, is necessary for and heavily recruited during singing. A second "anterior forebrain" pathway (AFP) receives input from HVC and ultimately projects onto RA. This includes the basal ganglia components of the song system (in Area X) (Carrillo & Doupe, 2004; Farries, Ding, & Perkel, 2005; Farries & Perkel, 2002) and represents a cortico-basal ganglia-thalamocortical pathway. AFP input to RA gives rise to variation in song production (Kao, Doupe, & Brainard, 2005; Olveczky, Andalman, & Fee, 2005; Scharff & Nottebohm, 1991; Sohrabji, Nordeen, & Nordeen, 1990). The AFP is the principle source of drive onto RA early in development, which is thought to explain the heightened variability of subsong (Aronov, Andalman, & Fee, 2008). The influence of the AFP wanes throughout development, and it has little direct influence on singing in adult birds. Nevertheless the AFP is involved in regulating the influence of auditory feedback (Brainard & Doupe, 2000), and so may contribute to sleep-related regulation of singing (Andalman & Fee, 2009). HVC receives auditory inputs from several sources that ultimately arise from higher-order auditory structures (reviewed above) as well as ascending input from the brainstem that is thought to contain feedback information (Ashmore, Renk, & Schmidt, 2008) (Fig. 6.3). RA projects to the brainstem and midbrain structures involved in syringeal and respiratory control, and these also provide feedback via the ascending input to HVC, completing the loop.

The physiology associated with song production, auditory feedback, song perception, and the development of these properties is complex and extensively analyzed (see reviews in Bolhuis & Gahr, 2006; Margoliash & Schmidt, 2010; Mooney, 2009). Here we emphasize the activity patterns of a few classes of neurons that have been the most carefully studied relative to sleep mechanisms of birdsong learning. In nucleus HVC of the motor pathway, there are two categories of projection neurons (and classes of neurons within each category): HVC neurons that project exclusively to RA (HVC-RAn) and HVC neurons that project exclusively to Area X (HVC-Xn). During singing in adult zebra finches HVC projection neurons emit precisely timed and structured bursts of a few (1–4) spikes at very high rate (burst duration 6–10 ms). Bursting is exceedingly sparse, with a given HVC-RAn emitting a single burst per motif and each HVC-Xn emitting 1–4 bursts per motif (Hahnloser, Kozhevnikov, & Fee, 2006; Kozhevnikov & Fee, 2007). In contrast, during singing the HVC interneurons (HVCi) fire tonically throughout song (Yu & Margoliash, 1996). As described below, a burst firing mode during sleep has been associated with neuronal replay.

Finally, during singing, RA projection neurons (RAn) also emit bursts characterized by a few spikes produced at high frequency. In contrast to the HVC-RAn neurons, RAn burst densely throughout many syllables of song, on average 12 bursts per motif, and they reliably emit each burst every time the bird sings the motif (Leonardo & Fee, 2005; Yu & Margoliash, 1996). Each burst occurs at the same point in time relative to the song acoustics and is thought to be associated with individual notes (components) of a syllable. Typically, each spike burst associated with a different part of a song has a different number of spikes and/or timing of spikes within the burst. The variation within bursts associated with one part of a song is far lower than the differences in bursts associated with other parts of a song. Thus, there are burst "types" that are unique, which has greatly facilitated analysis of neuronal replay during sleep.

NEURONAL REPLAY IN THE SONG SYSTEM

In neuronal rehearsal, activation across regions (and temporal relations in the activation patterns) observed offline, such as during sleep, approximates the patterns of activation observed during daytime behavior. In neuronal replay, the offline activity of neurons and populations of neurons produce veridical if approximate copies of the patterns observed during daytime behavior (Foster & Wilson, 2006; Louie & Wilson, 2001; Wilson & McNaughton, 1994). Veridical replay has been considered *sine qua non* for sleep-dependent processing. This is a limited view driven by the technical challenge of identifying offline structure in activity (typically bursting) without spike activity templates derived from online activity and by the conceptual challenge of interpreting offline activity that is not associated with an objective behavior. Neuronal activity during sleep, however, shows variability not observed during daytime activity, but this variation may carry information rather than noise—for example, activity whose patterns help regulate network plasticity by driving networks into nearby states or related dynamical paths. Furthermore, the activity during sleep may carry signals (information) about different modalities that are not represented in the daytime activity of the same neurons. This can arise via neuromodulatory mechanisms that can, for example, gate sensory information into a motor pathway. Such gating has been observed in the song system, where the responses of HVC and RA neurons to auditory stimuli are weaker or absent in awake compared to sleeping zebra finches (A. S. Dave, Yu, & Margoliash, 1998; Schmidt & Konishi, 1998). This gating also shows circadian variation. HVC neurons of juvenile zebra finches in the plastic song stage of sensorimotor development express auditory activity both day and night. However, these neurons respond more vigorously to playback of the tutor song during the day and more vigorously to playback of the bird's own song at night (Nick & Konishi, 2005). This suggests that different types of information are processed during singing and during offline replay of singing. Veridical playback is also observed in the song system (see below), but this result helps to emphasize that sleep-dependent processing is a far more complicated and multidimensional process than simple veridical playback.

There are extensive neuromodulatory pathways and mechanisms differentially activated in the song system depending on behavioral state (Cardin & Schmidt, 2004a, 2004b; Shea, Koch, Baleckaitis, Ramirez, & Margoliash, 2010; Shea & Margoliash, 2003). These features and the neurochemistry of the song system have been reviewed elsewhere (Ball & Balthazart, 2010; Castelino & Schmidt, 2010; Shea & Margoliash, 2010). The neuromodulatory pathways in birds help regulate state changes, hence sleep states, in much the same way they do in mammals. This is an integral part of sleep-dependent memory processing but will not be reviewed here.

NEURONAL REPLAY IN ADULT SONG BIRDS

Basic features. Neuronal replay of RAn in adult zebra finches is perhaps the most compelling example of replay at a single neuron level that has been reported to date (A. S. Dave & Margoliash, 2000). To demonstrate replay, it is necessary to directly compare single or population neuronal activity during waking and sleeping. Correlated bursting and other evidence suggest that replay is observed throughout the song system (A. S. Dave & Margoliash, 2000; Hahnloser, Kozhevnikov, & Fee, 2002; Kozhevnikov & Fee, 2007; Sutter & Margoliash, 1994), and this assumption underpins some analyses of HVC activity (Hahnloser et al., 2002). But to date, replay has only formally been demonstrated for RAn in adult zebra finches. The technical challenges of maintaining recordings during singing and then while an animal sleeps are substantial.

A related feature of zebra finch song system activity during sleep is that neurons express strong and highly selective response to a single stimulus—playback of the bird's own song (BOS) (Dave & Margoliash, 2000). Neurons give stronger responses to BOS than to conspecific songs or BOS presented in reverse (which preserves the overall spectral content).

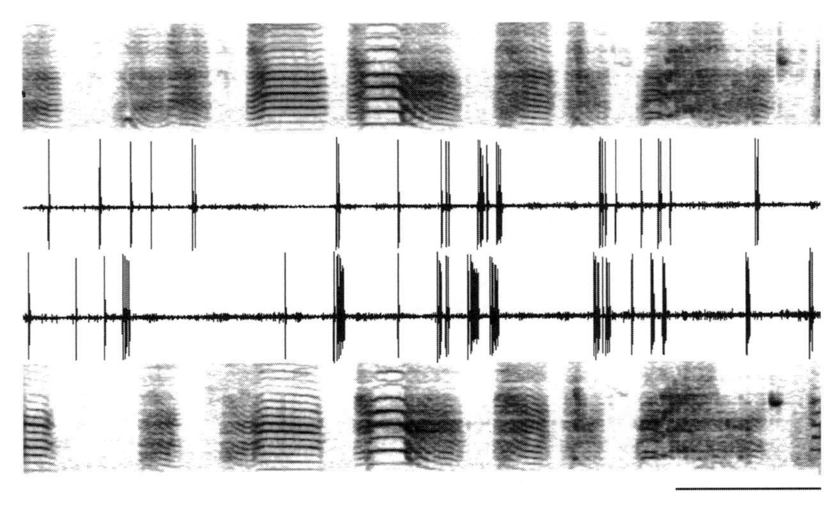

Figure 6.4 ***Neuronal replay and auditory responses in the song system.*** *Top two panels:* Sonograph of a rendition of the bird's own song and response of an RA neuron to playback of the song while the bird was sleeping. *Bottom two panels:* Activity of the same RA neuron during the day when the bird sang; the sonograph of the corresponding song is the *bottommost panel.* Note the pattern bursts are very similar for the right-hand side of the figure, where the syllables and timing of syllable sequences were the most similar. Scale bar is 200 ms. *From Dave (2001).*

The selective response to BOS arises from integration over the sequence of preceding syllables (up to many hundreds of ms) leading up to a nonlinear facilitated response expressed only for the correct sequence of syllables (Margoliash, 1983; Margoliash & Fortune, 1992). As with spontaneous replay, responses to BOS in sleeping birds release patterns of bursting that are related to the bursting observed during singing (Fig. 6.4). This indicates that information expressed in the bursting of RAn during sleep (and sleep bursting of other song system neurons) represents sensorimotor information. RAn activity during singing is premotor, preceding syllables by approximately 40 ms. RAn auditory responses show similar timing, hence the onset of a burst can precede the syllable the burst is associated with. This integration over prior syllables allows sensory and motor responses to be expressed in the same temporal framework, uniting the two.

RAn replay and its modification. The bursting pattern of an RAn during singing is precise and reliable. In a sleeping zebra finch, RAn occasionally emit individual bursts and trains of bursts. A burst was defined as continuous sequences of interspike intervals falling outside the normal (nonbursting) distribution. Considering only relatively long bursts ($\geq$8 spikes), approximately

7% of spikes occurred in bursts and approximately 15% of bursts matched a burst that the neuron had emitted during singing (A. S. Dave & Margoliash, 2000). These estimates represent a lower bound. Note that the definition is limited to considering only longer bursts and relies on the precise structure within individual bursts. Yet RAn occasionally emit multiple bursts in sequence whose relation to the sequence of bursts during singing is compelling even if one of the bursts in the sequence differs radically from that observed during singing. Moreover, bursting in the song system tends to recruit many cells synchronously. These considerations suggest that the actual numbers of RAn bursts related to song replay may be much higher than the original estimates we reported. Recording from a population of RAn during replay may resolve this issue.

The neuronal replay phenomenon in RA indicates that rather precise information—and perhaps precise variation in information—regarding singing is represented in the discharge of single neurons and populations of neurons during sleep. But how is the information used, if at all? The fundamental prediction of the sleep-learning hypothesis is that neuronal activity patterns should change over periods of sleep, and in a nonrandom fashion. A related prediction is that the observed changes should be adaptive with regard to learning. Given the reliability of RAn activity during singing, one conceptually simple if technically challenging approach to addressing this question is to compare the activity patterns of individual RAn recorded during singing before and after a period of sleep. We have recently reported this result, observing systematic changes in the burst discharge properties of RAn over periods of sleep (Rauske, Chi, Dave, & Margoliash, 2010). Of 115 spike bursts observed in 15 neurons, 33 bursts in 10 neurons showed changes in burst structure over sleep. In contrast, changes in burst structure during singing were about an order of magnitude less common (18 changed bursts of 551 bursts recorded during daytime singing). In almost all cases, the principal change over the sleep period was a decrease in the number of spikes in the burst (Fig. 6.5), thus the changes were not the result of random fluctuation around a mean number of spikes per burst. For altered bursts only, the interspike intervals increased so that the overall burst duration remained approximately constant.

Approximately half of all neurons exhibited one or more bursts that change when comparing the burst patterns over the two periods of singing. On average 20% of all bursts in RA emitted during singing changed after a night's sleep. Although these changes in burst structure are modest, even over one night and certainly over multiple nights, the accumulated changes would profoundly change a zebra finch's song if there was no compensatory response. Given that adult zebra finch song is highly stable, we conclude that spike loss during sleep is part of a mechanism to consistently modify song. We hypothesize this mechanism is involved in auditory

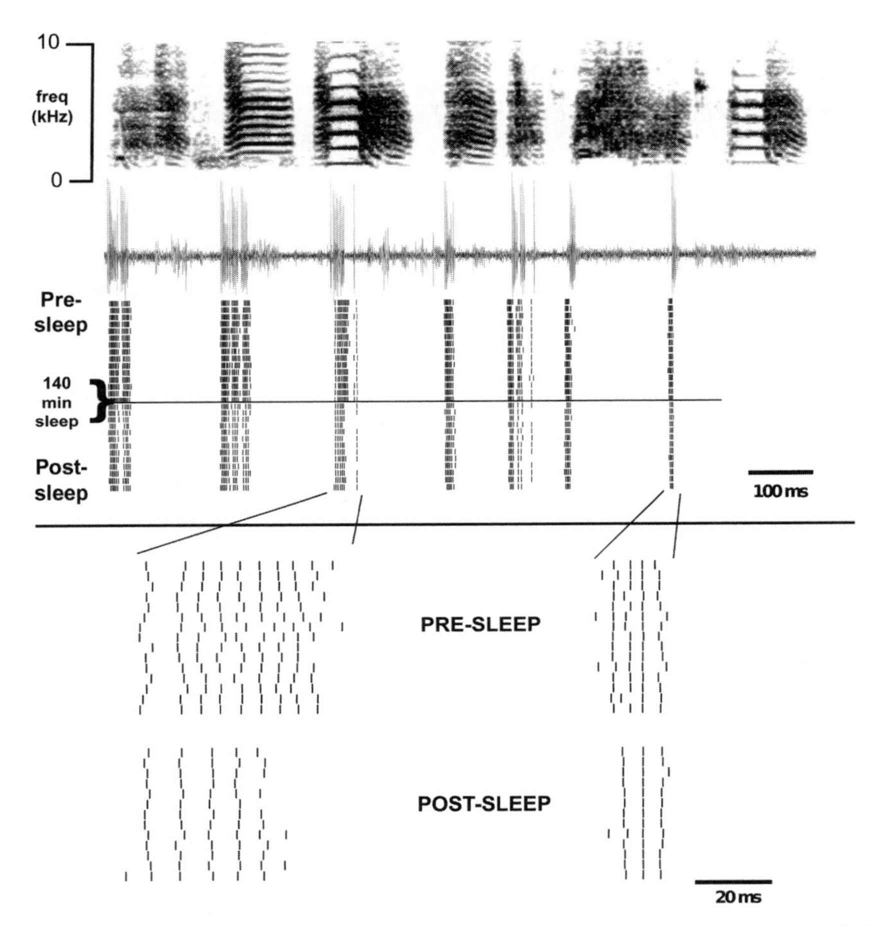

Figure 6.5 ***The structure of RA spike bursting during singing changes over a period of sleep.*** *Top panel:* Sonograph of a rendition of song the bird sang prior to sleep and a raw trace of an RA neuron's activity during that song. *Below* raster plots are shown of activity for 15 renditions of the song that were sang prior to a sleep period of 140 min and 13 renditions of the song that were sang after the sleep period. Note the timing of the burst sequences is preserved before and after sleep but there are slight differences in individual burst structure. *Bottom panel*: To help visualize the change in burst structure across sleep, two bursts are shown at higher temporal resolution. For both bursts, following sleep there are fewer spikes per burst and the interspike intervals have increased. *From Rauske (2005).*

feedback regulation of adult song (Nordeen & Nordeen, 1992) to maintain precision in singing.

Most neurons showed changes in a small number of bursts over sleep (typically zero, one or two), but one neuron showed changes in almost every burst (Fig. 6.5). The distribution of numbers of bursts changing per neuron was approximately exponential, although larger samples of neurons are required to confirm this result. This suggests that neurons expressing changes across sleep are selected randomly at a uniform rate. In neurons with two or more bursts that changed, however, the changes tended to cluster together (i.e., they were not distributed randomly across song).

Spike loss and synaptic homeostasis. If there is a strong tendency for bursts to loose spikes over periods of sleep, how do spikes get back into the system? There is a compelling neurogenesis phenomenon associated with HVC-RAn (Goldman & Nottebohm, 1983), in which new neurons are first incorporated into HVC local circuits (Paton & Nottebohm, 1984). They then project to and synapse in RA within one to two weeks (Alvarez-Buylla, Theelen, & Nottebohm, 1988). We speculate that newly incorporated HVC-RAn synapse onto those RAn that are expressing relatively few spikes in song. These RAn are weakly driven by local circuits, perhaps because they have unused synaptic space for which the new HVC-RAn can compete. To date we have not observed an RAn during sleep that suddenly and dramatically increased the number of bursts or the number of spikes within a burst—a plausible scenario arising from our speculation. But if our speculation is accurate, the RAn that would do so would be expressing very low levels of activity prior to the change in their status. Consequently, there would be a strong bias against detecting such silent neurons in extracellular recordings. Perhaps such changes could be observed in recordings of populations of RAn.

It appears that the loss of spikes within bursts of RAn is not first expressed during singing in the morning. Rather, changes in the structure of burst patterns are observed when emitted in ongoing discharge during sleep. The modified bursts show spike–timing structure that is more similar to the pattern of the corresponding burst that will be expressed during singing after sleep than the pattern of the corresponding burst that was expressed during singing before sleep. This was observed by using a course to fine temporal pattern filtering approach, first identifying sequences of bursts and then identifying the structure of a target burst within the sequence (Z. Chi, unpublished results). It remains to be determined if the modified burst pattern observed during sleep is really novel;

for example, the neuron could be cycling through a series of burst patterns (stable points in a network dynamics). However, the observation of a modified burst pattern during sleep prior to incorporation of that modified burst into singing after sleep is suggestive of a causal relation. The strong tendency of modified bursts to emit fewer spikes also argues against the caveat. We note that this description of spike loss over sleep may represent a specific instantiation of the proposed phenomenon of synaptic homeostasis, a sleep-related mechanism to regulate hyperexcitability of neuronal networks (Tononi & Cirelli, 2003). This helps to emphasize the point that the actual effects of proposed sleep mechanisms of neuronal activity have to be evaluated in the context of the specific systems that they operate.

Caveats and conclusions. A principal weakness that remains with our studies of adult replay is that we have not directly tied it to learning (in this case, auditory–feedback mediated song maintenance). To date we have only observed changes in bursting of individual neurons over periods of sleep. Singing in adult zebra finches shows circadian variation (Glaze & Troyer, 2006) that in juveniles is associated with sleep-related song learning (Derégnaucourt, Mitra, Feher, Pytte, & Tchernichovski, 2005), but that association has not been confirmed in adults. Singing in adult zebra finches also depends on auditory feedback, but the changes in song following adult deafening develop slowly over days and weeks (Leonardo & Konishi, 1999; Nordeen & Nordeen, 1992). The loss of spike bursts observed for RAn is presumably a mechanism to remove small variations in song, given that the songs zebra finch males direct towards females have much less variation than even their quite invariant undirected songs. We have yet to confirm this important prediction, however. This awaits development of a reliable technique for recording from populations of neurons and correlating changes in sleep bursts of the recorded population with song maintenance-related changes (e.g., by manipulating auditory feedback). In this important sense, both the neuronal replay work in bird song production and in the rat hippocampal replay system share a common and significant limitation.

These concerns notwithstanding, the replay phenomenon in RA offers significant potential for studying sleep and learning that has yet to be embraced. Given that the bird's song is known, recordings from a sufficient population of RA neurons should allow determination of the exact portion of song being replayed at each moment, without direct knowledge of the burst patterns the neurons express during singing. Such analysis could be bootstrapped by examining sequential burst structure in response to playback of the bird's own song. Such a development would be highly illuminating of many of the issues described above.

SLEEP AND SENSORIMOTOR SONG LEARNING: BEHAVIORAL EVIDENCE

We now turn to song developmental learning. There is only indirect evidence that the early perceptual learning experience involves sleep-dependent memory consolidation. When isolated birds are first exposed to the singing of an adult tutor, they sometimes respond by rapidly falling asleep, an observation first reported from the Tchernichovski lab (T. Lints, unpublished data), which we have confirmed (P. Adret, unpublished data). Certainly a relation between sleep and perceptual memory in juveniles would be consistent with studies of adults in humans and starlings.

The behavioral evidence supporting a role of sleep in the sensorimotor phase of song learning is far more secure, albeit to date this has been demonstrated only for a single species (zebra finch). To achieve this, it was necessary to develop techniques to analyze the large corpus of variable songs produced by each developing bird (Tchernichovski, Lints, Deregnaucourt, Cimenser, & Mitra, 2004; Tchernichovski, Mitra, Lints, & Nottebohm, 2001; Tchernichovski, Nottebohm, Ho, Pesaran, & Mitra, 2000). It was also necessary to develop a paradigm to bring the timing of developmental song learning under strong experimental control (Tchernichovski, Lints, Mitra, & Nottebohm, 1999; Tchernichovski & Nottebohm, 1998). This requires isolating the normally group-living juvenile birds from live tutors (adult males) to control for the variation in singing performance by tutors and social interactions between tutors and juveniles. In this approach, juvenile birds are raised by their mothers (who do not sing) until they are independent. They are then individually reared in a sound isolation acoustic chamber, which facilitates acquisition of high quality vocal recordings. Birds gain access to song through a form of instrumental conditioning, either pecking a key or pulling a string. Access to the song is limited by only rewarding a small number of pecks with song, a contingency that improves song copying (Tchernichovski et al., 1999).

A remarkable change in the singing behavior is observed under these conditions when a circa 40-day-old bird is first exposed to an adult conspecific song (Derégnaucourt et al., 2005). The young bird rapidly develops a circadian variation in singing behavior such that the songs produced early in the subjective day have less structure than songs sung later in the day. This pattern repeats each day so that songs produced in the afternoon are more complex and better copies of the tutor song than songs from the preceding or following morning. Notably, the circadian variation in singing does not emerge on the day of tutor song exposure but on the following day after a night of sleep.

The circadian pattern is adaptive and directly related to sleep (Derégnaucourt et al., 2005). Juveniles with a greater magnitude of circadian variation are those that eventually achieve the best copies of the tutor song. Juveniles prevented from singing in the morning will produce low structure songs in the afternoon when they are allowed to sing normally. Juveniles induced to sleep briefly during the day will sing a second bout of low structure songs upon awakening, followed by songs with increasing structure. These are forms of "awake deprivation" with a positive result that is much easier to interpret than the effects of sleep deprivation, the often-cited gold standard for sleep research.

The strength of these observations notwithstanding, a series of opportunities remain to more fully characterize the role of sleep in developmental song learning. The isolation rearing paradigm delays the onset of song copying later in development than normal. By this time the juveniles have already begun to sing but as birds that are isolated from tutor song exposure. Recent observations indicate that juveniles raised with live tutors develop a similar circadian variation, helping to address this point (M. Lusignan, unpublished observations). Juvenile zebra finches can show at least two different large-scale patterns of song learning (a syllable repetition or motif-centered pattern) (Liu, Gardner, & Nottebohm, 2004) but birds raised under instrumental conditioning of song exposure only express the motif-centered pattern (Tchernichovski et al., 2001). It would be valuable to compare sleep effects for both patterns of learning. The observed circadian variation is also somewhat fragile and shows significant individual variation, increasing the difficulty in studying it. Perhaps the biggest limitation to date, however, is that such studies have yet to be extended to other species, which limits confidence in the generality of the results. A challenge of the behavioral results is to explain why sleep (normally viewed as consolidating memories) should degrade song performance. Understanding how this arises requires a mechanistic explanation, to which we now turn.

NEURONAL REPLAY AND SONG DEVELOPMENT

Neuronal replay in the adult song system is expressed through high frequency spike bursting. Examining the emergence of bursting during development has shed additional light on neuronal replay and song learning (Shank & Margoliash, 2009). RAn were recorded on nights before and after the onset of instrumentally conditioned daily tutor song exposure. Prior to the onset of tutoring, RAn exhibited low spontaneous rates and little bursting during sleep, resulting in unimodal interspike interval (ISI) histograms. On the first night after tutor song exposure, however, RAn exhibited substantially modified ongoing discharge properties. There were more short-interval ISIs and many more sequences of short-interval ISIs organized into protobursts, which was characterized in the shape of the now bimodal ISI distributions. Note that the emergence of the sleep-related circadian pattern of singing behavior was observed on the day after the first day of tutor song exposure but prior to circadian singing patterns that begin the following morning, suggesting that the emergence of sleep bursting is part of the causal mechanism driving the circadian singing pattern.

In these experiments, most birds were exposed to only one of three tutor songs, resulting in three groups of birds. The average ISI histograms calculated for each bird were similar within each group and differed across groups. The shape of the ISI histogram is a dynamic property: in birds exposed to one tutor song for several days then switched to another, RAn recordings during sleep showed corresponding changes in ISI histograms. Given that the AFP, not HVC, provides the principal source of drive to RA in the first days after tutor song exposure when the bird is producing subsong (Aronov et al., 2008), this indicates that auditory song "template" information is expressed through the cortical–basal ganglia pathway in the song system.

It remains to be determined if there were systematic differences in the subsong of the different groups of birds, which would reflect the influence of the different tutor songs. The dynamic nature of the tutor song as represented in the ISI distributions during sleep, however, suggests that the information represented in the sleep discharges was directly related to the tutor song and not to singing. Furthermore, in the first nights after tutor song exposure, RAn do not respond to actual playback of the tutor song (or BOS). Thus, if RAn activity during sleep contributes to song development, it is probably not expressed through a replay mechanism as it is normally described.

Preliminary results indicate that RAn first exhibit auditory responses during sleep at the time the bird transitions from subsong to plastic song (M. Lusignan, unpublished results). This transition is also marked by HVC activity becoming dominant in driving RA. At this point, RAn neurons respond selectively to playback of BOS, which represents the emergence of neuronal replay as it is normally characterized. We speculate that sensorimotor mappings between HVC-Xn and HVC-RAn achieve some critical threshold, enabling populations of HVC-RAn to fire in a structured fashion. During sleep, this structured input to RA drives bursting and conveys auditory activity from HVC to RA.

OLD AND NEW IDEAS REGARDING SLEEP IN A NEW THEORY OF SONG LEARNING

The weight of the results in juvenile birds suggests that neuronal discharge during sleep contributes to developmental song learning. The first effects of exposure to a tutor song may include representational plasticity induced during sleep. This goes beyond the traditional view of how auditory memories influence song learning (Margoliash & Schmidt, 2010). In the new conception, there are distinct plastic processes, one associated with daytime singing and the other with sleep replay (Hinton, Dayan, Frey, & Neal, 1995). The distinction observed between the forms of auditory information expressed in daytime and sleep activity of HVCn is consistent with this hypothesis (Nick & Konishi, 2005). Early in development these two plastic processes are not well coordinated, resulting in the poorly structured songs observed during morning singing.

The principal feature of system consolidation theory is that information is transferred from local to global memories through long-term interactions (Buzsaki, 1998; Squire, Cohen, & Nadel, 1984). Consistent with this idea is the suggestion that song auditory memories are transferred through multiple structures including secondary auditory nuclei outside of the traditional song system, and through the AFP. System consolidation theory also is described in terms of a single mechanism or network structure, whereas the data from the song learning studies indicate that sleep-related information in networks changes over the time course of development. Finally, song learning has long been held up as a model for speech and language acquisition in humans (Doupe & Kuhl, 1999; Marler, 1970). The complex set of results obtained in zebra finches make basic predictions as to how humans should acquire and consolidate features of language (Margoliash, 2003).

SUMMARY AND FINAL CONCLUSIONS

We began by asking why a neuroscientist should study sleep in birds. One answer was grounded in an evolutionary perspective. If one seeks an understanding of sleep behavior and its ultimate functions, this requires a comparative approach. Complex mammalian-like sleep in some species of birds and shared organizational elements in neocortex and avian cortex represent part of a long lineage that includes shared traits across vertebrates and beyond. Examining sleep in the context of that lineage should help inform why sleep is observed so broadly in the animal kingdom. A second answer was grounded in a more ethological perspective. The study of animal behavior and its underlying physiology is facilitated by taking advantage of behavioral specializations in the target species. This neuroethological approach has guided our studies of sleep and learning in the two avian models we highlighted. The starling model we described, which relies on song recognition behavior, expresses a pattern of sleep-dependent memory benefits that is comparable to what is observed in humans. This will allow us to measure and manipulate memory consolidation while we examine its physiology in the starling auditory system. We contrasted this with the strengths and limitations in the prominent studies of sleep and learning in rodents.

Whereas the physiological analysis of sleep in relation to starling song perceptual learning is only beginning, the role of sleep for sensorimotor learning has profited by taking advantage of the intensively studied song learning and bird song system. We have observed a compelling neuronal replay phenomenon in adult zebra finches and that its emergence during development is related to significant song learning milestones. A promising avenue for future research is to investigate the role of replay in helping to shape the sleep-dependent vocal developmental trajectories. The novel hypotheses for sensorimotor learning these observations have suggested emphasize the potential for sleep research to uncover new physiological mechanisms.

We conclude by noting that sleep may have a myriad of effects in different species, in different physiological systems, and on different behaviors. Some overarching principles may apply very broadly, such as synaptic downscaling and system consolidation. Yet, there is no one thing that "sleep does." To understand the application of these principles, we need to look at specific behaviors and how those are affected by sleep. Songbirds in particular represent attractive model systems towards such goals.

REFERENCES

Adret-Hausberger, M., & Jenkins, P. F. (1988). Complex organization of the warbling song in the European starling *Sturnus vulgaris*. *Behaviour*, *107*, 138–156.

Alvarez-Buylla, A., Theelen, M., & Nottebohm, F. (1988). Birth of projection neurons in the higher vocal center of the canary forebrain before, during, and after song learning. *Proceedings of the National Academy of Sciences of the United States of America*, *85*(22), 8722–8726.

Andalman, A. S., & Fee, M. S. (2009). A basal ganglia-forebrain circuit in the songbird biases motor output to avoid vocal errors. *Proceedings of the National Academy of Sciences of the United States of America, 106*(30), 12518–12523.

Aronov, D., Andalman, A. S., & Fee, M. S. (2008). A specialized forebrain circuit for vocal babbling in the juvenile songbird. *Science, 320*(5876), 630–634.

Ashmore, R. C., Renk, J. A., & Schmidt, M. F. (2008). Bottom-up activation of the vocal motor forebrain by the respiratory brainstem. *Journal of Neuroscience, 28*(10), 2613–2623.

Ball, G. F., & Balthazart, J. (2010). Introduction to the chemical neuroanatomy of birdsong. *Journal of Chemical Neuroanatomy, 39*(2), 67–71.

Bolhuis, J. J., & Gahr, M. (2006). Neural mechanisms of birdsong memory. *Nature Reviews Neuroscience,* 7(5), 347–357.

Brainard, M. S., & Doupe, A. J. (2000). Alteration of auditory feedback causes both acute and lasting changes to Bengalese finch song. *Society Neuroscience Abstracts, 26,* 269.266

Brawn, T. P., Fenn, K. M., Nusbaum, H. C., & Margoliash, D. (2008). Consolidation of sensorimotor learning during sleep. *Learning & Memory, 15*(11), 815–819.

Brawn, T. P., Fenn, K. M., Nusbaum, H. C., & Margoliash, D. (2010). Consolidating the effects of waking and sleep on motor-sequence learning. *Journal of Neuroscience, 30*(42), 13977–13982.

Brawn, T. P., Nusbaum, H. C., & Margoliash, D. (2010). Sleep-dependent consolidation of auditory discrimination learning in adult starlings. *Journal of Neuroscience, 30*(2), 609–613.

Buzsaki, G. (1998). Memory consolidation during sleep: A neurophysiological perspective. *Journal of Sleep Research,* 7(Suppl. 1), 17–23.

Cardin, J. A., & Schmidt, M. F. (2004). Auditory responses in multiple sensorimotor song system nuclei are co-modulated by behavioral state. *Journal of Neurophysiology, 91*(5), 2148–2163.

Cardin, J. A., & Schmidt, M. F. (2004). Noradrenergic inputs mediate state dependence of auditory responses in the avian song system. *Journal of Neuroscience, 24*(35), 7745–7753.

Carrillo, G. D., & Doupe, A. J. (2004). Is the songbird Area X striatal, pallidal, or both? An anatomical study. *Journal of Comparative Neurology, 473*(3), 415–437.

Castelino, C. B., & Schmidt, M. F. (2010). What birdsong can teach us about the central noradrenergic system. *Journal of Chemical Neuroanatomy, 39*(2), 96–111.

Chaiken, M., Bohner, J., & Marler, P. (1993). Song acquisition in European starlings, Sturnus vulgaris: A comparison of the songs of live-tutored, tape-tutored, untutored, and wild-caught males. *Animal Behaviour, 46*(6), 1079–1090.

Cirelli, C., & Tononi, G. (2008). Is sleep essential? *PLoS Biology, 6*(8), e216.

Dave, A. S. (2001). *Mechanisms of sensorimotor vocal integration.* Chicago: University of Chicago.

Dave, A. S., & Margoliash, D. (2000). Song replay during sleep and computational rules for sensorimotor vocal learning. *Science, 290*(5492), 812–816.

Dave, A. S., Yu, A. C., & Margoliash, D. (1998). Behavioral state modulation of auditory activity in a vocal motor system. *Science, 282*(5397), 2250–2254.

Derégnaucourt, S., Mitra, P. P., Feher, O., Pytte, C., & Tchernichovski, O. (2005). How sleep affects the developmental learning of bird song. *Nature, 433*(7027), 710–716.

Dick, F., Lee, H. L., Nusbaum, H., & Price, C. J. (2011). Auditory-motor expertise alters "speech selectivity" in professional musicians and actors. *Cerebral Cortex, 21*(4), 938–948.

Diekelmann, S., & Born, J. (2010). The memory function of sleep. *Nature Reviews Neuroscience, 11*(2), 114–126.

Doupe, A. J., & Kuhl, P. K. (1999). Birdsong and human speech: Common themes and mechanisms. *Annual Review of Neuroscience, 22,* 567–631.

Dugas-Ford, J. (2009). *A comparative molecular study of the amniote dorsal telencephalon.* Chicago: University of Chicago.

Durand, S. E., Tepper, J. M., & Cheng, M. F. (1992). The shell region of the nucleus ovoidalis: A subdivision of the avian auditory thalamus. *The Journal of Comparative Neurology, 323*(4), 495–518.

Eens, M. (1997). Understanding the complex song of the European starling: An integrated ethological approach. *Advances in the Study of Behavior, 26*, 355–434.

Ego-Stengel, V., & Wilson, M. A. (2010). Disruption of ripple-associated hippocampal activity during rest impairs spatial learning in the rat. *Hippocampus, 20*(1), 1–10.

Eiland, M. M., Lyamin, O. I., & Siegel, J. M. (2001). State-related discharge of neurons in the brainstem of freely moving box turtles, terrapene carolina major. *Archives Italiennes de Biologie, 139*(1–2), 23–36.

Ellenbogen, J. M., Hu, P. T., Payne, J. D., Titone, D., & Walker, M. P. (2007). Human relational memory requires time and sleep. *Proceedings of the National Academy of Sciences of the United States of America, 104*(18), 7723–7728.

Ellenbogen, J. M., Hulbert, J. C., Jiang, Y., & Stickgold, R. (2009). The sleeping brain's influence on verbal memory: Boosting resistance to interference. *PLoS ONE, 4*(1), e4117.

Ellenbogen, J. M., Hulbert, J. C., Stickgold, R., Dinges, D. F., & Thompson-Schill, S. L. (2006). Interfering with theories of sleep and memory: Sleep, declarative memory, and associative interference. *Current Biology : CB, 16*(13), 1290–1294.

Euston, D. R., Tatsuno, M., & McNaughton, B. L. (2007). Fast-forward playback of recent memory sequences in prefrontal cortex during sleep. *Science, 318*(5853), 1147–1150.

Falls, J. B. (1982). Individual recognition by sound in birds. In D. E. Kroodsma, & E. H. Miller (Eds.), *Acoustic communication in birds, vol. 2, song learning and its consequences* (pp. 237–278). New York: Academic Press.

Farries, M. A., Ding, L., & Perkel, D. J. (2005). Evidence for "direct" and "indirect" pathways through the song system basal ganglia. *Journal of Comparative Neurology, 484*(1), 93–104.

Farries, M. A., & Perkel, D. J. (2002). A telencephalic nucleus essential for song learning contains neurons with physiological characteristics of both striatum and globus pallidus. *Journal of Neuroscience, 22*(9), 3776–3787.

Fenn, K. M., Nusbaum, H. C., & Margoliash, D. (2003). Consolidation during sleep of perceptual learning of spoken language. *Nature, 425*(6958), 614–616.

Ferrara, M., Iaria, G., Tempesta, D., Curcio, G., Moroni, F., Marzano, C., et al. (2008). Sleep to find your way: The role of sleep in the consolidation of memory for navigation in humans. *Hippocampus, 18*(8), 844–851.

Fischer, S., Nitschke, M. F., Melchert, U. H., Erdmann, C., & Born, J. (2005). Motor memory consolidation in sleep shapes more effective neuronal representations. *The Journal of Neuroscience : The official Journal of the Society for Neuroscience, 25*(49), 11248–11255.

Fortune, E. S., & Margoliash, D. (1992). Cytoarchitectonic organization and morphology of cells of the field L complex in male zebra finches (Taenopygia guttata). *The Journal of Comparative Neurology, 325*(3), 388–404.

Foster, D. J., & Wilson, M. A. (2006). Reverse replay of behavioural sequences in hippocampal place cells during the awake state. *Nature, 440*(7084), 680–683.

Frankland, P. W., & Bontempi, B. (2005). The organization of recent and remote memories. *Nature Reviews Neuroscience, 6*(2), 119–130.

Gais, S., Lucas, B., & Born, J. (2006). Sleep after learning aids memory recall. *Learning & Memory, 13*(3), 259–262.

Gais, S., Plihal, W., Wagner, U., & Born, J. (2000). Early sleep triggers memory for early visual discrimination skills. *Nature Neuroscience, 3*, 1335–1339.

Gentner, T. Q. (2004). Neural systems for individual song recognition in adult birds. *Annals of the New York Academy of Sciences, 1016*, 282–302.

Gentner, T. Q., & Hulse, S. H. (2000). Perceptual classification based on the component structure of song in European starlings. *Journal of the Acoustical Society of America, 107*(6), 3369–3381.

Gentner, T. Q., Hulse, S. H., Bentley, G. E., & Ball, G. F. (2000). Individual vocal recognition and the effect of partial lesions to HVc on discrimination, learning, and categorization of conspecific song in adult songbirds. *Journal of Neurobiology, 42*(1), 117–133.

Gentner, T. Q., & Margoliash, D. (2003). Neuronal populations and single cells representing learned auditory objects. *Nature, 424*(6949), 669–674.

Girardeau, G., Benchenane, K., Wiener, S. I., Buzsaki, G., & Zugaro, M. B. (2009). Selective suppression of hippocampal ripples impairs spatial memory. *Nature Neuroscience, 12*(10), 1222–1223.

Glaze, C. M., & Troyer, T. W. (2006). Temporal structure in zebra finch song: Implications for motor coding. *Journal of Neuroscience, 26*(3), 991–1005.

Goldman, S. A., & Nottebohm, F. (1983). Neuronal production, migration, and differentiation in a vocal control nucleus of the adult female canary brain. *Proceedings of the National Academy of Sciences of the United States of America, 80*(8), 2390–2394.

Gomez, R. L., Bootzin, R. R., & Nadel, L. (2006). Naps promote abstraction in language-learning infants. *Psychological Science, 17*(8), 670–674.

Hahnloser, R. H., Kozhevnikov, A. A., & Fee, M. S. (2002). An ultra-sparse code underlies the generation of neural sequences in a songbird. *Nature, 419*(6902), 65–70.

Hahnloser, R. H., Kozhevnikov, A. A., & Fee, M. S. (2006). Sleep-related neural activity in a premotor and a basal-ganglia pathway of the songbird. *Journal of Neurophysiology, 96*(2), 794–812.

Hennevin, E., Huetz, C., & Edeline, J. M. (2007). Neural representations during sleep: From sensory processing to memory traces. *Neurobiol Learning & Memory, 87*(3), 416–440.

Hinton, G. E., Dayan, P., Frey, B. J., & Neal, R. M. (1995). The "wake-sleep" algorithm for unsupervised neural networks. *Science, 268*(5214), 1158–1161.

Hu, P., Stylos-Allan, M., & Walker, M. P. (2006). Sleep facilitates consolidation of emotional declarative memory. *Psychological Science, 17*(10), 891–898.

Hupbach, A., Gomez, R. L., Bootzin, R. R., & Nadel, L. (2009). Nap-dependent learning in infants. *Developmental Science, 12*(6), 1007–1012.

Jarvis, E. D., Gunturkun, O., Bruce, L., Csillag, A., Karten, H., Kuenzel, W., et al. (2005). Avian brains and a new understanding of vertebrate brain evolution. *Nature Reviews Neuroscience, 6*(2), 151–159.

Jeanne, J. M., Thompson, J. V., Sharpee, T. O., & Gentner, T. Q. (2011). Emergence of learned categorical representations within an auditory forebrain circuit. *Journal of Neuroscience, 31*(7), 2595–2606.

Ji, D., & Wilson, M. A. (2007). Coordinated memory replay in the visual cortex and hippocampus during sleep. *Nature Neuroscience, 10*(1), 100–107.

Jones, S. G., Vyazovskiy, V. V., Cirelli, C., Tononi, G., & Benca, R. M. (2008). Homeostatic regulation of sleep in the white-crowned sparrow (Zonotrichia leucophrys gambelii). *BMC Neuroscience, 9*, 47.

Kao, M. H., Doupe, A. J., & Brainard, M. S. (2005). Contributions of an avian basal ganglia-forebrain circuit to real-time modulation of song. *Nature, 433*(7026), 638–643.

Karten, H. J. (1967). The organization of the ascending auditory pathway in the pigeon (Columba livia). I. Diencephalic projections of the inferior colliculus (nucleus mesencephali lateralis, pars dorsalis). *Brain Research, 6*(3), 409–427.

Karten, H. J. (1968). The ascending auditory pathway in the pigeon (Columba livia). II. Telencephalic projections of the nucleus ovoidalis thalami. *Brain Research, 11*(1), 134–153.

Karten, H. J. (1997). Evolutionary developmental biology meets the brain: The origins of mammalian cortex. *Proceedings of the National Academy of Sciences of the United States of America, 94*(7), 2800–2804.

Konishi, M. (1978). Auditory environment and vocal development in birds. In R. D. Walk, & H. L. J. Pick (Eds.), *Perception and experience* (pp. 105–118). New York: Plenum.

Korman, M., Doyon, J., Doljansky, J., Carrier, J., Dagan, Y., & Karni, A. (2007). Daytime sleep condenses the time course of motor memory consolidation. *Nature Neuroscience, 10*(9), 1206–1213.

Kozhevnikov, A. A., & Fee, M. S. (2007). Singing-related activity of identified HVC neurons in the zebra finch. *Journal of Neurophysiology, 97*(6), 4271–4283.

Kroodsma, D. E. (1982). Learning and the ontogeny of sound signals in birds. In D. E. Kroodsma, & E. H. Miller (Eds.), *Acoustic communication in birds, vol. 2, song learning and its consequences* (pp. 1–23). New York: Academic Press.

Kroodsma, D. E., & Baylis, J. R. (1982). Appendix: A world survey of evidence for vocal learning in birds. In D. E. Kroodsma, & E. H. Miller (Eds.), *Acoustic Communication in Birds, vol. 2, song learning and its consequences* (pp. 311–337). New York: Academic Press.

Kroodsma, D. E., & Konishi, M. (1991). A suboscine bird, eastern phoebe (Sayornis-phoebe), develops normal song without auditory feedback. *Animal Behaviour, 42*(3), 477–488.

Kroodsma, D. E., & Miller, E. H. (1982). *Acoustic communication in birds. vol. 2. song learning and its consequences*. New York: Academic Press.

Kudrimoti, H. S., Barnes, C. A., & McNaughton, B. L. (1999). Reactivation of hippocampal cell assemblies: Effects of behavioral state, experience, and EEG dynamics. *Journal of Neuroscience, 19*(10), 4090–4101.

Lee, A. K., & Wilson, M. A. (2002). Memory of sequential experience in the hippocampus during slow wave sleep. *Neuron, 36*(6), 1183–1194.

Leonardo, A., & Fee, M. S. (2005). Ensemble coding of vocal control in birdsong. *Journal of Neuroscience, 25*(3), 652–661.

Leonardo, A., & Konishi, M. (1999). Decrystallization of adult birdsong by perturbation of auditory feedback. *Nature, 399*(6735), 466–470.

Liu, W. C., Gardner, T. J., & Nottebohm, F. (2004). Juvenile zebra finches can use multiple strategies to learn the same song. *Proceedings of the National Academy of Sciences of the United States of America, 101*(52), 18177–18182.

Louie, K., & Wilson, M. A. (2001). Temporally structured replay of awake hippocampal ensemble activity during rapid eye movement sleep. *Neuron, 29*(1), 145–156.

Low, P. S., Shank, S. S., Sejnowski, T. J., & Margoliash, D. (2008). Mammalian-like features of sleep structure in zebra finches. *Proceedings of the National Academy of Sciences of the United States of America, 105*(26), 9081–9086.

Margoliash, D. (1983). Acoustic parameters underlying the responses of song-specific neurons in the white-crowned sparrow. *Journal of Neuroscience, 3*(5), 1039–1057.

Margoliash, D. (2002). Evaluating theories of bird song learning: Implications for future directions. *Journal of Comparative Physiology A, 188*(11–12), 851–866.

Margoliash, D. (2003). Offline learning and the role of autogenous speech: New suggestions from birdsong research. *Speech Communication, 41*, 165–178.

Margoliash, D., & Fortune, E. S. (1992). Temporal and harmonic combination-sensitive neurons in the zebra finch's HVc. *Journal of Neuroscience, 12*(11), 4309–4326.

Margoliash, D., & Schmidt, M. F. (2010). Sleep, off-line processing, and vocal learning. *Brain and Language, 115*(1), 45–58.

Marler, P. (1970). Birdsong and speech development: Could there be parallels? *American Scientist, 58*(6), 669–673.

Marler, P. (1997). Three models of song learning: Evidence from behavior. *Journal of Neurobiology, 33*, 501–516.

Marshall, L., & Born, J. (2007). The contribution of sleep to hippocampus-dependent memory consolidation. *Trends in Cognitive Sciences, 11*(10), 442–450.

Mednick, S., Nakayama, K., & Stickgold, R. (2003). Sleep-dependent learning: A nap is as good as a night. *Nature Neuroscience, 6*(7), 697–698.

Mooney, R. (2009). Neural mechanisms for learned birdsong. *Learning & Memory, 16*(11), 655–669.

Muller, C. M., & Leppelsack, H. J. (1985). Feature extraction and tonotopic organization in the avian auditory forebrain. *Experimental Brain Research, 59*(3), 587–599.

Nick, T. A., & Konishi, M. (2005). Neural song preference during vocal learning in the zebra finch depends on age and state. *Journal of Neurobiology, 62*(2), 231–242.

Nishida, M., & Walker, M. P. (2007). Daytime naps, motor memory consolidation and regionally specific sleep spindles. *PLoS ONE, 2*(4), e341.

Nordeen, K.W., & Nordeen, E.J. (1992). Auditory feedback is necessary for the maintenance of stereotyped song in adult zebra finches. *Behavioral & Neural Biology*, *57*(1), 58–66.

O'Neill, J., Pleydell-Bouverie, B., Dupret, D., & Csicsvari, J. (2010). Play it again: Reactivation of waking experience and memory. *Trends in Neurosciences*, *33*(5), 220–229.

Olveczky, B. P., Andalman, A. S., & Fee, M. S. (2005). Vocal experimentation in the juvenile songbird requires a basal ganglia circuit. *PLoS Biology*, *3*(5), e153.

Paton, J. A., & Nottebohm, F. (1984). Neurons born in adult brain are recruited into functional circuits. *Science*, *225*, 1046–1048.

Payne, J. D., Stickgold, R., Swanberg, K., & Kensinger, E. A. (2008). Sleep preferentially enhances memory for emotional components of scenes. *Psychological Science*, *19*(8), 781–788.

Peyrache, A., Khamassi, M., Benchenane, K., Wiener, S. I., & Battaglia, F. P. (2009). Replay of rule-learning related neural patterns in the prefrontal cortex during sleep. *Nature Neuroscience*, *12*(7), 919–926.

Poe, G. R. (2010). Cognitive neuroscience of sleep. In G. A. Kerkhof & H. P. A. van Dongen (Eds.), *Human sleep and cognition* (Vol. 185, pp. 1–19). Oxford: Elsevier Science.

Racsmany, M., Conway, M. A., & Demeter, G. (2010). Consolidation of episodic memories during sleep: Long-term effects of retrieval practice. *Psychological Science*, *21*(1), 80–85.

Ramadan, W., Eschenko, O., & Sara, S. J. (2009). Hippocampal sharp wave/ripples during sleep for consolidation of associative memory. *PLoS ONE*, *4*(8), e6697.

Rasch, B., & Born, J. (2007). Maintaining memories by reactivation. *Current Opinion in Neurobiology*, *17*(6), 698–703.

Rattenborg, N. C. (2006). Evolution of slow-wave sleep and palliopallial connectivity in mammals and birds: A hypothesis. *Brain Research Bulletin*, *69*(1), 20–29.

Rattenborg, N. C., Martinez-Gonzalez, D., Roth, T. C., II, & Pravosudov, V. V. (2011). Hippocampal memory consolidation during sleep: A comparison of mammals and birds. *Biological Reviews of the Cambridge Philosophical Society*, *86*(3), 658–691.

Rauske, P. L. (2005). The modulation of sensorimotor activity by sleep. Chicago: University of Chicago.

Rauske, P. L., Chi, Z., Dave, A. S., & Margoliash, D. (2010). Neuronal stability and drift across periods of sleep: Premotor activity patterns in a vocal control nucleus of adult zebra finches. *Journal of Neuroscience*, *30*(7), 2783–2794.

Reiner, A., Perkel, D. J., Bruce, L. L., Butler, A. B., Csillag, A., Kuenzel, W., et al. (2004). Revised nomenclature for avian telencephalon and some related brainstem nuclei. *Journal of Comparative Neurology*, *473*(3), 377–414.

Robertson, E. M., Pascual-Leone, A., & Press, D. Z. (2004). Awareness modifies the skill-learning benefits of sleep. *Current Biology : CB*, *14*(3), 208–212.

Saranathan, V., Hamilton, D., Powell, G.V., Kroodsma, D. E., & Prum, R. O. (2007). Genetic evidence supports song learning in the three-wattled bellbird Procnias tricarunculata (Cotingidae). *Molecular Ecology*, *16*(17), 3689–3702.

Scharff, C., & Nottebohm, F. (1991). A comparative study of the behavioral deficits following lesions of various parts of the zebra finch song system: Implications for vocal learning. *The Journal of Neuroscience*, *11*(9), 2896–2913.

Schmidt, M. F., & Konishi, M. (1998). Gating of auditory responses in the vocal control system of awake songbirds. *Nature Neuroscience*, *1*(6), 513–518.

Scullin, M. K., & McDaniel, M. A. (2010). Remembering to execute a goal: Sleep on it!. *Psychological Science*, *21*, 1028–1035.

Sen, K., Theunissen, F. E., & Doupe, A. J. (2001). Feature analysis of natural sounds in the songbird auditory forebrain. *Journal of Neurophysiology*, *86*(3), 1445–1458.

Shank, S. S., & Margoliash, D. (2009). Sleep and sensorimotor integration during early vocal learning in a songbird. *Nature*, *458*(7234), 73–77.

Shea, S. D., Koch, H., Baleckaitis, D., Ramirez, J. M., & Margoliash, D. (2010). Neuron-specific cholinergic modulation of a forebrain song control nucleus. *Journal of Neurophysiology*, *103*(2), 733–745.

Shea, S. D., & Margoliash, D. (2003). Basal forebrain cholinergic modulation of auditory activity in the zebra finch song system. *Neuron*, *40*(6), 1213–1226.

Shea, S. D., & Margoliash, D. (2010). Behavioral state-dependent reconfiguration of song-related network activity and cholinergic systems. *Journal of Chemical Neuroanatomy*, *39*(2), 132–140.

Siegel, J. M. (2008). Do all animals sleep? *Trends in Neurosciences*, *31*(4), 208–213.

Simpson, H. B., & Vicario, D. S. (1990). Brain pathways for learned and unlearned vocalizations differ in zebra finches. *The Journal of Neuroscience*, *10*(5), 1541–1556.

Sohrabji, F., Nordeen, E. J., & Nordeen, K. W. (1990). Selective impairment of song learning following lesions of a forebrain nucleus in the juvenile zebra finch. *Behavioral and Neural Biology*, *53*(1), 51–63.

Squire, L. R., Cohen, N. J., & Nadel, L. (1984). The medial temporal region and memory consolidation: A new hypothesis. In H. Weingartner, & E. Parker (Eds.), *Memory consolidation* (pp. 185–210). Hillsdale, NJ: Lawrence Erlbaum.

Stickgold, R., James, L., & Hobson, J. A. (2000). Visual discrimination learning requires sleep after training. *Nature Neuroscience*, *3*(12), 1237–1238.

Stoddard, P. K. (1996). Vocal recognition of neighbors by territorial passerines. In D. E. Kroodsma, & E. H. Miller (Eds.), *Ecology and evolution of acoustic communication in birds* (pp. 356–374). Cornell, Ithaca: Cromstock.

Strait, D. L., Chan, K., Ashley, R., & Kraus, N. (2012). Specialization among the specialized: Auditory brainstem function is tuned in to timbre. *Cortex; a Journal Devoted to the Study of the Nervous System and Behavior*, *48*(3), 360–362.

Suga, N., & Ma, X. (2003). Multiparametric corticofugal modulation and plasticity in the auditory system. *Nature Reviews. Neuroscience*, *4*(10), 783–794.

Sutter, M. L., & Margoliash, D. (1994). Global synchronous response to autogenous song in zebra finch HVc. *Journal of Neurophysiology*, *72*(5), 2105–2123.

Szymczak, J. T. (1987). Daily distribution of sleep states in the rook *Corvus frugilegus*. *Journal of Comparative Physiology [A]*, *161*(2), 321–327.

Talamini, L. M., Nieuwenhuis, I. L., Takashima, A., & Jensen, O. (2008). Sleep directly following learning benefits consolidation of spatial associative memory. *Learning & Memory*, *15*(4), 233–237.

Tarullo, A. R., Balsam, P. D., & Fifer, W. P. (2011). Sleep and infant learning. *Infant and Child Development*, *20*(1), 35–46.

Tchernichovski, O., Lints, T., Mitra, P. P., & Nottebohm, F. (1999). Vocal imitation in zebra finches is inversely related to model abundance. *Proceedings of the National Academy of Sciences of the United States of America*, *96*(22), 12901–12904.

Tchernichovski, O., Lints, T. J., Deregnaucourt, S., Cimenser, A., & Mitra, P. P. (2004). Studying the song development process: Rationale and methods. *Annals of the New York Academy of Sciences*, *1016*, 348–363.

Tchernichovski, O., Mitra, P. P., Lints, T., & Nottebohm, F. (2001). Dynamics of the vocal imitation process: How a zebra finch learns its song. *Science*, *291*(5513), 2564–2569.

Tchernichovski, O., & Nottebohm, F. (1998). Social inhibition of song imitation among sibling male zebra finches. *Proceedings of the National Academy of Sciences of the United States of America*, *95*(15), 8951–8956.

Tchernichovski, O., Nottebohm, F., Ho, C. E., Pesaran, B., & Mitra, P. P. (2000). A procedure for an automated measurement of song similarity. *Animal Behaviour*, *59*(6), 1167–1176.

Theunissen, F. E., Amin, N., Shaevitz, S. S., Woolley, S. M., Fremouw, T., & Hauber, M. E. (2004). Song selectivity in the song system and in the auditory forebrain. *Annals of the New York Academy of Sciences*, *1016*, 222–245.

Thompson, J. V., & Gentner, T. Q. (2010). Song recognition learning and stimulus-specific weakening of neural responses in the avian auditory forebrain. *Journal of Neurophysiology*, *103*(4), 1785–1797.

Tobler, I., & Borbely, A. A. (1988). Sleep and EEG spectra in the pigeon (Columba livia) under baseline conditions and after sleep deprivation. *Journal of Comparative Physiology A*, *163*, 729–738.

Tononi, G., & Cirelli, C. (2003). Sleep and synaptic homeostasis: A hypothesis. *Brain Research Bulletin*, *62*(2), 143–150.

Tucker, M. A., Hirota, Y., Wamsley, E. J., Lau, H., Chaklader, A., & Fishbein, W. (2006). A daytime nap containing solely non-REM sleep enhances declarative but not procedural memory. *Neurobiology of Learning and Memory*, *86*(2), 241–247.

Ulinski, P. S. (1984). Design features in vertebrate sensory systems. *American Zoologist*, *24*, 717–731.

Vates, G. E., Broome, B. M., Mello, C. V., & Nottebohm, F. (1996). Auditory pathways of caudal telencephalon and their relation to the song system of adult male zebra finches. *Journal of Comparative Neurology*, *366*(4), 613–642.

Walker, M. P., Brakefield, T., Morgan, A., Hobson, J. A., & Stickgold, R. (2002). Practice with sleep makes perfect: Sleep-dependent motor skill learning. *Neuron*, *35*(1), 205–211.

Walker, M. P., & Stickgold, R. (2006). Sleep, memory, and plasticity. *Annual Review of Psychology*, *57*, 139–166.

Wang, Y., Brzozowska-Prechtl, A., & Karten, H. J. (2010). Laminar and columnar auditory cortex in avian brain. *Proceedings of the National Academy of Sciences of the United States of America*, *107*(28), 12676–12681.

Wild, J. M., Karten, H. J., & Frost, B. J. (1993). Connections of the auditory forebrain in the pigeon (Columba livia). *Journal of Comparative Neurology*, *337*(1), 32–62.

Wilson, M. A., & McNaughton, B. L. (1994). Reactivation of hippocampal ensemble memories during sleep [see comments]. *Science*, *265*(5172), 676–679.

Yu, A. C., & Margoliash, D. (1996). Temporal hierarchical control of singing in birds. *Science*, *273*(5283), 1871–1875.

CHAPTER 7

Phasic Pontine-Wave (P-Wave) Generation: Cellular-Molecular-Network Mechanism and Functional Significance

Subimal Datta
Laboratory of Sleep and Cognitive Neuroscience, Departments of Psychiatry and Neurology and Program in Neuroscience, Boston University School of Medicine

INTRODUCTION

Sleep is a highly evolved global behavioral state in mammals. Sleep provides an exceptional opportunity to study the brain-based physical and physiological foundation of cognitive and homeostatic regulatory processes. The basic stages of sleep are introduced first in this chapter, followed by the basic elements of memory, and finally, a description of how these two fields are integrated.

Stages of sleep. Normally, when we first enter the sleep state, it is via quiet (non-rapid-eye-movement; NREM) sleep, which is a state that, behaviorally, is not very dramatic. We simply lie still, our eyes drift slowly back and forth, and at indeterminate intervals, we shift our sleep position. Upon first falling asleep, individuals may progressively lose awareness of the outside world and experience microhallucinations and illusions of movement of the body in space. During NREM sleep there are notable decreases in body temperature, blood pressure, heart rate, and respiratory rate. These decreases are accompanied by routine increases in the production of antibodies and pulsatile release of growth and sex hormones from the pituitary gland. NREM sleep is characterized by a change in the EEG from a low-amplitude, high-frequency pattern to one that is high amplitude, low frequency. The degree to which the EEG is progressively synchronized (that is, of high amplitude and low frequency) can be subdivided into four stages in humans: Stage one sleep (NREM-I) is characterized by relatively low-amplitude (<50 μV), high-frequency (4–7 cycles per second; Hz) theta activity and vertex sharp waves in the EEG. Stage two sleep (NREM-II) is characterized by the appearance of distinctive sleep spindles (lasting between 0.5–1.0 sec, with peak amplitudes of 100 μV

Sleep and Brain Activity
DOI: http://dx.doi.org/10.1016/B978-0-12-384995-3.00007-1

and composed of augmenting and decrementing waves at a frequency of 12–14 Hz) and K-complex (a negative sharp wave followed immediately by a slower positive component) waveforms in the EEG. Stage three sleep (NREM-III) is characterized by the addition of high-amplitude (>100 μV) slow waves (1–4 Hz), but with no more than 50% of the EEG record occupied by these slow waves. In stage four sleep (NREM-IV), the EEG record is dominated by these high-amplitude (150–250 μV) slow waves (1–4 Hz). The NREM-III and NREM-IV sleep stages are now considered to be a single sleep stage known as "slow-wave sleep" (SWS). Throughout the progression of NREM sleep (including SWS), as the EEG frequency decreases and the amplitude increases, muscle tone progressively declines and may be lost altogether in most of the somatic musculature. The slow rolling eye movements that first replaced the rapid saccadic eye movements of waking gradually subside, with the eyes finally assuming a divergent upward gaze. After varying amounts of time (depending upon the size of the animal and its brain), the progressive set of changes in the EEG reverses itself and the EEG resumes the low-amplitude, fast character previously seen in waking. Instead of waking, however, behavioral sleep persists, and this sleep phase is REM sleep. REM sleep is characterized by a constellation of events that includes the following: an activated pattern of cortical EEG activity; marked atonia of the postural muscles; rapid eye movements; a theta rhythm within the hippocampus; field potentials in the pons (P-wave); lateral geniculate nucleus and occipital cortex (ponto-geniculo-occipital [PGO]) spikes; myoclonic twitches, most apparent in the facial and distal limb musculature; pronounced fluctuations in cardiorespiratory rhythms and core body temperature; and finally, penile erection and clitoral tumescence. With a basic understanding of the stages of sleep now established, this chapter will move on to an introduction to memory processing, which is highly dependent on specific brain activity during sleep.

Stages of memory formation. The development and maturation of memory is a complex process that occurs in several distinct stages over time. The two major stages of memory formation are (1) acquisition of information (learning or encoding), and (2) consolidation of memory trace. Currently, there is a large body of evidence that has demonstrated that the different stages of memory development are influenced by specific stages of sleep. The first stage, acquisition, is triggered by engaging with an object or performing an action. The initial memory that then forms or becomes encoded leads to the formation of a representation of the object or action within the brain. This initial encoding of a memory is a relatively rapid process that requires only a few milliseconds. At this stage, memory remains in a short-term store that has very limited capacity and, in the absence of rehearsal, persists for only minutes at most. However, if this encoded information persists in the form of reverberating activity in neuronal circuits, then another process transforms this short-term memory into an intermediate form. This intermediate form of memory is relatively more stable than the short-term memory and can last for several hours. However, at this stage, memory still remains sensitive to interference from competing or disrupting factors. This susceptibility is overcome through a process of consolidation. Memory consolidation was originally defined as a process whereby a memory trace, through the simple passage of time, becomes increasingly resistant to interference from competing or disrupting factors in the absence of further practice. However, the past 50 years of sleep and memory research have revealed that in addition to the passage of time, an adequate amount of sleep during this time is also required for memory to consolidate successfully. At the end of the consolidation stage, a memory has become stable and resistant to even extreme disruptions, such as electroconvulsive shock or application of neuronal gene and protein activation inhibitors. Memory consolidation itself is not a single-step process, but rather a multistep process that occurs exclusively during periods of sleep (Datta, 2010). Operationally, the cascading memory consolidation process can be divided into four stages: (1) search and read out of the intermediate form of memory, (2) elimination of unnecessary and/or redundant memory, (3) strengthening of cognitively relevant memory, and (4) transfer of stable memory to long-term storage. All of these processes occur over time, automatically, outside of awareness and without intent. Thus, they are distinct from the changes that result from conscious reminiscing or intentional rehearsal. Additionally, there is now a clear consensus that P-wave activity in REM sleep is critically involved in the last two steps of the memory consolidation process.

DESCRIPTION OF PGO/P-WAVES

Prominent phasic events of REM sleep include characteristic field potentials in the pontine tegmentum, which begin just prior to the onset of REM sleep and continue through its duration (Jouvet et al., 1959; Brooks and Bizzi, 1963; Datta & Hobson, 1994, 1995; Datta et al., 1998). These field potentials have been recorded in both the lateral geniculate body (LGB) and the occipital cortex of the cat (Mikiten et al., 1961; Mouret et al., 1963). Since, in the cat, these field potentials originate in the pons (P) and then propagate to the geniculate (G) and occipital cortex (O), they are called

PGO waves (Bizzi & Brooks, 1963; Brooks & Bizzi, 1963). Subsequent studies found that PGO waves in the cat could also be recorded at points throughout the extent of the thalamus and cortex. However, such PGO waves reach their highest amplitude in the LGB, primary visual cortex, and association visual cortex (reviewed in Datta, 1997). In addition to the pons, thalamus, and cortex, phasic potentials have been recorded in both the oculomotor nuclei (Brooks & Bizzi, 1963) and the cerebellum of the cat (Jouvet et al., 1965). Phasic potentials of pontine origin have also been recorded in the amygdala, cingulate gyrus, and hippocampus, which suggests that PGO waves also occur in the limbic system (Calvo & Fernandez-Guardiola, 1984). More importantly, all of these studies that have mapped PGO waves in the cat have demonstrated that the pons is the primary site of origin for PGO wave activity (reviewed in Datta, 1995, 1997). In addition to cats, PGO waves have also been documented and studied in other mammalian species including nonhuman primates, humans, and rodents. In nonhuman primates, PGO wave-like phasic field potentials have been recorded from the LGB and pons of macaques (Cohen & Feldman, 1968; Feldman & Cohen, 1968) and in the LGB of baboons (Vuillon-Cacciuttolo & Seri, 1978). In humans, phasic potentials have been recorded in the striate cortex during REM sleep (Salzarulo et al., 1975). Such striate field potentials are probably cortical components of state-specific phasic potentials of pontine origin. The observation of phasic scalp potentials associated with eye movements during REM sleep originally suggested that PGO wave-like activity may also be present in humans (McCarley et al., 1983; Miyauchi et al., 1987). Indeed, PGO waves have recently been recorded in the human pons, occurring immediately before and during REM sleep (Lim et al., 2007).

Based on recordings of PGO waves in the cat, initial attempts to record similar potentials in the LGB of the rat were unsuccessful (Gottesmann, 1969; Stern et al., 1974). Subsequent studies have recorded PGO-like waves in the pons of the rat that are equivalent to those in the pons of the cat (Gottesmann, 1969; Farber et al., 1980; Kaufman, 1983; Sanford et al., 1995; Datta et al., 1998, 1999). These initial failures led to the conclusion that state-specific pontine phasic waves in rats do not excite LGB neurons to produce the geniculate components of PGO waves (Datta, 1995). More recently, the absence of PGO wave-like activity in the rat LGB was shown to be due to a lack of afferent inputs from P-wave generating cells to the LGB (Datta et al., 1998). This field potential in the rat is therefore called a pontine-wave (P-wave), since it does not activate the geniculate nucleus (Datta et al., 1999, Datta 2000).

The waveform, amplitude, and frequency characteristics of PGO waves recorded from the pons, geniculate, and occipital cortex have been examined most intensively in the cat (reviewed in Datta, 1997). PGO waves are biphasic in shape with a duration of 60–120 ms and an amplitude between 200–300 μV (Datta & Hobson, 1994). The P-wave in the rat is equivalent to the pontine component of the PGO wave in the cat (Datta & Hobson, 1994; Datta et al., 1998, 1999), with similar duration (see Fig. 7.1) (75–100 msec) and amplitude (100–150 μV) (Datta et al., 1998). PGO/P-waves can occur either as singlets or as clusters containing a variable number of waves (3–5 waves/burst) at a density range of 30–60 spikes/min during REM sleep (Datta & Hobson, 2000). Singlet PGO/P-waves, known as Type I waves, occur commonly in NREM sleep and are independent of eye movement; conversely, clusters of PGO waves, known as Type II waves, are associated with eye movement bursts and are typically indicative of REM sleep (Morrison & Pompeiano, 1966). In fact, Type II PGO wave activity accounts for 55–65% of the total number of PGO waves recorded during REM sleep (Datta et al., 1992; Datta & Hobson, 2000).

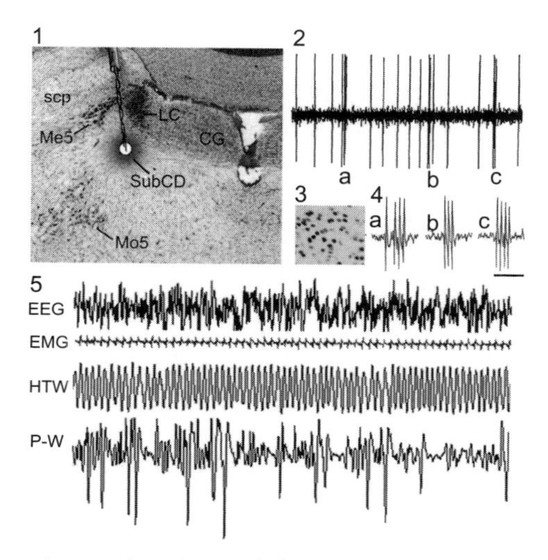

Figure 7.1 The location and activity of the pontine-wave (P-wave) generator. (1) Coronal section through the pons showing the P-wave generator (SubCD) of the rat in relation to neighboring areas of the brain. The figure also shows a bipolar recording electrode targeted directly into the P-wave generator to record P-wave activity. (2) A train of action potentials from a P-wave generating neuron showing recurrent high-frequency bursts in the background of tonic action potentials. (3) Photomicrograph of the P-wave generator of the rat showing pCREB-ir nuclei after cholinergic activation. (4) The three high-frequency bursts (a, b, and c) seen in (2), displayed on an expanded time scale (time scale = 10 ms). (5) Sample polygraphic appearance of REM sleep (trace duration = 10 sec) in rats showing low-voltage, high-frequency waves recorded from the frontal cortex (EEG); neck muscle atonia (EMG); theta-waves in the hippocampal EEG (HTW); and P-waves in the pontine EEG (P-W). Unpublished results of S. Datta.

CELLULAR AND MOLECULAR CHARACTERISTICS OF THE PGO/P-WAVE GENERATOR

Early transection and PGO wave recording studies in the cat indicated that the PGO wave generator is located within the pons (Bizzi & Brooks, 1963; Jouvet et al., 1965; Gottesmann, 1969; Datta, 1997). Subsequently, a number of single cell activity recordings in and around the PPT and laterodorsal tegmentum (LDT) observed a small population of neurons (about 3–5%) that discharged in bursts (3–5 spikes/burst) immediately preceding individual LGB PGO waves (McCarley et al., 1978; Steriade et al., 1990a,b). Based on this observation, these cells were originally believed to be PGO-wave generating neurons (McCarley et al., 1978; Steriade et al., 1990b). Recent studies in the cat, however, clearly indicate that the burst cells in the PPT/LDT are not PGO-wave generating neurons (reviewed in Datta, 1995). Instead, these cells, called transferring neurons, are responsible for conveying information from the pontine PGO wave generator to the forebrain (Datta, 1997). In the rat, because P-wave generating cells transmit P-wave information directly to the forebrain (Datta et al., 1998), these transferring neurons are absent (Datta & Siwek, 2002).

Utilizing chemical microstimulation, cell-specific lesions, and single cell recording techniques, the P-wave generator in the cat was localized within the caudolateral peribrachial (C-PBL) area (Datta et al., 1992; Datta & Hobson, 1994, 1995). Subsequently, using similar experimental techniques, the P-wave generator in the rat was localized within the dorsal subcoeruleus nucleus (SubCD) (Datta et al., 1998; 1999). In humans, as in the cat, the PGO wave generator is located in the C-PBL (Lim et al., 2007). Immunohistochemical identification of cholinergic and glutamatergic cells in the brainstem indicates that PGO-wave generating cells in the cat are capable of synthesizing both acetylcholine and glutamate (Quattrochi et al., 1998), thus these cells could be labeled as both cholinergic and glutamatergic. However, P-wave generating cells in the rat have been identified by specific monoclonal antibodies as glutamatergic, but not cholinergic (Datta, 2006).

Since the P-wave generator is also involved in sensorimotor integration (Morrison & Bowker, 1975), the differences in its anatomical location and neurotransmitter identity between the rat and cat may provide species-specific advantages. Specifically, in prey animals (i.e., the rat), the P-wave generator is anatomically closer to the locus coeruleus (LC). This shorter distance is advantageous during REM sleep (when animals are naturally paralyzed due to muscle atonia) because it permits quick communication with the LC for a flight response, which facilitates escape from predators. This rapid flight response is vital for the survival of prey animals. In contrast, the predatory mammalian (i.e., the cat) PGO wave generator is farther from the LC, and instead, closer to the PPT. Since predators rarely face the threat of predation, there is no advantage to having a quick arousal response to any nonthreatening type of noise during REM sleep. In fact, frequent interruptions could actually harm a predatory animal by preventing the necessary functions (i.e., cognitive) of REM sleep. Thus, for these types of interruptions, the P-wave generator signals the cholinergic PPT to intensify REM sleep, rather than wake up the animal, by activating the LC.

Single cell recording studies have shown that P-wave generating neurons discharge high-frequency spike bursts (>500 Hz, 3–5 spikes/burst) in the background of tonically increased firing rates (30–40 Hz) during the P-wave related transitional state between SWS and REM sleep (tS-R) and REM sleep (Datta & Hobson, 1994; Datta, 1997). Normally, the glutamatergic P-wave generating cells remain silent during wake (W) and SWS (Datta & Hobson, 1994). A neuroanatomical pathway tracing study demonstrated that functionally identified P-wave generator cells in the rat project to the dorsal hippocampus (DH), amygdala, entorhinal cortex, visual cortex, and many other regions of the brain involved in cognitive functions (see Fig 7.2) (Datta et al., 1998). Similar studies have also demonstrated that the P-wave generator in both the cat and rat receives afferent projections from the raphe nuclei (RN) and locus coeruleus (LC) nuclei (Quattrochi et al., 1998; Datta et al., 1999). It has been demonstrated that cholinergic activation of the P-wave generator increases glutamate release in the DH (Datta, 2006). In addition, P-wave activity has a positive influence on hippocampal theta-wave activity in the DH (Karashima et al., 2002, 2005). Most recently, we have demonstrated that activation of the P-wave generator increases (1) phosphorylation of cAMP response element-binding protein (CREB), (2) activity-regulated cytoskeletal-associated protein (Arc), and (3) brain-derived neurotrophic factor (BDNF), as well as the messenger ribonucleic acids (mRNAs) of Arc, BDNF, and early growth response-1 (Egr-1) in the DH and amygdala (see Fig. 7.2) (Saha & Datta, 2005; Ulloor & Datta, 2005; Datta et al., 2008, 2009).

MECHANISMS OF P-WAVE ACTIVITY GENERATION

As mentioned in the previous section, experimental evidence has demonstrated that P-wave activity of REM sleep is generated by the activation of a distinct cell group, located in the SubCD in the rat and C-PBL in the cat (reviewed in Datta, 1995; Datta & McLean, 2007). It should be emphasized

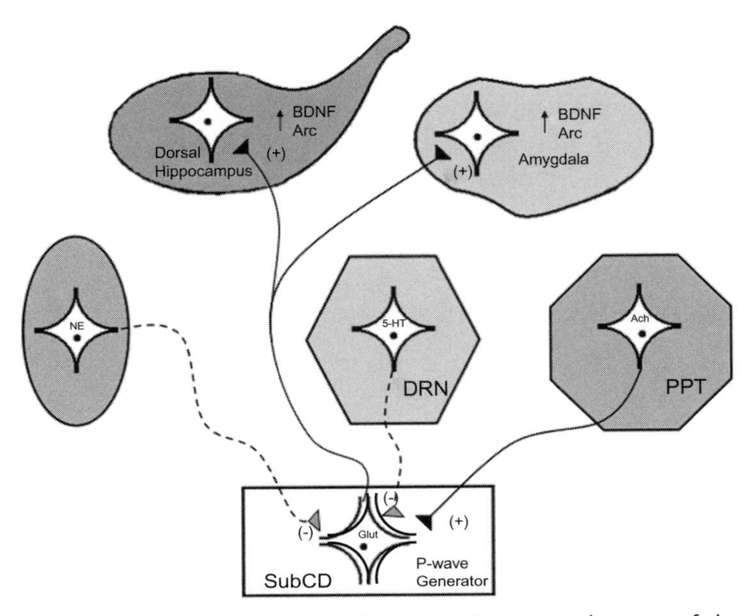

Figure 7.2 A summary diagram illustrating some important elements of the circuit regulating pontine-wave (P-wave) generator activity and targets of P-wave generating cells in the forebrain. The P-wave generator in the rat is located in the dorsal part of the subcoeruleus nucleus (SubCD), and the neurochemical phenotype of the P-wave generating neurons is glutamatergic (Glut). These P-wave generating cells receive afferent inputs from the cholinergic cells (Ach) in the pedunculopontine tegmentum (PPT), norepinephrinergic cells (NE) in the locus coeruleus (LC), and serotonergic cells (5-HT) in the dorsal raphe nucleus (DRN). Activation of PPT Ach cells excites (+) P-wave generating cells by releasing acetylcholine into the P-wave generator. Conversely, activation of LC NE cells and DRN 5-HT cells inhibits (-) P-wave generating cells by releasing NE and 5-HT into the P-wave generator. Thus, P-wave generating cells are activated in the presence of high levels of Ach and low levels of NE and 5-HT in the P-wave generator. Also, activation of P-wave generating cells increases glutamate release in the dorsal hippocampus and amygdala. Increased glutamate release activates dorsal hippocampus and amygdala cells via NMDA receptors, which ultimately increases expression of plasticity-related genes, BDNF and Arc.

here that this particular cell group simply represents the executive neurons for the P-wave. Turn-on or turn-off conditions of P-wave generating executive neurons are regulated by the ratios of available aminergic and cholinergic neurotransmitters within the P-wave generator. The source of aminergic neurotransmitters is the LC and DRN, while cholinergic neurotransmitters originate from the PPT (Fig. 7.2). The activity of both aminergic and cholinergic cells is approximately equal during wakefulness, and the onset of SWS results in an equal reduction in activity. Therefore, the ratio of aminergic to cholinergic neurotransmitters in the P-wave generator is proportionate during wakefulness and through SWS. During REM sleep, however, aminergic cell activities are markedly reduced or absent and cholinergic cell activities are comparatively high (Datta et al., 2009b). The level of cholinergic cell activity during REM sleep is roughly 35% less than that of wakefulness. Thus, when a hypothetical ratio of aminergic and cholinergic neurotransmitters is 1:1, the P-wave generator remains in turned-off condition; however, when this ratio is 0:0.65, the generator is turned on to express P-wave activity (Datta & Siwek, 2002). Besides cholinergic and aminergic neurotransmitters, the inhibitory neurotransmitter GABA is also involved in the regulation of P-wave generating cells' activity, especially in the expression of high-frequency bursts.

EVIDENCE TO LINK THE P-WAVE GENERATOR WITH MEMORY CONSOLIDATION

Physiological evidence. Long-term potentiation (LTP) of synaptic transmission is widely considered to be a model of activity-dependent synaptic plasticity that could be involved in certain forms of learning and memory (Bliss & Collingridge 1993; Datta, 2006). It is well documented that REM sleep increases following learning trials and that deprivation of REM sleep soon after learning trials causes a subsequent decrease in performance of the learned task (Karni et al. 1994; Datta & Patterson, 2003; Datta et al., 2004). Associated with these changes in REM sleep are changes in the efficacy of synaptic transmission in the brain, manifested as LTP (for references see Datta, 2006). LTP is significant in that it is thought to be the physiological substrate of learning and memory at the level of the hippocampus and amygdala (Bliss & Collingridge, 1993). The standard protocols used by most researchers to induce LTP in the hippocampus, amygdala, neocortex, and many other areas of the brain are (1) high-frequency stimulation in which several hundred pulses at frequencies of 250–400 Hz are given; and (2) short high-frequency (>200 Hz) bursts of stimuli with an interburst interval of ~200 msec, called theta-patterned stimulation (for references see Datta & Patterson, 2003; Datta, 2006). In an experimental situation, the high-frequency electrical stimulation of an afferent pathway is key for induction of LTP. However, during REM sleep, the physiological source of this presynaptic high-frequency stimulation is unclear. Therefore, the identification of this source would be a significant contribution to the current body of knowledge about the physiological substrates of learning and memory.

For REM sleep-dependent memory processing and learning, the source of the LTP-inducing high-frequency stimulus must come from the REM sleep sign generating structures of the brainstem. Over the past 25 years, a number of laboratories have recorded the single cell activity patterns of several different REM sleep sign generating structures in rats, cats, and nonhuman primates (for reviews see Datta, 1995, 1997; Datta & MacLean, 2007). Depending on the specific REM sleep sign generating structure, the neuronal activity patterns of these singular cells are classified as tonic single-spike type, bursting type, or both tonic and bursting type. The only type of cell within the REM sleep sign generating structures that fires as a high-frequency burst, similar to the high-frequency stimulus required for the generation of LTP, is located within the P-wave generator (Datta, 1997). These P-wave generating neurons discharge high-frequency (>500 Hz) spike bursts (3–5 spikes/burst) on the background of tonically increased firing rates (30–40 Hz) during the P-wave related states of tS-R and REM sleep (Datta & Hobson, 1994; Datta, 1997). High-frequency bursting patterns of these P-wave generating cells support the idea that the P-wave generator may be the source of electrical stimulus for the induction of physiological LTP. Experimental evidence suggests that the activation of P-wave generating cells is capable of inducing LTP. Microinjection of the cholinergic agonist carbachol into the P-wave generator activates P-wave generating cells (Datta et al., 1991, 1998). Cholinergic activation of the P-wave generator in the cat markedly increases P-wave activity and REM sleep (Datta et al., 1991, 1992). This cholinergic stimulation-induced potentiation of P-wave density and REM sleep lasts for about 7–10 days. This long-lasting increase in P-wave density and REM sleep is a physiological sign of synaptic, as well as intracellular, plasticity. Activation of the P-wave generator facilitates hippocampal theta activity (Karashima et al., 2002, 2005; Datta, 2006). Physiological evidence suggests that the hippocampal theta rhythm favors induction of LTP in the hippocampus, as well as in many different parts of the cerebral cortex (for references see Poe et al., 2000; Pavlides and Ribeiro, 2003; Booth and Poe, 2006). Thus, the collection of P-wave generating cells is not only capable of inducing physiological LTP, but also represents the only group of cells in the REM sleep generating network that is capable of inducing this type of physiological plasticity.

Anatomical evidence. If the P-wave generator is a presynaptic input for the induction of synaptic plasticity, a prerequisite for learning and memory processing, P-wave generating cells would be expected to send anatomical connections to the forebrain structures involved in memory processing. To test this hypothesis, the anterograde tracer biotinylated dextran amine (BDA) was microinjected into the physiologically identified cholinoceptive pontine P-wave generating site of rats to identify brain structures receiving efferent projections from those P-wave generating sites (Datta et al., 1998). In all cases, small volume injections of BDA in the cholinoceptive P-wave generating sites resulted in anterograde labeling of fibers and terminals in many regions of the brain. The most important output structures of those P-wave generating cells were the occipital cortex, entorhinal cortex, piriform cortex, amygdala, hippocampus, and many other thalamic, hypothalamic, and brainstem nuclei that participate in the generation of REM sleep (Datta 1995, 1997; Datta et al., 1998). All of these forebrain structures are also well known to be involved in memory processing (for references see Datta & Patterson, 2003). More recently, it has been demonstrated that these functionally identified P-wave generating cells are glutamatergic and stimulation of these cells releases glutamate in the DH (Datta, 2006). These monosynaptic axonal connections between P-wave generating glutamatergic cells and forebrain structures provide anatomical evidence that P-wave generating cells have the necessary anatomical substrate to be the presynaptic input for the induction of synaptic plasticity, a required process for learning and memory processing.

Behavioral evidence. Several studies indicate that rapid eye movements may represent the element of REM sleep that is crucial for memory consolidation (Verschoor & Holdstock 1984; Mandai et al., 1989; Smith & Weeden 1990; Smith & Lapp, 1991). For example, when a background clicking noise was presented during acquisition of a learned skill, presentation of the same auditory stimulus during subsequent eye movements during REM sleep (cueing) was correlated with a 23% improvement on retest performance one week later. The same cueing applied during non-eye-movement REM sleep episodes correlated with only an 8.8% retest improvement. It has been hypothesized, therefore, that the eye movements (or at least that segment of REM sleep in which they occur) are selectively important in REM sleep-dependent memory consolidation (Smith & Weeden, 1990). Visual learning tests in human volunteers showed that in addition to increases in percentage of REM sleep, the percentage of eye bursts during post-training REM sleep increased (Verschoor & Holdstock, 1984). Researchers hypothesize that these augmented eye bursts represent the scanning of visual stimuli encountered during the learning task, as part of the process of sorting, organizing, and consolidating daily input

(Verschoor & Holdstock, 1984). A study of Morse language learning in humans provides further evidence for an eye movement role in learning and memory processing during REM sleep. After a 90-minute Morse language learning session immediately prior to bedtime, subjects who had the greatest success had the densest rapid eye movements (Mandai et al., 1989). It is well established that the occurrence and direction of rapid eye movements during REM sleep depends exclusively on the excitation of P-wave generating cells (Datta & Hobson, 1994). Therefore, the studies described above indirectly suggest that the excitation of P-wave generating cells may be involved in REM sleep-dependent memory consolidation. The following paragraph describes some of the behavioral studies that have tested directly the relationship between P-wave generator activity and memory consolidation.

Using two different types of learning paradigms—two-way active avoidance (TWAA) and the Morris water maze (spatial learning)—studies have shown that learning training increases REM sleep and P-wave activity (Datta, 2000, 2006). More importantly, the results of such studies have shown that the increase in P-wave density during post-training REM sleep episodes is positively correlated with the effective consolidation, retention, and recall of the learning task. Together, the results of these studies indicate that P-wave generator activation may have a positive influence in the REM sleep-dependent memory processing of TWAA and spatial navigational learning.

In another behavioral study, we have demonstrated that supplemental activation of the P-wave generator, to a level greater than the normal post-training increased level, boosts retention of TWAA learning (Mavanji & Datta, 2003). The evidence from this study suggests that P-wave generator activation during REM sleep may enhance consolidation and integration of memories, resulting in improved performance on a recently learned task. Subsequently, another study has shown that activation of the P-wave generator prevents the memory-impairing effects of post-training REM sleep deprivation (Datta et al., 2004). The results of this study further substantiate the idea activation of the P-wave generator during REM sleep enhances the physiological process of memory processing, which naturally occurs during post-training REM sleep. Finally, another study has shown that selective elimination of cell bodies from the P-wave generator prevents retention of TWAA learning memory (Mavanji et al., 2004). The results of this study also have shown that lesions in the P-wave generator eliminate P-waves during REM sleep without changing the amount of time spent in W, SWS, or REM sleep. These findings provide direct evidence that P-wave generating cells are crucial for normal REM sleep-dependent memory processing.

Biochemical/molecular evidence. A number of studies have shown that the afferent path for DH reactivation-dependent LTP and/or memory formation is glutamatergic, and transmission at these synapses involves NMDA receptors on the postsynaptic side (Morris et al., 1986; Zanatta et al., 1996; Packard & Teather, 1997; Steward & Worley, 2001). As mentioned earlier, the P-wave generator directly projects to the DH, amygdala, and many other forebrain structures that are involved in memory processing (Datta et al., 1998). More importantly, we have shown that P-wave generating cells are glutamatergic and activation of P-wave generating cells increases glutamate release in the DH (Datta, 2006). Additionally, our behavioral studies have shown that learning training increases P-wave activity and activation of the P-wave generator during a post-training period improves memory (Datta, 2000, 2006; Mavanji & Datta, 2003; Datta et al., 2004). We have also demonstrated that the elimination of P-wave generating cells prevents retention of memory (Mavanji et al., 2004). Collectively, the results of these studies suggest that P-wave generating cells are one of the major sources of glutamate for postsynaptic NMDA receptor activation–mediated memory processing in the DH.

A number of studies have suggested that neuronal activation-induced stimulation of the cAMP and/or Ca^{++}-PKA-CREB pathway is involved in the induction of a variety of gene expressions and ultimately in the protein synthesis of long-term memory formation (Kandel & Schwartz, 1982; Abel et al., 1997; Datta et al., 2009a). Using molecular and cellular techniques, we have shown that TWAA learning training causes P-wave generator activation and spatiotemporal phosphorylation of CREB (pCREB) in the DH and amygdala (Saha & Datta, 2005). Similarly, we have also demonstrated that TWAA learning training increases pCREB, BDNF, and Arc proteins in the DH and amygdala (Ulloor & Datta, 2005). The results of this study show that the increase in P-wave activity during the post-training 3-hour recording session is positively correlated with the increased levels of pCREB, BDNF, and Arc in the DH. This suggests that memory processing of TWAA learning involves excitation of P-wave generating cells and increased expression of pCREB, Arc, and BDNF proteins in the DH and amygdala. Finally, using a combination of cell-specific localized lesions and molecular techniques, we have shown that elimination of P-wave-generating cells abolishes P-wave activity and TWAA learning training trials-induced expression of pCREB and Arc proteins and Arc, BDNF, and Egr-1 mRNAs in the DH and amygdala (Datta et al., 2008). More recently, it has been demonstrated that P-wave generator activation-dependent TWAA memory processing involves the intracellular PKA signaling system in the DH (Datta et al., 2009a). This study has shown that P-wave generator activation-mediated PKA activation is necessary for the expression of TWAA learning training-induced BDNF gene expression in the DH. Collectively, these cellular and molecular studies have shown that TWAA memory processing-specific gene activation and protein synthesis in the DH and amygdala are initiated by the activation of the P-wave generator. These studies also suggest that the P-wave generator activation is the primary mechanism for the REM sleep-dependent memory consolidation process.

CONCLUSIONS

In this chapter, I have discussed some of the compelling evidence that I believe to be significant for our understanding of the functional significance of P-wave generator activity in both REM sleep and REM sleep-dependent memory processing. These findings are the following: (1) Both TWAA and Morris water maze spatial navigation learning training increase REM sleep and P-wave activity during the subsequent sleep period. Improved performance in both TWAA and Morris water maze spatial navigation learning is proportional to the increase in P-wave density during the REM sleep episodes following these training trials (Datta, 2000, 2006). (2) After TWAA training trials, immediate supplemental activation of the P-wave generator, to a level above the normal post-training increased level, significantly increases retention of learning in later testing (Mavanji & Datta, 2003). (3) Activation of the P-wave generator prevents the TWAA memory-impairing effects of post-training REM sleep deprivation (Datta et al., 2004). (4) Elimination of P-waves by selective elimination of P-wave generating cells prevents retention of TWAA learning in the test trials (Mavanji et al., 2004). (5) We have shown that P-wave generating cells are glutamatergic, which project directly to a number of forebrain regions, including the DH and amygdala (Datta et al., 1998; Datta, 2006). Efferents from the P-wave generator project to areas that are directly involved in learning and memory processing. Activation of the P-wave generator increases glutamate release and theta wave frequency in the DH; both of these conditions have a positive influence on memory processing (Datta, 2006). (6) REM sleep-dependent TWAA memory processing depends on the P-wave generator activation-mediated interaction with the DH-CA3 region (Datta et al., 2005). (7) Chemical activation of the P-wave generator and/or TWAA learning training increases the phosphorylation of transcription factor CREB and expression of immediate early genes Arc, BDNF, and Egr-1 in the DH, amygdala, and cerebral cortex (Saha & Datta, 2005; Ulloor & Datta, 2005; Datta et al., 2008).(8) P-wave generator activation-mediated TWAA memory processing involves PKA activation and PKA activation-mediated BDNF expression in the DH-CA3 (Datta et al., 2009a). These findings substantiate the idea that P-wave generator activation during post-training REM sleep may be critical for REM sleep-dependent memory processing of two-way active avoidance and spatial learning.

At present, our understanding of sleep-dependent memory processing mechanisms remains incomplete. Nevertheless, based on the existing findings, we suggest that training paradigms cause an increase in homeostatic demand for the activation of P-wave generating cells in the brainstem, which ultimately increases the total duration of P-wave related states, tS-R and REM sleep. Activation of P-wave generating cells during post-learning-training tS-R and REM sleep provides a glutamatergic-activating stimulus to the hippocampus and amygdala, which leads to the physiological reactivation and neuronal activation-dependent gene expression and protein synthesis processes that are necessary for long-term neuronal plasticity and memory formation.

ACKNOWLEDGEMENTS

This work was supported by National Institutes of Health (USA) Research Grants NS 34004 and MH 59839.

REFERENCES

Abel, T., Nguyen, P.V., Barad, M., Deuel, T. A., Kandel, E. R., & Bourtchouladze, R. (1997). Genetic demonstration of a role for PKA in the late phase of LTP and in hippocampus-based long-term memory. *Cell, 88*, 615–626.

Bizzi, E., & Brooks, D. C. (1963). Functional connections between pontine reticular formation and lateral geniculate nucleus during sleep sleep. *Archives Italiennes De Biologie, 101*, 666–680.

Bliss, T.V., & Collingridge, G. L. (1993). A synaptic model of memory: Long-term potentiation in the hippocampus. *Nature, 361*, 31–39.

Booth, V., & Poe, G. R. (2006). Input source and strength influences overall firing phase of model hippocampal CA1 pyramidal cells during theta: Elevance to REM sleep reactivation and memory consolidation. *Hippocampus, 16*, 161–173.

Brooks, D. C., & Bizzi, E. (1963). Brain stem electrical activity during deep sleep. *Archives Italiennes De Biologie, 101*, 648–665.

Calvo, J. M., & Fernandez-Guardiola, A. (1984). Phasic activity of the basolateral amygdala, cingulate gyrus and hippocampus during REM sleep in the cat. *Sleep*, 7, 202–210.

Cohen, J. M., & Feldman, M. (1968). Relationship to electrical activity in the pontine reticular formation and lateral geniculate body to rapid eye movements. *Journal of Neurophysiology*, *31*, 807–817.

Datta, S. (1995). Neuronal activity in the peribrachial area: Relationship to behavioral state control. *Neuroscience and Biobehavioral Reviews*, *19*, 67–84.

Datta, S. (1997). Cellular basis of Pontine Ponto-geniculo-occipital wave generation and modulation. *Cellular and Molecular Neurobiology*, *17*, 341–365.

Datta, S. (2000). Avoidance task training potentiates phasic pontine-wave density in the rat: A mechanism for sleep-dependent plasticity. *The Journal of Neuroscience*, *20*, 8607–8613.

Datta, S. (2006). Activation of phasic pontine-wave generator: A mechanism for sleep-dependent memory processing. *Sleep and Biological Rhythms*, *4*, 16–26.

Datta, S. (2010). Sleep: Learning and Memory. In G. F. Koob, M. Le Moal, & R. F. Thompson (Eds.), *Encyclopedia of behavioral neuroscience* (Vol. 3, pp. 218–226). Oxford: Academic Press.

Datta, S., & Hobson, J. A. (1994). Neuronal activity in the caudo-lateral peribrachial pons: Relationship to PGO waves and rapid eye movements. *Journal of Neurophysiology*, *71*, 95–109.

Datta, S., & Hobson, J. A. (1995). Suppression of ponto-geniculo-occipital waves by neurotoxic lesions of pontine caudo-lateral peribrachial cells. *Neuroscience*, *67*, 703–712.

Datta, S., & Hobson, J. A. (2000). The rat as an experimental model for sleep neurophysiology. *Behavioral Neuroscience*, *114*, 1239–1244.

Datta, S., & Maclean, R. R. (2007). Neurobiological mechanisms for the regulation of mammalian sleep-wake behavior: reinterpretation of historical evidence and inclusion of contemporary cellular and molecular evidence. *Neuroscience and Biobehavioral Reviews*, *31*, 775–824.

Datta, S., & Patterson, E. H. (2003). Activation of phasic pontine wave (P-wave): A mechanism of learning and memory processing. In J. Maquet, R. Stickgold, & C. Smith (Eds.), *Sleep and brain plasticity* (pp. 135–156). Oxford: Oxford University Press.

Datta, S., & Siwek, D. F. (2002). Single cell activity patterns of pedunculopontine tegmentum neurons across the sleep-wake cycle in the freely moving rats. *Journal of Neuroscience Research*, *70*, 611–621.

Datta, S., Calvo, J. M., Quattrochi, J. J., & Hobson, J. A. (1991). Long-term enhancement of REM sleep following cholinergic stimulation. *Neuroreport*, *2*, 619–622.

Datta, S., Calvo, J. M., & Quatrochi, J. (1992). Cholinergic microstimulation of the peribrachial nucleus in the cat. I. immediate and prolonged increases in ponto-geniculo-occipital waves. *Archives Italiennes De Biologie*, *130*, 263–284.

Datta, S., Siwek, D. F., Patterson, E. H., & Cipolloni, P. B. (1998). Localization of pontine PGO wave generation sites and their anatomical projctions in the rat. *Synapse*, *30*, 409–423.

Datta, S., Patterson, E. H., & Siwek, D. F. (1999). Brainstem afferents of the cholinoceptive pontine wave generation sites in the rat. *Sleep Research Online*, *2*, 79–82.

Datta, S., Mavanji, V., Ulloor, J., & Patterson, E. H. (2004). Activation of phasic pontine-wave generator prevents rapid eye movement sleep deprivation-induced learning impairment in the rat: a mechanism for sleep-dependent plasticity. *Journal of Neuroscience*, *24*, 1416–1427.

Datta, S., Saha, S., Prutzman, S. L., Mullins, O. J., & Mavanji, V. (2005). Pontine-wave generator activation-dependent memory processing of avoidance learning involves the dorsal hippocampus in the rat. *Journal of Neuroscience Research*, *80*, 727–737.

Datta, S., Li, G., & Auerbach, S. (2008). Activation of phasic pontine-wave generator in the rat: a mechanism for expression of plasticity-related genes and proteins in the dorsal hippocampus and amygdala. *The European Journal of Neuroscience*, *27*, 1876–1892.

Datta, S., Siwek, D. F., & Huang, M. P. (2009). Improvement of two-way active avoidance memory requires protein kinase a activation and brain-derived neurotrophic factor expression in the dorsal hippocampus. *Journal of Molecular Neuroscience*, *38*, 257–264.

Datta, S., Siwek, D. F., & Stack, E. C. (2009). Identification of cholinergic and non-cholinergic neurons in the pons expressing phosphorylated cyclic AMP response element-binding protein as a function of rapid eye movement sleep. *Neuroscience*, *163*, 397–414.

Farber, J., Marks, G. A., & Roffwarg, H. P. (1980). Rapid eye movement sleep PGO-type waves are present in the dorsal pons of the albino rat. *Science*, *209*, 615–617.

Feldman, M., & Cohen, B. (1968). Electrical activity in the lateral geniculate body of the alert monkey associated with eye movements. *Journal of Neurophysiology*, *31*, 455–466.

Gottesmann, C. (1969). Etude sur les activites electrophysiologiques phasiques chez la rat. *Physiology & Behavior*, *4*, 495–504.

Jouvet, M., Jeannerod, M., & Delorme, F. (1965). Organization of the system responsible for phase activity during paradoxal sleep. *Comptes Rendus Des Seances De La Societe De Biologie Et De Ses Filiales*, *159*, 1599–1604.

Jouvet, M., Michel, F., & Courjon, J. (1959). L'activite electrique du rhinencephale au cours du sommeil chez le chat. *Comptes Rendus Social Biologies*, *153*, 101–105.

Kandel, E. R., & Schwartz, J. H. (1982). Molecular biology of learning: modulation of transmitter release. *Science*, *218*, 433–443.

Karashima, A., Nakamura, K., Sato, N., Nakao, M., Katayama, N., & Yamamoto, M. (2002). Phase-locking of spontaneous and elicited ponto-geniculo-occipital waves is associated with acceleration of hippocampal theta waves during rapid eye movement sleep in cats. *Brain Research*, *958*, 347–358.

Karashima, A., Nakao, M., Katayama, N., & Honda, K. (2005). Instantaneous acceleration and amplification of hippocampal theta wave coincident with phasic pontine activities during REM sleep. *Brain Research*, *1051*, 50–56.

Kaufman, L. S. (1983). Parachlorophenylalanine does not affect pontine-geniculate-occipital waves in rats despite significant effects on other sleep-waking parameters. *Experimental Neurology*, *80*, 410–417.

Karni, A., Tanne, D., Rubenstein, B. S., Askenasy, J. J., & Sagi, D. (1994). Dependence on REM sleep of overnight improvement of a perceptual skill. *Science*, *265*, 679–682.

Lim, Y. G., Kim, K. K., & Park, K. S. (2007). ECG recording on a bed during sleep without direct skin-contact. *IEEE Transactions on Bio-Medical Engineering*, *54*, 718–725.

Mandai, O., Guerrien, A., Sockeel, P., Dujardin, K., & Leconte, P. (1989). REM sleep modifications following a Morse code learning session in humans. *Physiology & Behavior*, *46*, 639–642.

Mavanji, V., & Datta, S. (2003). Activation of the phasic pontine-wave generator enhances improvement of learning performance: a mechanism for sleep-dependent plasticity. *The European Journal of Neuroscience*, *17*, 359–370.

Mavanji, V., Ulloor, J., Saha, S., & Datta, S. (2004). Neurotoxic lesions of phasic pontine-wave generator cells impair retention of 2-way active avoidance memory. *Sleep*, *27*, 1282–1292.

McCarley, R. W., Nelson, J. P., & Hobson, J. A. (1978). Ponto-geniculo-occipital (PGO) burst neurons: correlative evidence for neuronal generators of PGO waves. *Science*, *201*, 269–272.

McCarley, R. W., Winkelman, J. W., & Duffy, F. H. (1983). Human cerebral potentials associated with REM sleep rapid eye movements: links to PGO waves and waking potentials. *Brain Research*, *274*, 359–364.

Mikiten, T. M., Niebyl, P. H., & Hendley, C. D. (1961). EEG desynchronization during behavioral sleep associated with spike discharges from the thalamus of the cat. *Federation Proceedings*, *20*, 327.

Miyauchi, S., Takino, R., Fukuda, H., & Torii, S. (1987). Electrophysiological evidence for dreaming: human cerebral potentials associated with rapid eye movement during REM sleep. *Electroencephalography and Clinical Neurophysiology*, *66*, 383–390.

Morris, R. G., Anderson, E., Lynch, G. S., & Baudry, M. (1986). Selective impairment of learning and blockade of long-term potentiation by an N-methyl-D-aspartate receptor antagonist, AP5. *Nature*, *319*, 774–776.

Morrison, A. R., & Bowker, R. M. (1975). The biological significance of PGO spikes in the sleeping cat. *Acta Neurobiologiae Experimentalis (Wars)*, *35*, 821–840.

Morrison, A. R., & Pompeiano, O. (1966). Vestibular influences during sleep. IV. Functional relations between vestibular nuclei and lateral geniculate nucleus during desynchronized sleep. *Archives Italiennes De Biologie*, *104*, 425–458.

Mouret, J., Jeannerod, M., & Jouvet, M. (1963). L'active electrique du systeme visuel au cours de la phase paradoxale du sommeil chez le chat. *Journal of Physiology (Paris)*, *55*, 305–306.

Packard, M. G., & Teather, L. A. (1997). Posttraining injections of MK-801 produce a time-dependent impairment of memory in two water maze tasks. *Neurobiology of Learning and Memory*, *68*, 42–50.

Pavlides, C., & Ribeiro, S. (2003). Recent evidence of memory processing in sleep. In J. Maquet, R. Stickgold, & C. Smith (Eds.), *Sleep and brain plasticity* (pp. 327–362). Oxford: Oxford University Press.

Poe, G. R., Nitz, D. A., McNaughton, B. L., & Barnes, C. A. (2000). Experience-dependent phase-reversal of hippocampal neuron firing during REM sleep. *Brain Research*, *855*, 176–180.

Quattrochi, J., Datta, S., & Hobson, J. A. (1998). Cholinergic and non-cholinergic afferents of the caudolateral parabrachial nucleus: a role in the long-term enhancement of rapid eye movement sleep. *Neuroscience*, *83*, 1123–1136.

Saha, S., & Datta, S. (2005). Two-way active avoidance training-specific increases in phosphorylated cAMP response element-binding protein in the dorsal hippocampus, amygdala, and hypothalamus. *The European Journal of Neuroscience*, *21*, 3403–3414.

Salzarulo, P., Pelloni, G., & Lairy, G. C. (1975). [Electrophysiologic semiology of daytime sleep in 7 to 9-year-old children]. *Electroencephalography and Clinical Neurophysiology*, *38*, 473–494.

Sanford, L. D., Tejani-Butt, S. M., Ross, R. J., & Morrison, A. R. (1995). Amygdaloid control of alerting and behavioral arousal in rats: involvement of serotonergic mechanisms. *Archives Italiennes De Biologie*, *134*, 81–99.

Smith, C., & Lapp, L. (1991). Increases in number of REMS and REM density in humans following an intensive learning period. *Sleep*, *14*, 325–330.

Smith, C., & Weeden, K. (1990). Post training REMs coincident auditory stimulation enhances memory in humans. *Psychiatric Journal of the University of Ottawa*, *15*, 85–90.

Steriade, M., Datta, S., Pare, D., Oakson, G., & Curro Dossi, R. C. (1990). Neuronal activities in brain-stem cholinergic nuclei related to tonic activation processes in thalamocortical systems. *The Journal of Neuroscience*, *10*, 2541–2559.

Steriade, M., Pare, D., Datta, S., Oakson, G., & Curro Dossi, R. (1990). Different cellular types in mesopontine cholinergic nuclei related to ponto-geniculo-occipital waves. *The Journal of Neuroscience*, *10*, 2560–2579.

Stern, W. C., Forbes, W. B., & Morgane, P. J. (1974). Absence of ponto-geniculo-occipital (PGO) spikes in rats. *Physiology & Behavior*, *12*, 293–295.

Steward, O., & Worley, P. F. (2001). Selective targeting of newly synthesized Arc mRNA to active synapses requires NMDA receptor activation. *Neuron*, *30*, 227–240.

Ulloor, J., & Datta, S. (2005). Spatio-temporal activation of cyclic AMP response element-binding protein, activity-regulated cytoskeletal-associated protein and brain-derived nerve growth factor: a mechanism for pontine-wave generator activation-dependent two-way active-avoidance memory processing in the rat. *Journal of Neurochemistry*, *95*, 418–428.

Verschoor, G. J., & Holdstock, T. L. (1984). REM bursts and REM sleep following visual and auditory learning. *South African Journal of Psychology*, *14*, 69–74.

Vuillon-Cacciuttolo, G., & Seri, B. (1978). Effects of optic nerve section in baboons on the geniculate and cortical spike activity during various states of vigilance. *Electroencephalography and Clinical Neurophysiology*, *44*, 754–768.

Zanatta, M. S., Schaeffer, E., Schmitz, P. K., Medina, J. H., Quevedo, J., Quillfeldt, J. A., et al. (1996). Sequential involvement of NMDA receptor-dependent processes in hippocampus, amygdala, entorhinal cortex and parietal cortex in memory processing. *Behavioural Pharmacology*, *7*, 341–345.

CHAPTER 8

Neural Correlates of Human Sleep and Sleep-Dependent Memory Processing

Christelle Meyer, Vincenzo Muto, Mathieu Jaspar, Caroline Kussé, Ariane Foret, Laura Mascetti and Pierre Maquet
Cyclotron Research Centre, University of Liège, Belgium

INTRODUCTION

The brain is able to generate three distinct functional modes that are associated with specific neural firing patterns, oscillatory modes, and neuromodulatory contexts: wakefulness, non- rapid-eye-movement (NREM) sleep, and REM sleep. The mechanisms underlying these vigilance states are known with increasing detail through neurophysiological and molecular studies conducted in animals. Sleep being conserved across species, similar mechanisms should also underlie the organization of human brain function during sleep and wakefulness. The characterization of human brain function during sleep thus appears as an important step in the translational effort that aims at understanding the mechanisms underlying human sleep. Unfortunately, the access to human brain function is limited. With the exception of depth recordings in selected patients with brain disorders (e.g., intractable epilepsy or Parkinson disease), functional neuroimaging techniques provide the most straightforward access to regional human brain function at a macroscopic level. Various imaging techniques are available. Some techniques measure slow metabolic or hemodynamic signals at spatial resolution of a few millimeters: single photon emission computed tomography (SPECT), positron emission tomography (PET), and functional magnetic resonance imaging (fMRI). In contrast, electroencephalography (EEG) and magnetoencephalography (MEG) follow electromagnetic signals sampled at high temporal resolution but with a limited spatial resolution. These various techniques provide complementary views of human brain function and a comprehensive characterization of human sleep ultimately requires their integration with the mechanistic data revealed by animal research.

Sleep and Brain Activity
DOI: http://dx.doi.org/10.1016/B978-0-12-384995-3.00008-3

This chapter focuses on the modifications of brain activity reported during normal human sleep. First, we focus on the changes in regional cerebral metabolism and hemodynamics during NREM and REM sleep. In particular, we describe novel data combining EEG and fMRI recordings, which characterize changes in brain activity associated with slow waves and spindles, as well as the modification in functional brain connectivity during NREM sleep. Then we review the influence of previous waking experience on brain activity during sleep, and discuss the implication of sleep in memory consolidation.

NREM SLEEP

During NREM sleep, neuronal activity is organized by a slow rhythm (<1 Hz). Intracellular recordings showed that the neuronal membrane potential alternates between a depolarized state, associated with sustained firing, and a hyperpolarized phase, during which most neurons remain silent (Steriade, Nunez, & Amzica, 1993b). These variations in membrane potential occur locally in close synchrony within large neural populations, which therefore alternate between activated ("ON" state) and quiet periods ("OFF" state) (Vyazovskiy et al., 2009). EEG slow waves recorded on the scalp occur in synchrony with these neural events, the peak negativity being considered as corresponding to the neural "OFF" state (Molle, Marshall, Gais, & Born, 2002). On scalp EEG in humans, slow waves predominate over frontal areas and appear as traveling waves: each single wave is initiated at a definite cortical site and propagates following a given trajectory on the scalp (Massimini, Huber, Ferrarelli, Hill, & Tononi, 2004).

The slow oscillation temporally organizes other sleep rhythms such as spindles or hippocampal sharp waves and ripples. Spindles result from cyclic inhibition of thalamocortical neurons by reticular thalamic neurons. Post-inhibitory rebound spike bursts in thalamocortical cells entrain cortical populations in spindle oscillations (Steriade & McCarley, 2005). Corticothalamic neurons in turn synchronize thalamic spindling activity (Contreras, Destexhe, Sejnowski, & Steriade, 1996). Spindles preferentially occur during the neuronal "UP" state (Steriade, Nunez, & Amzica, 1993a) or in humans, during the positive phase of the EEG slow waves (Molle et al., 2002). In humans, two kinds of spindles are observed on EEG recordings (De Gennaro & Ferrara, 2003). Slow spindles (typically <13 Hz) predominate over frontal areas, whereas fast spindles (typically >13 Hz) prevail over centroparietal regions. The two spindles types have been associated with several functional differences (De Gennaro & Ferrara, 2003).

The hippocampal activity during NREM sleep is characterized by sharp waves and ripples, which are synchronous to the cortical slow oscillation (Clemens et al., 2007; Isomura et al., 2006; Ji & Wilson, 2007; Molle, Yeshenko, Marshall, Sara, & Born, 2006), although it is not yet clear which oscillation is driving the other (Molle & Born, 2009; Tononi, Massimini, & Riedner, 2006).

Global Changes in Brain Energy Metabolism

On average, neural firing activity is decreased during NREM sleep, relative to wakefulness or REM sleep (Vyazovskiy et al., 2009). Accordingly, measures of brain energy metabolism by tracer techniques showed that it is decreased in NREM sleep, relative to wakefulness, in cats (Ramm & Frost, 1983, 1986), monkeys (Kennedy et al., 1982) and humans (Buchsbaum et al., 1989; Madsen, Schmidt, Wildschiodtz, et al., 1991; Maquet et al., 1990). These early data have been reviewed elsewhere (Madsen & Vorstrup, 1991; Maquet, 2000). Quantitatively, cerebral glucose metabolic rates are 11% lower in stage 2 sleep (Maquet et al., 1992) and about 40% lower in deep NREM sleep (Buchsbaum et al., 1989; Maquet et al., 1990), relative to resting wakefulness. Likewise, cerebral oxygen utilization decreases by 5% (Madsen, Schmidt, Holm et al., 1991) to 7% (Takahashi, 1989) during light NREM sleep and by 25% in deep NREM sleep (Madsen, Schmidt, Wildschiodtz et al., 1991), as compared to resting wakefulness. Cerebral blood flow measurements showed the same general decrease in NREM sleep in humans (Braun et al., 1997; Kajimura et al., 1999; Madsen, Schmidt, Wildschiodtz, et al., 1991; Takahashi, 1989). In deep NREM sleep, the most deactivated areas were located in the dorsal pons and mesencephalon, cerebellum, thalami, basal ganglia, basal forebrain/hypothalamus, mesial frontal areas, and the precuneus (Andersson et al., 1998; Braun et al., 1997; Kajimura et al., 1999; Maquet et al., 1997). Regional decreases in blood flow in the brainstem, thalamus, and basal forebrain were interpreted as the diminished activity of activating structures maintaining wakefulness. The decreases in cortical blood flow were considered as identifying the areas where the slow oscillation consistently drives the activity of a large fraction of the local neural population. In keeping with this interpretation, blood flow in the medial prefrontal cortex was also inversely proportional to the EEG power in the delta frequency band (1.5–4 Hz), suggesting that local blood flow decreases as the synchronization of neural firing increases in this area (Dang-Vu et al., 2005). Tracer techniques like 2-deoxyglucose autoradiography in animals (Sokoloff, 1981), or positron emission tomography in humans (Phelps & Mazziotta, 1985) require long uptake periods (in the order of one to tens of minutes), several order of magnitude larger than the dynamics of neural activity during slow oscillation. In this respect, they can be thought of as estimating the average metabolic consequences of the alternating "ON" and "OFF" states.

Neural Correlates of Spindles and Slow Waves

These decreased average metabolic demands should by no means suggest that NREM sleep is a resting period for the brain; the slow oscillation is associated with substantial neural activity. Accordingly, with the advent of simultaneous recordings of functional magnetic resonance imaging and EEG signals, it became possible to characterize consistent changes in regional brain activity associated with the NREM sleep neural transients such as slow waves or spindles.

In contrast to EEG recordings, which insisted upon the variability between slow waves (Massimini et al., 2004), EEG/fMRI data identified the regions that are systematically recruited when a slow wave is recorded on the EEG. During NREM sleep, significant *increases* in activity were associated with slow waves in several cortical areas including inferior frontal, medial prefrontal, precuneus, and posterior cingulate cortex (Dang-Vu et al., 2008). These results were confirmed by reconstruction of electric sources of slow waves, as recorded by high-density EEG: slow waves were consistently associated with currents in the medial frontal gyrus, the inferior frontal gyrus, the anterior cingulate, the precuneus, and the posterior cingulate cortex (Murphy et al., 2009). This peculiar response pattern that primarily involves the medial aspect of the frontal, cingulate, and parietal cortices is thought to reflect the propagation of slow waves through major connectivity pathways of the human brain (Hagmann et al., 2008).

In addition, EEG/fMRI data further showed that the largest slow waves (>140 μV) were associated with significant activity in the parahippocampal gyrus, cerebellum, and pontine tegmentum (Dang-Vu et al., 2008). This finding indicates that slow waves reflect the synchronous firing of distributed cortical populations but also recruit mesiotemporal cortex and subcortical structures. In keeping with this view, depth recording in epileptic patients showed that gamma oscillations, a major determinant of fMRI signal (Logothetis, Pauls, Augath, Trinath, & Oeltermann, 2001), are recorded in synchrony with cortical slow oscillation at approximately the same time in many different cortical areas, the most prominent source being the parahippocampal gyrus (Le Van Quyen et al., 2010). The recruitment of the pons was unexpected, because the slow oscillation is interrupted by the stimulation of brainstem structures promoting wakefulness (Steriade, Amzica, & Nunez, 1993). However, there is now evidence that some brainstem populations involved in sleep/wake regulation fire in synchrony with the cortical slow oscillation (Eschenko, Magri, Panzeri, & Sara, 2012; Mena-Segovia, Sims, Magill, & Bolam, 2008). Collectively, EEG/fMRI data show that slow waves are associated with a transient increase in activity, not only in a distributed set of highly connected cortical areas but also in mesiotemporal regions and even in brainstem structures.

EEG/fMRI data also showed that spindles were associated with increased activity in the thalami, anterior cingulate, insular, and superior temporal cortices (Schabus et al., 2007). In addition, fast spindles recruited a set of cortical regions involved in sensorimotor processing (precentral and postcentral gyri, supplementary motor area, and neighboring midcingulate cortex), as well as the mesial frontal cortex and hippocampus. In contrast, slow spindles were associated with increased activity in the superior frontal gyrus (Schabus et al., 2007). These results were consistent with EEG source reconstruction reporting a spindle source in medial prefrontal cortex for slow spindles, and in the precuneus for fast spindles (Anderer et al., 2001). Moreover, EEG/fMRI data further showed that functional connectivity between the hippocampal formation and the neocortical regions recruited by fast spindles was increased in light NREM sleep, during which spindles predominate, relative to wakefulness (Andrade et al., 2011). These results suggest that fast spindles are associated with a synchronous activity in distributed hippocampocortical circuits, a condition suspected to favor exchange of information within these networks.

However, there seems to be more than two spindle types in humans. MEG recordings in humans identified multiple asynchronous neural generators during sleep spindles (Dehghani, Cash, Rossetti, Chen, & Halgren, 2010), a finding recently corroborated by intracerebral recordings in epileptic patients (Nir et al., 2011). Why MEG identifies multiples spindle sources is currently not fully understood. It was suggested that MEG preferentially recruit the recruitment of the focal core thalamocortical system, whereas EEG would be more sensitive to the distributed matrix thalamic system (Dehghani et al., 2010). In this view, the functional impact of spindles on information processing during sleep might imply multiple thalamocortical loops, in addition to the set of brain areas recruited by the two spindle classes identified on scalp EEG recordings.

In summary, EEG/fMRI data identified regional brain responses consistently associated with slow waves and spindles. Both slow waves and spindles result in increased activity in a distributed set of cortical and subcortical areas, suggesting that these oscillations transiently synchronize large-scale brain networks and potentially affect information processing during sleep. These consistent regionally specific response patterns complement the characterization of their variability illustrated by electrophysiological recordings. It is also worth observing that the description of the neural correlates of human NREM sleep rhythms is still fragmentary and other oscillations remain to be fully characterized. For instance, preliminary results showed that infraslow EEG oscillations (<0.1 Hz) during NREM sleep appear to be associated with activity increases in subcortical structures and negative responses in the cortex (Picchioni et al., 2011).

Functional Connectivity

Interactions between neural populations or, at the macroscopic level, between brain areas constitute a fundamental aspect of brain function. Functional connectivity is usually assessed by temporal covariations of activity between brain areas (Friston, Frith, Liddle, & Frackowiak, 1993). During resting wakefulness, spontaneous fluctuations of regional brain activity occur simultaneously in several distributed sets of brain areas. Several such functional networks are reliably identified across subjects (Damoiseaux et al., 2006). One of these networks is composed of brain areas that are consistently more active during resting wakefulness than when the subject is engaged in a cognitive task (Raichle et al., 2001). This network, which is usually referred to as the default mode network, includes medial anterior areas (medial prefrontal cortex, anterior cingulate cortex) and posterior regions (precuneus, posterior cingulate cortex, inferior parietal lobule).

During light NREM sleep, functional connectivity does not change in primary sensory areas and in the default mode network, relative to wakefulness (Larson-Prior et al., 2009). The connectivity between the intraparietal sulcus and frontal eye fields, two regions involved in attention, is even increased (Larson-Prior et al., 2009). In contrast, during deep NREM sleep, the connectivity between frontal and posterior components of the default mode network significantly decreases (Horovitz et al., 2009). These findings are consistent with a comprehensive analysis of functional connectivity using graph theory, which showed that changes in functional connectivity were specific to sleep stages (Spoormaker et al., 2010). Thalamocortical connectivity was significantly reduced at the transition from wakefulness to light NREM sleep. In contrast, the connectivity between cortical areas was maintained or even increased during light sleep, but broke down during deep NREM sleep. The reduction in corticocortical connectivity was more pronounced for long than short connections, which resulted in increased local clustering in deep NREM sleep. Although the functional significance of these changes in functional connectivity are not yet fully understood, they imply that information processing during deep NREM sleep substantially differs from wakefulness. The results suggest that brain function during deep NREM sleep is organized in multiple segregated functional systems, a pattern which echoes the local expression of slow waves and spindles.

REM SLEEP

REM sleep is characterized by tonic features that persist throughout REM sleep, such as fast, low amplitude EEG oscillations, muscle atonia or, in animals, rhythmic hippocampal theta rhythm. It is also associated with phasic features, that occur episodically, such as rapid eye movements, muscle twitches, autonomous instability or, in animals, pontine waves.

REM sleep is associated with an intense neuronal activity, similar to waking levels (Steriade & McCarley, 2005). Accordingly, brain glucose metabolism (Maquet et al., 1990) and oxygen utilization (Madsen, Schmidt, Wildschiodtz et al., 1991) are elevated during REM sleep and reach levels comparable to wakefulness. However, the spatial distribution of brain activity during REM sleep and wakefulness differ considerably. Cerebral blood flow measurements using PET characterized the distribution of regional brain activity during REM sleep. In keeping with animal data (Lydic et al., 1991; Ramm & Frost, 1986), REM sleep was associated with a high activity in the brainstem and thalamic nuclei as well as in limbic and paralimbic areas: the amygdala, the hippocampal formation, and the anterior cingulate, orbitofrontal, and insular cortices (Braun et al., 1997; Maquet et al., 1996; Nofzinger, Mintun, Wiseman, Kupfer, & Moore, 1997). Temporal and occipital cortices (Braun et al., 1997) as well as motor and premotor areas (Maquet et al., 2000) were also shown to be very active during human REM sleep. These activity increases contrasted with the relative quiescence of the associative frontal and parietal cortices (Braun et al., 1997; Maquet et al., 1996; Maquet et al., 2005).

Functional brain connectivity is also modified during REM sleep. For instance, REM sleep was associated with selective activation of extrastriate visual cortices, which was correlated with decreases in the striate cortex (Braun et al., 1998). Likewise, the functional relationship between the amygdala and the temporal and occipital cortices was enhanced during REM sleep relative to wakefulness or NREM sleep (Maquet & Phillips, 1998).

The mechanisms explaining this peculiar distribution of cortical activity are not yet fully understood. They might be intimately related to the activity of brain structures that generate REM sleep. For instance, the precoeruleus area in rat, a pontine REM-on area, projects to the medial septum and might influence hippocampal theta EEG during REM sleep (Lu, Sherman, Devor, & Saper, 2006). Changes in neuromodulation, predominantly cholinergic during REM sleep (Steriade & McCarley, 2005), could also participate in modifying the distribution of forebrain activity although to our knowledge, this suggestion has not been experimentally tested.

The neural correlates of phasic REM sleep have seldom been investigated in humans. Several observations suggest the presence of pontine waves during human REM sleep. Intracerebral recordings in the striate cortex of epileptic patients showed monophasic or diphasic potentials during REM sleep, appearing in isolation or in bursts (Salzarulo, Lairy, Bancaud, & Munari, 1975). Phasic potentials suggestive of PGO waves were also recorded from the pedunculopontine nucleus (Lim et al., 2007) and subthalamic nucleus (Fernandez-Mendoza et al., 2009) in patients with Parkinson disease before and during REM sleep. In normal subjects, the evidence is naturally more speculative. Transient occipital and/or parietal potentials time-locked to rapid eye movements were observed on EEG recordings in normal volunteers during REM sleep (McCarley, Winkelman, & Duffy, 1983). MEG recordings revealed similar potentials during REM sleep and their magnetic source was localized in the brainstem, thalamus, hippocampus, and occipital cortex (Inoué, Saha, & Musha, 1999). The sequence of activation of these magnetic sources suggested that the activation of the frontal eye field and the pons precedes the recruitment of limbic and paralimbic areas (orbitofrontal cortex, amygdala, parahippocampal gyrus) (Ioannides et al., 2004). Using PET and cerebral blood flow measurements, it was shown that the activity in the right geniculate body and the primary occipital cortex increases in proportion to the density of eye movements to a larger extent during REM sleep than during wakefulness (Peigneux et al., 2001), a result confirmed by fMRI studies (Hong et al., 2009; Miyauchi, Misaki, Kan, Fukunaga, & Koike, 2009; Wehrle et al., 2005).

The variability of respiratory and heart rates is another phasic aspect of REM sleep. A preliminary PET study showed that the variability of heart rate was more tightly correlated with the activity in the extended amygdala during REM sleep than during wakefulness (Desseilles et al., 2006). In addition, the functional connectivity between the amygdala and the insular cortex, a region involved in cardiovascular regulation during wakefulness (Critchley, Corfield, Chandler, Mathias, & Dolan, 2000), was weaker during REM sleep than during wakefulness. These results suggest a functional reorganization of central cardiovascular regulation during REM sleep that deserves further investigation.

FUNCTIONAL NEUROIMAGING OF MEMORY PROCESSING DURING SLEEP

Due to his permanent interaction with the environment, the individual continuously acquires new memories. Initially labile, these representations are further processed and transformed into more stable ones, which are progressively incorporated into long-term memories. This process, which is referred to as memory consolidation, involves changes in brain structure and function at both synaptic and systems levels. Sleep was shown to actively promote memory consolidation. It is associated with improved retention of declarative memories (Gais, Lucas, & Born, 2006) and enhanced performance in procedural learning tasks (Gais, Plihal, Wagner, & Born, 2000; Stickgold, James, & Hobson, 2000; Walker, Brakefield, Morgan, Hobson, & Stickgold, 2002). It also renders memories more resistant to interference (Ellenbogen, Hulbert, Stickgold, Dinges, & Thompson-Schill, 2006). The mechanisms underlying sleep-dependent memory consolidation are currently conceived within two conceptual frameworks. The first one suggests that sleep locally and ubiquitously maintains synaptic homeostasis; the second assumes that sleep promotes the systems-level reorganization of memories within hippocampo-neocortical networks.

Sleep and Synaptic Homeostasis

This theory claims that wakefulness is associated with a net increase in synaptic strength in the brain, which would become energetically unsustainable in the long term (Tononi & Cirelli, 2003, 2006). Due to this progressive synaptic potentiation, sleep pressure would increase monotonically in the brain with time spent awake, or more precisely, in proportion to neural activity accrued locally during wakefulness. During NREM sleep, slow waves would be associated with a gradual downscaling of the average brain synaptic strength to a baseline level, a process beneficial for learning and memory because it ultimately increases the signal-to-noise ratio related to the learned material. In support of this hypothesis, a local increase in slow wave activity (SWA) is locally enhanced following increased neural activity during waking: vibratory stimulation of the hand (Kattler, Dijk, & Borbely, 1994), training to a visuomotor adaptation task (Huber, Ghilardi, Massimini, & Tononi, 2004), transcranial magnetic stimulation of the motor cortex (Huber, Esser, et al., 2007), or after spike timing-dependent activity is elicited during waking by transcranial paired associative stimulation (Huber et al., 2008). In contrast, arm immobilization results in a decrease in SWA over controlateral sensorimotor areas during subsequent NREM sleep (Huber et al., 2006).

This hypothesis is currently supported by a number of animal data. Slow wave activity increases in rats exposed to enriched environment in relation to release of BDNF (brain-derived neurotrophic factor), a neurotrophin involved in synaptic potentiation (Huber, Tononi, & Cirelli, 2007). At the cellular level, multiunit recordings showed that the mean firing rate in the cerebral cortex increases after periods of wakefulness and decreases after periods of sleep, consistent with a net change in synaptic strength (Vyazovskiy et al., 2009). Changes in firing patterns in NREM sleep correlate with changes in slow-wave activity (Vyazovskiy et al., 2009). The slope and amplitude of cortical evoked responses, taken as markers of local synaptic strength, increase after wakefulness and decrease after sleep in proportion to changes in SWA (Vyazovskiy, Cirelli, Pfister-Genskow, Faraguna, & Tononi, 2008). The level of several molecular markers of synaptic potentiation are elevated in the cortex after a period of wakefulness and low after sleep, consistent with synaptic potentiation during wakefulness and depression during sleep (Vyazovskiy et al., 2008). Finally, there is evidence in flies that synapse size or number increases after a few hours of wake and decreases during sleep (Bushey, Tononi, & Cirelli, 2011).

Sleep-Dependent Systems-Level Consolidation of Hippocampal-Dependent Memories

Organisms are continuously exposed to novel pieces of information and have to flexibly retain this new information while preserving the knowledge, concepts, and skills gradually forged by earlier experience. To resolve this conflict, it has been assumed that the brain resorts to complementary learning systems with different dynamics (Marr, 1970, 1971; McClelland, McNaughton, & O'Reilly, 1995). On the one hand, novel information would quickly induce substantial changes in synaptic strength in the hippocampus. By contrast, in the cortex, it would result in limited synaptic changes that would not be sufficient to allow for the reliable reinstatement of the specific response pattern associated with a recent memory. Multiple repetitions of similar phases of information processing would be necessary for synaptic changes to accumulate thereby gradually reinforcing this novel representation and integrating it into the corpus of long-term representations stored predominantly in the cortex. Spontaneously reinstatements of hippocampal and neocortical activity associated with newly encoded representations would participate in this process and progressively strengthen corticocortical connections, which eventually buttress long-term memories (Marr, 1970, 1971; McClelland et al., 1995). These so-called "reactivations" can occur both during wakefulness and sleep. However, sleep appears as a particularly favorable period for memory reactivation, since the brain is less responsive than during wakefulness to environmental stimuli, which might potentially interfere with the learned material.

Reactivations have indeed been observed at the cellular level during both NREM and REM sleep in rodents. Neural firing patterns recorded during wakefulness are spontaneously repeated during sleep in the hippocampus (e.g., Louie & Wilson, 2001; Nadasdy, Hirase, Czurko, Csicsvari, & Buzsaki, 1999), which coincide with sharp waves and ripples (Kudrimoti, Barnes, & McNaughton, 1999). The selective elimination of hippocampal ripples during post-training consolidation periods impairs spatial memory tasks (Girardeau, Benchenane, Wiener, Buzsaki, & Zugaro, 2009). Replay of neural firing patterns is also observed in various brain structures such as the neocortex (Ji & Wilson, 2007; Peyrache, Khamassi, Benchenane, Wiener, & Battaglia, 2009; Ribeiro et al., 2004), or the striatum (Pennartz et al., 2004). Importantly, during NREM sleep, reactivations in the neocortex appear in close temporal synchrony with hippocampal sharp waves (Ji & Wilson, 2007; Peyrache et al., 2009), within 100 ms after the hippocampal cells (Wierzynski, Lubenov, Gu, & Siapas, 2009). In particular, the interactions between the hippocampus and the medial prefrontal cortex seem to play a key role in memory consolidation of hippocampal-dependent memories. The medial prefrontal cortex shows compressed replays of neural activity patterns during sleep (Euston, Tatsuno, & McNaughton, 2007), which coincide with hippocampal sharp waves and are selectively induced by the acquisition of novel information (Peyrache et al., 2009). Collectively, these findings illustrate learning-dependent hippocampo-neocortical interactions during post-training sleep, consistent with the hypothesis of a systems-level memory consolidation.

Some results suggest that at the macroscopic brain systems level, similar reactivations occur during NREM sleep in humans. After the exploration of a virtual tridimensional maze, the activity is enhanced during NREM sleep in occipital, parietal, and mesiotemporal areas (Peigneux et al., 2004). Moreover, the increase in hippocampal activity is linearly related to the individual gain in the ability to navigate in the maze the next day, suggesting that the changes in hippocampal activity during NREMS relates to the offline processing of topographical memory. No such reactivation was observed during REM sleep. In order to experimentally induce learning-related reactivations, olfactory cues were associated with the encoding of object locations during a source memory task. Reexposure to conditioned cues during NREM sleep improved memory retention and increased hippocampal activity, as assessed by fMRI (Rasch, Buchel, Gais, & Born, 2007). Again, conditioned cues were ineffective if delivered during REM sleep or wakefulness. Finally, reactivating memories by conditioned cues during NREM sleep not only is associated with significant hippocampal and neocortical responses, but it also increases their subsequent resistance to interference (Diekelmann, Buchel, Born, & Rasch, 2011). These findings support the view that the hippocampal activity during NREM sleep results in a strengthening and stabilization of recent memories.

Conversely, hindering the offline memory processing by sleep deprivation modifies the neural correlates of subsequent retrieval. In a within-subject cross-over design, recall of word-pair associates was assessed using fMRI 48 hours after encoding (Gais et al., 2007). In one condition, sleep was allowed in as usual during the two post-encoding nights. In the other condition, the volunteers were totally sleep deprived on the first post-encoding night. Hippocampal responses were significantly larger during recall of words learned before sleep than before sleep deprivation. In addition, sleep enhanced the functional connectivity between the hippocampus and the mPFC during recall. Six months later, memory recall more strongly recruited the medial prefrontal and occipital cortex for words that were encoded before sleep than before sleep deprivation. These results confirm earlier experiments in humans showing that over the course of three months, hippocampal activity during memory retrieval gradually decreases whereas activity in a ventral medial prefrontal region increases (Takashima et al., 2006). However, they further show that sleep after encoding leads to a long-lasting reorganization in memories in the cortex. Sleep deprivation had similar effects on emotional memory, although the recruitment of mesial prefrontal cortex could be observed as early as 72 hours after encoding. In a between-subjects design, episodic recognition of emotional and neutral stimuli was tested 72 hours after encoding, with or without total sleep deprivation during the first post-encoding night (Sterpenich et al.,

2007). Successful recollection of emotional stimuli elicited larger responses in the hippocampus and the medial prefrontal cortex in the sleep group than in the sleep deprived group. In addition, the functional connectivity between hippocampus and medial prefrontal cortex was enhanced during recollection of emotional items after sleep. Six months later, recollection was associated with significantly larger responses in subjects allowed to sleep than in sleep-deprived subjects, in a set of cortical areas, including the medial prefrontal cortex, the precuneus, and the occipital cortex (Sterpenich et al., 2009). Moreover, the functional connectivity was enhanced between the medial prefrontal cortex and the precuneus. These results confirm that sleep during the first night after encoding profoundly influences the long-term organization of memories in cortical networks.

In sum, NREM sleep is associated with a strengthening of memories that become resistant to interference. Functional neuroimaging in humans shows that hippocampal activity increases during post-training NREM sleep, suggesting that the hippocampus, and its interactions with cortical areas, takes part in memory consolidation. When these processes are disturbed by sleep deprivation, responses at retrieval are altered in the hippocampus and, in the long term, in cortical networks which involve the medial prefrontal cortex.

Early Memory Structuring during Encoding

Given the central role of the hippocampus in the consolidation of declarative memories, one may wonder if the hippocampal activity during encoding is a critical factor in making a memory trace susceptible to sleep-dependent consolidation. Some functional neuroimaging data support this view. In a study testing the effect of directed forgetting, volunteers were asked to learn a series of words (Rauchs et al., 2011). Each word was followed by an instruction indicating whether the item was to be remembered (TBR item) or forgotten (TBF item). During the following night, half of the volunteers were allowed to sleep whereas the others were totally sleep deprived. Three days after encoding, memory for TBR and TBF items was probed using a recognition task during which subjects had to categorize each word presented as previously learned or not. The key finding was a larger response in the hippocampus during encoding of TBR items that were later remembered compared with TBR items that were ultimately forgotten. No such difference was detected for the TBF items indicating that the hippocampal recruitment during encoding identified memories that would be ultimately consolidated. In addition, the increase in hippocampal response was observed only in the volunteers allowed to sleep and not in the sleep deprived group, indicating that memory consolidation was sleep-dependent. These findings suggest that the recruitment of the hippocampus during encoding foreshadows subsequent sleep-dependent memory consolidation.

Sleep-Dependent Consolidation of Procedural Motor Memories

Sleep promotes the consolidation of motor memories. In particular, during motor sequence learning sleep has been associated with spontaneous gains in performance (Walker et al., 2002) and increased resistance to interference (Korman et al., 2007). The neural correlates of this sleep-dependent motor consolidation have not yet been systematically characterized.

The activity during REM sleep after motor sequence learning is enhanced in various brain areas (Maquet et al., 2000). Normal volunteers were trained to a probabilistic serial reaction task. In this task, participants have to press as fast and as accurately as possible on the key corresponding to a stimulus appearing at one out of six possible screen positions. Unknown to subjects, the sequence of stimulus positions was generated by a probabilistic finite-state grammar. In comparison to control volunteers who were not trained to the task, the activity in premotor and occipital cortices, thalamus, and upper brainstem was increased during REM sleep in trained participants. In addition, the functional connectivity between premotor cortex and posterior parietal cortex and presupplementary motor area was also enhanced during REM sleep after training (Laureys et al., 2001). These changes in regional activity during REM sleep were observed only if the learned material was structured by hidden rules imposed by the probabilistic grammar and not when it was random, suggesting that they were related to the processing of the underlying higher-order sequential structure (Peigneux et al., 2003).

Other attempts to assess the effects of sleep on procedural motor learning characterized the neural correlates of the finger tapping task. In this task, volunteers have to repeat an explicitly known five-element finger sequence as rapidly and as accurately as possible with their nondominant hand. In two studies, volunteers were either trained in the morning or in the evening, and tested 12 hours later, in the morning (after sleep) or in the evening, after an equivalent period of wakefulness. Increased responses were observed at retest after sleep in the ventral striatum (Debas et al., 2010), or in the right primary motor cortex, medial prefrontal lobe, hippocampus. and left cerebellum (Walker, Stickgold, Alsop, Gaab, & Schlaug, 2005). In another study, volunteers were scanned during initial training and 2 days later, at retesting, with either sleep or sleep deprivation during the first post-training night (Fischer, Nitschke, Melchert, Erdmann, & Born, 2005). In these conditions, sleep was associated with reduced brain responses in prefrontal, premotor, and primary motor cortical areas. The reasons for these discrepancies are unclear but they probably arise from differences in experimental design (day/evening versus sleep/sleep deprivation), which might be associated with different levels of local sleep pressure. They might also characterize the evolution of motor memory trace at different stages of their consolidation.

The detailed mechanisms underlying sleep-dependent consolidation of motor memories are still unsettled. Behavioral evidence indicates that it might vary according to different experimental factors. For instance, the consolidation of dynamic visuomotor adaptation would differ from sequence learning (Debas et al., 2010); implicit and explicit learning would result in different time courses of consolidation (Robertson, Pascual-Leone, & Press, 2004); and the complexity of the learned material might influence memory consolidation (Kuriyama, Stickgold, & Walker, 2004). Likewise, procedural motor memory has been associated to various aspects of sleep: sleep in the second half of the night (Plihal & Born, 1997), sleep spindles (Fogel & Smith, 2006; Morin et al., 2008), or REM sleep (Fogel, Smith, & Cote, 2007). Both synaptic (Huber et al., 2004) and system-level consolidation (Debas et al., 2010) have been involved in motor memory consolidation. Further research is obviously required to gain a thorough understanding of motor memory consolidation during sleep.

CONCLUSIONS

Over the years, functional neuroimaging has revealed several important aspects of human sleep. After demonstrating the profound differences in brain energy metabolism and hemodynamics between wakefulness and sleep, functional data have showed the dynamic fluctuations of regional brain activity resulting from sleep-specific oscillations. Slow waves and spindles have been shown to result in transient synchronous activity in distributed brain areas. These activity patterns certainly constrain the way the brain processes information during sleep. This was also suggested by profound changes in functional integration taking place across brain areas during sleep. Finally, functional neuroimaging has shown the significant influence of waking experience on regional brain function during sleep. During NREM sleep, data are consistent with both a ubiquitous use-dependent increase in slow oscillation but also with a reorganization of declarative memories within hippocampal-neocortical circuits.

ACKNOWLEDGEMENTS

Personal research reported in this review was supported by the Belgian Fonds National de la Recherche Scientifique (FNRS), Fondation Médicale Reine Elisabeth (FMRE), Research Fund of the University of Liège, and "Interuniversity Attraction Poles Programme—Belgian State—Belgian Science Policy."

REFERENCES

Anderer, P., Klosch, G., Gruber, G., Trenker, E., Pascual-Marqui, R. D., Zeitlhofer, J., et al. (2001). Low-resolution brain electromagnetic tomography revealed simultaneously active frontal and parietal sleep spindle sources in the human cortex. *Neuroscience*, *103*(3), 581–592.

Andersson, J. L., Onoe, H., Hetta, J., Lidstrom, K., Valind, S., Lilja, A., et al. (1998). Brain networks affected by synchronized sleep visualized by positron emission tomography. *Journal of Cerebral Blood Flow and Metabolism*, *18*(7), 701–715.

Andrade, K. C., Spoormaker, V. I., Dresler, M., Wehrle, R., Holsboer, F., Samann, P. G., et al. (2011). Sleep Spindles and Hippocampal Functional Connectivity in Human NREM Sleep. *The Journal of Neuroscience*, *31*(28), 10331–10339.

Braun, A. R., Balkin, T. J., Wesensten, N. J., Gwadry, F., Carson, R. E., Varga, M., et al. (1998). Dissociated pattern of activity in visual cortices and their projections during human rapid eye movement sleep. *Science*, *279*(5347), 91–95.

Braun, A. R., Balkin, T. J., Wesenten, N. J., Carson, R. E., Varga, M., Baldwin, P., et al. (1997). Regional cerebral blood flow throughout the sleep-wake cycle. An H2(15)O PET study. *Brain*, *120*(Pt 7), 1173–1197.

Buchsbaum, M. S., Gillin, J. C., Wu, J., Hazlett, E., Sicotte, N., Dupont, R. M., et al. (1989). Regional cerebral glucose metabolic rate in human sleep assessed by positron emission tomography. *Life Sciences*, *45*(15), 1349–1356.

Bushey, D., Tononi, G., & Cirelli, C. (2011). Sleep and synaptic homeostasis: Structural evidence in Drosophila. *Science*, *332*(6037), 1576–1581.

Clemens, Z., Molle, M., Eross, L., Barsi, P., Halasz, P., & Born, J. (2007). Temporal coupling of parahippocampal ripples, sleep spindles and slow oscillations in humans. *Brain*, *130*(Pt 11), 2868–2878.

Contreras, D., Destexhe, A., Sejnowski, T. J., & Steriade, M. (1996). Control of spatiotemporal coherence of a thalamic oscillation by corticothalamic feedback. *Science*, *274*(5288), 771–774.

Critchley, H. D., Corfield, D. R., Chandler, M. P., Mathias, C. J., & Dolan, R. J. (2000). Cerebral correlates of autonomic cardiovascular arousal: A functional neuroimaging investigation in humans. *The Journal of Physiology*, *523*(*Pt 1*), 259–270.

Damoiseaux, J. S., Rombouts, S. A., Barkhof, F., Scheltens, P., Stam, C. J., Smith, S. M., et al. (2006). Consistent resting-state networks across healthy subjects. *Proceedings of the National Academy of Sciences of the United States of America*, *103*(37), 13848–13853.

Dang-Vu, T. T., Desseilles, M., Laureys, S., Degueldre, C., Perrin, F., Phillips, C., et al. (2005). Cerebral correlates of delta waves during non-REM sleep revisited. *Neuroimage*, *28*(1), 14–21.

Dang-Vu, T. T., Schabus, M., Desseilles, M., Albouy, G., Boly, M., Darsaud, A., et al. (2008). Spontaneous neural activity during human slow wave sleep. *Proceedings of the National Academy of Sciences of the United States of America*, *105*(39), 15160–15165.

De Gennaro, L., & Ferrara, M. (2003). Sleep spindles: An overview. *Sleep Medicine Reviews*, 7(5), 423–440.

Debas, K., Carrier, J., Orban, P., Barakat, M., Lungu, O., Vandewalle, G., et al. (2010). Brain plasticity related to the consolidation of motor sequence learning and motor adaptation. *Proceedings of the National Academy of Sciences of the United States of America*, *107*(41), 17839–17844.

Dehghani, N., Cash, S. S., Rossetti, A. O., Chen, C. C., & Halgren, E. (2010). Magnetoencephalography demonstrates multiple asynchronous generators during human sleep spindles. *Journal of Neurophysiology*, *104*(1), 179–188.

Desseilles, M., Dang Vu, T., Laureys, S., Peigneux, P., Degueldre, C., Phillips, C., et al. (2006). A prominent role for amygdaloid complexes in the Variability in Heart Rate (VHR) during Rapid Eye Movement (REM) sleep relative to wakefulness. *Neuroimage*, *32*, 1008–1015.

Diekelmann, S., Buchel, C., Born, J., & Rasch, B. (2011). Labile or stable: Opposing consequences for memory when reactivated during waking and sleep. *Nature Neuroscience.*, *14*(3), 381–386.

Ellenbogen, J. M., Hulbert, J. C., Stickgold, R., Dinges, D. F., & Thompson-Schill, S. L. (2006). Interfering with theories of sleep and memory: Sleep, declarative memory, and associative interference. *Current Biology*, *16*(13), 1290–1294.

Eschenko, O., Magri, C., Panzeri, S., & Sara, S. J. (2012). Noradrenergic neurons of the locus coeruleus are phase locked to cortical up-down states during sleep. *Cereb Cortex*.

Euston, D. R., Tatsuno, M., & McNaughton, B. L. (2007). Fast-forward playback of recent memory sequences in prefrontal cortex during sleep. *Science*, *318*(5853), 1147–1150.

Fernandez-Mendoza, J., Lozano, B., Seijo, F., Santamarta-Liebana, E., Ramos-Platon, M. J., Vela-Bueno, A., et al. (2009). Evidence of subthalamic PGO-like waves during REM sleep in humans: A deep brain polysomnographic study. *Sleep*, *32*(9), 1117–1126.

Fischer, S., Nitschke, M. F., Melchert, U. H., Erdmann, C., & Born, J. (2005). Motor memory consolidation in sleep shapes more effective neuronal representations. *The Journal of Neuroscience*, *25*(49), 11248–11255.

Fogel, S. M., & Smith, C. T. (2006). Learning-dependent changes in sleep spindles and stage 2 sleep. *Journal of Sleep Research*, *15*(3), 250–255.

Fogel, S. M., Smith, C. T., & Cote, K. A. (2007). Dissociable learning-dependent changes in REM and non-REM sleep in declarative and procedural memory systems. *Behavioural Brain Research*, *180*(1), 48–61.

Friston, K. J., Frith, C. D., Liddle, P. F., & Frackowiak, R. S. (1993). Functional connectivity: The principal-component analysis of large (PET) data sets. *Journal of Cerebral Blood Flow and Metabolism*, *13*(1), 5–14.

Gais, S., Albouy, G., Boly, M., Dang-Vu, T. T., Darsaud, A., Desseilles, M., et al. (2007). Sleep transforms the cerebral trace of declarative memories. *Proceedings of the National Academy of Sciences of the United States of America*, *104*(47), 18778–18783.

Gais, S., Lucas, B., & Born, J. (2006). Sleep after learning aids memory recall. *Learning & Memory*, *13*(3), 259–262.

Gais, S., Plihal, W., Wagner, U., & Born, J. (2000). Early sleep triggers memory for early visual discrimination skills. *Nature Neuroscience*, *3*(12), 1335–1339.

Girardeau, G., Benchenane, K., Wiener, S. I., Buzsaki, G., & Zugaro, M. B. (2009). Selective suppression of hippocampal ripples impairs spatial memory. *Nature Neuroscience*, *12*(10), 1222–1223.

Hagmann, P., Cammoun, L., Gigandet, X., Meuli, R., Honey, C. J., Wedeen, V. J., et al. (2008). Mapping the structural core of human cerebral cortex. *PLoS Biology*, *6*(7), e159.

Hong, C. C., Harris, J. C., Pearlson, G. D., Kim, J. S., Calhoun, V. D., Fallon, J. H., et al. (2009). fMRI evidence for multisensory recruitment associated with rapid eye movements during sleep. *Human Brain Mapping*, *30*(5), 1705–1722.

Horovitz, S. G., Braun, A. R., Carr, W. S., Picchioni, D., Balkin, T. J., Fukunaga, M., et al. (2009). Decoupling of the brain's default mode network during deep sleep. *Proceedings of the National Academy of Sciences of the United States of America*, *106*(27), 11376–11381.

Huber, R., Esser, S. K., Ferrarelli, F., Massimini, M., Peterson, M. J., & Tononi, G. (2007). TMS-Induced cortical potentiation during wakefulness locally increases slow wave activity during sleep. *PLoS ONE*, *2*, e276.

Huber, R., Ghilardi, M. F., Massimini, M., Ferrarelli, F., Riedner, B. A., Peterson, M. J., et al. (2006). Arm immobilization causes cortical plastic changes and locally decreases sleep slow wave activity. *Nature Neuroscience*, *9*(9), 1169–1176.

Huber, R., Ghilardi, M. F., Massimini, M., & Tononi, G. (2004). Local sleep and learning. *Nature*, *430*(6995), 78–81.

Huber, R., Maatta, S., Esser, S. K., Sarasso, S., Ferrarelli, F., Watson, A., et al. (2008). Measures of cortical plasticity after transcranial paired associative stimulation predict changes in electroencephalogram slow-wave activity during subsequent sleep. *The Journal of Neuroscience*, *28*(31), 7911–7918.

Huber, R., Tononi, G., & Cirelli, C. (2007). Exploratory behavior, cortical BDNF expression, and sleep homeostasis. *Sleep*, *30*(2), 129–139.

Inoué, S., Saha, U. K., & Musha, T. (1999). Spatio-temporal distribution of neuronal activities and REM sleep. In B. N. Mallick, & S. Inoue (Eds.), *Rapid eye movement sleep* (pp. 214–220). New Dehli: Narosa Publishing.

Ioannides, A. A., Corsi-Cabrera, M., Fenwick, P. B., del Rio Portilla, Y., Laskaris, N. A., Khurshudyan, A., et al. (2004). MEG tomography of human cortex and brainstem activity in waking and REM sleep saccades. *Cerebral Cortex*, *14*(1), 56–72.

Isomura, Y., Sirota, A., Ozen, S., Montgomery, S., Mizuseki, K., Henze, D. A., et al. (2006). Integration and segregation of activity in entorhinal-hippocampal subregions by neocortical slow oscillations. *Neuron*, *52*(5), 871–882.

Ji, D., & Wilson, M. A. (2007). Coordinated memory replay in the visual cortex and hippocampus during sleep. *Nature Neuroscience*, *10*(1), 100–107.

Kajimura, N., Uchiyama, M., Takayama, Y., Uchida, S., Uema, T., Kato, M., et al. (1999). Activity of midbrain reticular formation and neocortex during the progression of human non-rapid eye movement sleep. *The Journal of Neuroscience*, *19*(22), 10065–10073.

Kattler, H., Dijk, D. J., & Borbely, A. A. (1994). Effect of unilateral somatosensory stimulation prior to sleep on the sleep EEG in humans. *Journal of Sleep Research*, *3*(3), 159–164.

Kennedy, C., Gillin, J. C., Mendelson, W., Suda, S., Miyaoka, M., Ito, M., et al. (1982). Local cerebral glucose utilization in non-rapid eye movement sleep. *Nature*, *297*(5864), 325–327.

Korman, M., Doyon, J., Doljansky, J., Carrier, J., Dagan, Y., & Karni, A. (2007). Daytime sleep condenses the time course of motor memory consolidation. *Nature Neuroscience*, *10*(9), 1206–1213.

Kudrimoti, H. S., Barnes, C. A., & McNaughton, B. L. (1999). Reactivation of hippocampal cell assemblies: Effects of behavioral state, experience, and EEG dynamics. *The Journal of Neuroscience*, *19*(10), 4090–4101.

Kuriyama, K., Stickgold, R., & Walker, M. P. (2004). Sleep-dependent learning and motor-skill complexity. *Learning & Memory*, *11*(6), 705–713.

Larson-Prior, L. J., Zempel, J. M., Nolan, T. S., Prior, F. W., Snyder, A. Z., & Raichle, M. E. (2009). Cortical network functional connectivity in the descent to sleep. *Proceedings of the National Academy of Sciences of the United States of America*, *106*(11), 4489–4494.

Laureys, S., Peigneux, P., Phillips, C., Fuchs, S., Degueldre, C., Aerts, J., et al. (2001). Experience-dependent changes in cerebral functional connectivity during human rapid eye movement sleep. *Neuroscience*, *105*(3), 521–525.

Le Van Quyen, M., Staba, R., Bragin, A., Dickson, C., Valderrama, M., Fried, I., et al. (2010). Large-scale microelectrode recordings of high-frequency gamma oscillations in human cortex during sleep. *The Journal of Neuroscience*, *30*(23), 7770–7782.

Lim, A. S., Lozano, A. M., Moro, E., Hamani, C., Hutchison, W. D., Dostrovsky, J. O., et al. (2007). Characterization of REM-sleep associated ponto-geniculo-occipital waves in the human pons. *Sleep*, *30*(7), 823–827.

Logothetis, N. K., Pauls, J., Augath, M., Trinath, T., & Oeltermann, A. (2001). Neurophysiological investigation of the basis of the fMRI signal. *Nature*, *412*(6843), 150–157.

Louie, K., & Wilson, M. A. (2001). Temporally structured replay of awake hippocampal ensemble activity during rapid eye movement sleep. *Neuron*, *29*(1), 145–156.

Lu, J., Sherman, D., Devor, M., & Saper, C. B. (2006). A putative flip-flop switch for control of REM sleep. *Nature*, *441*(7093), 589–594.

Lydic, R., Baghdoyan, H. A., Hibbard, L., Bonyak, E.V., DeJoseph, M. R., & Hawkins, R. A. (1991). Regional brain glucose metabolism is altered during rapid eye movement sleep in the cat: A preliminary study. *The Journal of Comparative Neurology*, *304*(4), 517–529.

Madsen, P. L., Schmidt, J. F., Holm, S., Vorstrup, S., Lassen, N. A., & Wildschiodtz, G. (1991). Cerebral oxygen metabolism and cerebral blood flow in man during light sleep (stage 2). *Brain Research*, *557*(1–2), 217–220.

Madsen, P. L., Schmidt, J. F., Wildschiodtz, G., Friberg, L., Holm, S., Vorstrup, S., et al. (1991). Cerebral O2 metabolism and cerebral blood flow in humans during deep and rapid-eye-movement sleep. *Journal of Applied Physiology*, *70*(6), 2597–2601.

Madsen, P. L., & Vorstrup, S. (1991). Cerebral blood flow and metabolism during sleep. *Cerebrovascular and Brain Metabolism Reviews*, *3*(4), 281–296.

Maquet, P. (2000). Functional neuroimaging of normal human sleep by positron emission tomography. *Journal of Sleep Research*, *9*(3), 207–231.

Maquet, P., Degueldre, C., Delfiore, G., Aerts, J., Peters, J. M., Luxen, A., et al. (1997). Functional neuroanatomy of human slow wave sleep. *The Journal of Neuroscience*, *17*(8), 2807–2812.

Maquet, P., Dive, D., Salmon, E., Sadzot, B., Franco, G., Poirrier, R., et al. (1992). Cerebral glucose utilization during stage 2 sleep in man. *Brain Research*, *571*(1), 149–153.

Maquet, P., Dive, D., Salmon, E., Sadzot, B., Franco, G., Poirrier, R., et al. (1990). Cerebral glucose utilization during sleep-wake cycle in man determined by positron emission tomography and [18F]2-fluoro-2-deoxy-D-glucose method. *Brain Research*, *513*(1), 136–143.

Maquet, P., Laureys, S., Peigneux, P., Fuchs, S., Petiau, C., Phillips, C., et al. (2000). Experience-dependent changes in cerebral activation during human REM sleep. *Nature Neuroscience*, *3*(8), 831–836.

Maquet, P., Peters, J., Aerts, J., Delfiore, G., Degueldre, C., Luxen, A., et al. (1996). Functional neuroanatomy of human rapid-eye-movement sleep and dreaming. *Nature*, *383*(6596), 163–166.

Maquet, P., & Phillips, C. (1998). Functional brain imaging of human sleep. *Journal of Sleep Research*, 7(Suppl. 1), 42–47.

Maquet, P., Ruby, P., Maudoux, A., Albouy, G., Sterpenich, V., Dang-Vu, T., et al. (2005). Human cognition during REM sleep and the activity profile within the frontal and parietal cortices: A reappraisal of functional neuroimaging data. In S. Laureys (Ed.), *Progress in brain research* (Vol. 150, pp. 219–227). Amsterdam: Elsevier.

Marr, D. (1970). A theory for cerebral neocortex. *Proceedings of the Royal Society London B: Biological Sciences*, *176*(43), 161–234.

Marr, D. (1971). Simple memory: A theory for archicortex. *Philosophical Transactions of the Royal Society of London Series B, Biological Sciences*, *262*(841), 23–81.

Massimini, M., Huber, R., Ferrarelli, F., Hill, S., & Tononi, G. (2004). The sleep slow oscillation as a traveling wave. *The Journal of Neuroscience*, *24*(31), 6862–6870.

McCarley, R. W., Winkelman, J. W., & Duffy, F. H. (1983). Human cerebral potentials associated with REM sleep rapid eye movements: Links to PGO waves and waking potentials. *Brain Research*, *274*(2), 359–364.

McClelland, J. L., McNaughton, B. L., & O'Reilly, R. C. (1995). Why there are complementary learning systems in the hippocampus and neocortex: Insights from the successes and failures of connectionist models of learning and memory. *Psychological Review*, *102*(3), 419–457.

Mena-Segovia, J., Sims, H. M., Magill, P. J., & Bolam, J. P. (2008). Cholinergic brainstem neurons modulate cortical gamma activity during slow oscillations. *The Journal of Physiology*, *586*(Pt 12), 2947–2960.

Miyauchi, S., Misaki, M., Kan, S., Fukunaga, T., & Koike, T. (2009). Human brain activity time-locked to rapid eye movements during REM sleep. *Experimental Brain Research, 192*(4), 657–667.

Molle, M., & Born, J. (2009). Hippocampus whispering in deep sleep to prefrontal cortex—for good memories? *Neuron, 61*(4), 496–498.

Molle, M., Marshall, L., Gais, S., & Born, J. (2002). Grouping of spindle activity during slow oscillations in human non-rapid eye movement sleep. *The Journal of Neuroscience, 22*(24), 10941–10947.

Molle, M., Yeshenko, O., Marshall, L., Sara, S. J., & Born, J. (2006). Hippocampal sharp wave-ripples linked to slow oscillations in rat slow-wave sleep. *Journal of Neurophysiology, 96*(1), 62–70.

Morin, A., Doyon, J., Dostie, V., Barakat, M., Hadj Tahar, A., Korman, M., et al. (2008). Motor sequence learning increases sleep spindles and fast frequencies in post-training sleep. *Sleep, 31*(8), 1149–1156.

Murphy, M., Riedner, B. A., Huber, R., Massimini, M., Ferrarelli, F., & Tononi, G. (2009). Source modeling sleep slow waves. *Proceedings of the National Academy of Sciences of the United States of America, 106*(5), 1608–1613.

Nadasdy, Z., Hirase, H., Czurko, A., Csicsvari, J., & Buzsaki, G. (1999). Replay and time compression of recurring spike sequences in the hippocampus. *The Journal of Neuroscience, 19*(21), 9497–9507.

Nir, Y., Staba, R. J., Andrillon, T., Vyazovskiy, V. V., Cirelli, C., Fried, I., et al. (2011). Regional slow waves and spindles in human sleep. *Neuron, 70*(1), 153–169.

Nofzinger, E. A., Mintun, M. A., Wiseman, M., Kupfer, D. J., & Moore, R. Y. (1997). Forebrain activation in REM sleep: An FDG PET study. *Brain Research, 770*(1–2), 192–201.

Peigneux, P., Laureys, S., Fuchs, S., Collette, F., Perrin, F., Reggers, J., et al. (2004). Are spatial memories strengthened in the human hippocampus during slow wave sleep? *Neuron, 44*(3), 535–545.

Peigneux, P., Laureys, S., Fuchs, S., Delbeuck, X., Degueldre, C., Aerts, J., et al. (2001). Generation of rapid eye movements during paradoxical sleep in humans. *Neuroimage, 14*(3), 701–708.

Peigneux, P., Laureys, S., Fuchs, S., Destrebecqz, A., Collette, F., Delbeuck, X., et al. (2003). Learned material content and acquisition level modulate cerebral reactivation during posttraining rapid-eye-movements sleep. *Neuroimage, 20*(1), 125–134.

Pennartz, C. M., Lee, E., Verheul, J., Lipa, P., Barnes, C. A., & McNaughton, B. L. (2004). The ventral striatum in off-line processing: Ensemble reactivation during sleep and modulation by hippocampal ripples. *The Journal of Neuroscience, 24*(29), 6446–6456.

Peyrache, A., Khamassi, M., Benchenane, K., Wiener, S. I., & Battaglia, F. P. (2009). Replay of rule-learning related neural patterns in the prefrontal cortex during sleep. *Nature Neuroscience, 12*(7), 919–926.

Phelps, M. E., & Mazziotta, J. C. (1985). Positron emission tomography: Human brain function and biochemistry. *Science, 228*(4701), 799–809.

Picchioni, D., Horovitz, S. G., Fukunaga, M., Carr, W. S., Meltzer, J. A., Balkin, T. J., et al. (2011). Infraslow EEG oscillations organize large-scale cortical-subcortical interactions during sleep: A combined EEG/fMRI study. *Brain Research, 1374*, 63–72.

Plihal, W., & Born, J. (1997). Effects of early and late nocturnal sleep on declarative and procedural memory. *Journal of Cognitive Neuroscience, 9*, 534–547.

Raichle, M. E., MacLeod, A. M., Snyder, A. Z., Powers, W. J., Gusnard, D. A., & Shulman, G. L. (2001). A default mode of brain function. *Proceedings of the National Academy of Sciences of the United States of America, 98*(2), 676–682.

Ramm, P., & Frost, B. J. (1983). Regional metabolic activity in the rat brain during sleep-wake activity. *Sleep, 6*(3), 196–216.

Ramm, P., & Frost, B. J. (1986). Cerebral and local cerebral metabolism in the cat during slow wave and REM sleep. *Brain Research, 365*(1), 112–124.

Rasch, B., Buchel, C., Gais, S., & Born, J. (2007). Odor cues during slow-wave sleep prompt declarative memory consolidation. *Science*, *315*(5817), 1426–1429.

Rauchs, G., Feyers, D., Landeau, B., Bastin, C., Luxen, A., Maquet, P., et al. (2011). Sleep contributes to the strengthening of some memories over others, depending on hippocampal activity at learning. *The Journal of Neuroscience, 31*(7), 2563–2568.

Ribeiro, S., Gervasoni, D., Soares, E. S., Zhou, Y., Lin, S. C., Pantoja, J., et al. (2004). Long-lasting novelty-induced neuronal reverberation during slow-wave sleep in multiple forebrain areas. *PLoS Biology*, *2*(1), E24.

Robertson, E. M., Pascual-Leone, A., & Press, D. Z. (2004). Awareness modifies the skill-learning benefits of sleep. *Current Biology*, *14*(3), 208–212.

Salzarulo, P., Lairy, G. C., Bancaud, J., & Munari, C. (1975). Direct depth recording of the striate cortex during REM sleep in man: Are there PGO potentials? *EEG Clinical Neurophysiology*, *38*, 199–202.

Schabus, M., Dang-Vu, T. T., Albouy, G., Balteau, E., Boly, M., Carrier, J., et al. (2007). Hemodynamic cerebral correlates of sleep spindles during human non-rapid eye movement sleep. *Proceedings of the National Academy of Sciences of the United States of America*, *104*(32), 13164–13169.

Sokoloff, L. (1981). Localization of functional activity in the central nervous system by measurement of glucose utilization with radioactive deoxyglucose. *Journal of Cerebral Blood Flow and Metabolism*, *1*(1), 7–36.

Spoormaker, V. I., Schroter, M. S., Gleiser, P. M., Andrade, K. C., Dresler, M., Wehrle, R., et al. (2010). Development of a large-scale functional brain network during human non-rapid eye movement sleep. *The Journal of Neuroscience*, *30*(34), 11379–11387.

Steriade, M., Amzica, F., & Nunez, A. (1993). Cholinergic and noradrenergic modulation of the slow (approximately 0.3 Hz) oscillation in neocortical cells. *Journal of Neurophysiology*, *70*(4), 1385–1400.

Steriade, M., & McCarley, R. W. (2005). *Brain control of wakefulness and sleep*. New York: Kluwer Academic.

Steriade, M., Nunez, A., & Amzica, F. (1993). Intracellular analysis of relations between the slow (<1 Hz) neocortical oscillation and other sleep rhythms of the electroencephalogram. *The Journal of Neuroscience*, *13*(8), 3266–3283.

Steriade, M., Nunez, A., & Amzica, F. (1993). A novel slow (<1 Hz) oscillation of neocortical neurons in vivo: Depolarizing and hyperpolarizing components. *The Journal of Neuroscience*, *13*(8), 3252–3265.

Sterpenich, V., Albouy, G., Boly, M., Vandewalle, G., Darsaud, A., Balteau, E., et al. (2007). Sleep-related hippocampo-cortical interplay during emotional memory recollection. *PLoS Biology*, *5*(11), e282.

Sterpenich, V., Albouy, G., Darsaud, A., Schmidt, C., Vandewalle, G., Dang Vu, T. T., et al. (2009). Sleep promotes the neural reorganization of remote emotional memory. *The Journal of Neuroscience*, *29*(16), 5143–5152.

Stickgold, R., James, L., & Hobson, J. A. (2000). Visual discrimination learning requires sleep after training. *Nature Neuroscience*, *3*(12), 1237–1238.

Takahashi, K. (1989). [Regional cerebral blood flow and oxygen consumption during normal human sleep]. *No To Shinkei*, *41*(9), 919–925.

Takashima, A., Petersson, K. M., Rutters, F., Tendolkar, I., Jensen, O., Zwarts, M. J., et al. (2006). From the Cover: Declarative memory consolidation in humans: A prospective functional magnetic resonance imaging study. *Proceedings of the National Academy of Sciences of the United States of America*, *103*(3), 756–761.

Tononi, G., & Cirelli, C. (2003). Sleep and synaptic homeostasis: A hypothesis. *Brain Research Bulletin*, *62*(2), 143–150.

Tononi, G., & Cirelli, C. (2006). Sleep function and synaptic homeostasis. *Sleep Medicine Reviews*, *10*(1), 49–62.

Tononi, G., Massimini, M., & Riedner, B. A. (2006). Sleepy dialogues between cortex and hippocampus: Who talks to whom? *Neuron*, *52*(5), 748–749.

Vyazovskiy, V. V., Cirelli, C., Pfister-Genskow, M., Faraguna, U., & Tononi, G. (2008). Molecular and electrophysiological evidence for net synaptic potentiation in wake and depression in sleep. *Nature Neuroscience*, *11*(2), 200–208.

Vyazovskiy, V. V., Olcese, U., Lazimy, Y. M., Faraguna, U., Esser, S. K., Williams, J. C., et al. (2009). Cortical firing and sleep homeostasis. *Neuron*, *63*(6), 865–878.

Walker, M. P., Brakefield, T., Morgan, A., Hobson, J. A., & Stickgold, R. (2002). Practice with sleep makes perfect: Sleep-dependent motor skill learning. *Neuron*, *35*(1), 205–211.

Walker, M. P., Stickgold, R., Alsop, D., Gaab, N., & Schlaug, G. (2005). Sleep-dependent motor memory plasticity in the human brain. *Neuroscience*, *133*(4), 911–917.

Wehrle, R., Czisch, M., Kaufmann, C., Wetter, T. C., Holsboer, F., Auer, D. P., et al. (2005). Rapid eye movement-related brain activation in human sleep: A functional magnetic resonance imaging study. *Neuroreport*, *16*(8), 853–857.

Wierzynski, C. M., Lubenov, E. V., Gu, M., & Siapas, A. G. (2009). State-dependent spike-timing relationships between hippocampal and prefrontal circuits during sleep. *Neuron*, *61*(4), 587–596.

Sleep EEG Rhythms and System Consolidation of Memory

Gordon B. Feld[1,2] and Jan Born[1,2]

[1]University of Tübingen, Department of Medical Psychology and Behavioral Neurobiology, [2]University of Lübeck, Department of Neuroendocrinology

MEMORY AND SLEEP

The Two-Stage Model of Memory Formation

The adaptive capability of individuals is greatly enhanced by the brain's plasticity, i.e., the ability of the brain to acquire new information and new skills. However, for proper functioning the brain at the same time must preserve a certain degree of stability in memory, i.e., to keep a certain amount of stable knowledge about rules and regularities in the environment. This plasticity-stability dilemma asks how the brain achieves the task of providing stability, i.e., protecting already acquired traces from being continuously overwritten by new information, and, concurrently, plasticity that allows for the fast assimilation of adaptive information in the same storage system (Carpenter & Grossberg, 1988). The two-stage model of memory formation offers a solution to this problem as it enables the fast acquisition of new traces into memory and their subsequent integration within long-term memories without deleting them (Marr, 1971; McClelland, McNaughton, & O'Reilly, 1995). The model is a general basis of the standard theory of system memory consolidation (termed "standard consolidation theory") and more recent developments thereof (Nadel, Samsonovich, Ryan, & Moscovitch, 2000; Winocur & Moscovitch, 2011).

In this section we introduce the model and throughout the chapter we will add sleep's participation in this memory process. The model assumes that long-term memories are produced by the brain in a two-stage process. The first involves the storage of information into a readily established but transient memory representation, which constitutes a memory store able to learn at a fast rate, but which must only hold this information for a limited duration. However, this temporary store is expanded by a long-term store, which learns at a slower rate, which is, accordingly, also able to maintain the trace for a much longer time. At first, new information is encoded as

Sleep and Brain Activity
DOI: http://dx.doi.org/10.1016/B978-0-12-384995-3.00009-5

a representation in the temporary memory system, which is associated to the specific information in the long-term store, thereby binding this pattern of information with similar old memories. The repeated reactivation of the trace stored in the temporary store in conjunction with older already established memories during subsequent periods of consolidation in a gradual process updates the long-term representation and thereby strengthens connections related to the new pattern of information and integrates it into the preexisting knowledge networks. This process also allows the extraction of relevant and invariant features of the new trace, as overlapping amounts of information are reactivated several times (Lewis & Durrant, 2011). Irrelevant features however may be erased due to only marginal reactivation.

The two-stage model has been applied to systems involving learning and subsequent consolidation. However, the declarative memory system has gained the most attention in this respect. Here, the role of the fast-learning, transient memory store is accredited to the hippocampus, whereas the slow learning long-term store resides in the neural network of the neocortex. Through repeated reactivation of the hippocampal representation the traces in the neocortex are strengthened and over a course of days to years may thereby lose their dependence on the hippocampal structure entirely (Frankland & Bontempi, 2005; McClelland et al., 1995; Zola-Morgan & Squire, 1990). The declarative memory system can be subdivided into two classes of memory (see Box 9.1 on declarative and procedural memory). Episodic memories include the representation of an event in a temporal and spatial context, thereby setting an episode. Semantic memories are independent of the hippocampus and can be viewed as memories that only contain the informational content of the event, while all context information is lost. Loss of the hippocampus leaves semantic memories more or less intact but heavily compromises associated episodic context (Nadel et al., 2000). In essence, semantic information is considered to be deduced from a series of episodes through consolidation, in the course of which most of the individual episodes are lost. Declarative memory is commonly distinguished from procedural memory for skills and habits that do not essentially rely on the hippocampus (see Box 9.1). However, both memories may not be as independent as originally conceptualized, as procedural skill representations, like semantic memory, may originate from a series of overlapping episodes of practicing a skill (e.g., Albouy et al., 2008; Packard & McGaugh, 1996; Schendan, Searl, Melrose, & Stern, 2003).

Memory formation according to the two-stage model represents a system consolidation process, in such a sense that the model assumes that

Box 9.1 Declarative and Procedural Memory

Memory can be subdivided into different systems, depending on the type of the stored information and the involved brain regions. One division is that into declarative and nondeclarative memory with the procedural memory representing the most important nondeclarative memory system (Squire, 1992; Winocur, Moscovitch, & Bontempi, 2010). Declarative memory is further divided into memory for facts (semantic) and memory for events (episodic), and is defined anatomically by its dependence on the hippocampus. Declarative memories are encoded fast, but thereafter also decay rapidly unless they are consolidated. Semantic memories can be considered to result from repeated episodes including overlapping items. For example, the knowledge that Paris is the capital of France results from repeated episodes including this association. In a process of semantization the episodic context (source), i.e., when and where this association was learned, is forgotten, whereas the fact (item) remains stored in memory.

Procedural memory is that of perceptual and motor skills and is learned through repeated practice and, henceforth, remains relatively stable. For this, motor skill memories rely on corticostriatal and corticocerebellar loops. Whereas declarative memories are encoded and retrieved explicitly, i.e., with the person's awareness of what information is learned and recollected, procedural memories can be acquired both explicitly and implicitly, i.e., without awareness. Normally, when a skill is learned, at first explicit attention and binding adherence to task rules is predominant; later as the task becomes automated attention and thus awareness towards the singular elements fades. It is important to note that, although traditionally procedural tasks are thought not to rely on the hippocampus, fMRI studies have shown hippocampal activation during explicit and implicit motor skill learning on a serial reaction time task (Schendan et al., 2003). This speaks at least initially for dependence on the hippocampus. Explicit acquisition of procedural skills involves in addition increased prefrontal cortex activation. Similar to semantic memory, procedural memory can be considered to originate from the repeated episodes of practicing a skill, including overlapping sequential elements. In an extraction process a procedural representation is formed whereby the representation loses its initially strong dependency on the context in which the skill was acquired (e.g., Packard & McGaugh, 1996).

different neuronal networks are used for storing the information temporarily and for the long term. System consolidation is commonly distinguished from synaptic consolidation, which implies the strengthening of memory representations at the synaptic level, i.e., within localized synaptic networks (Dudai, 2004). System consolidation is believed to take place during offline periods of the brain, as the reactivation and redistribution of fresh memories heavily depends on the same neuronal resources as the processing of external stimuli, rendering these processes mutually exclusive. We will present here data indicating that system consolidation is mainly associated with slow wave sleep (SWS). Synaptic consolidation, on the other hand, may be more effective during REM sleep, to strengthen representations after redistribution during SWS, and during waking.

Sleep and Memory Consolidation—Preferential Consolidation of Explicit Memory

Sleep has been shown to benefit declarative and procedural memory (Diekelmann & Born, 2010). However, the specific impact of sleep on different aspects and forms of memory is dependent on multiple variables. We first review important psychological factors and modulators of sleep-dependent memory consolidation and then focus on the qualitative transformation of memory achieved during sleep.

Declarative memories are encoded and recollected explicitly, i.e., with conscious awareness, whereas procedural memories can be acquired also implicitly, i.e., without awareness of what is learned. There is evidence that sleep-dependent memory consolidation favors explicitly encoded information (see Box 9.1), as after sleep performance on a procedural serial reaction time task was more robust if sequences were trained explicitly than implicitly (Robertson, Pascual-Leone, & Press, 2004). In cases where all the information was encoded explicitly, memory consolidation favors strengthening of weak traces over strong traces (Drosopoulos, Schulze, Fischer, & Born, 2007). The degree to which a memory gains access to sleep's beneficial processes is also dependent on the task difficulty and the emotional tone of the event composing the task with the more demanding and more emotional tasks showing a greater benefit from sleep (Kuriyama, Stickgold, & Walker, 2004; Wagner, Hallschmid, Rasch, & Born, 2006).

The above-mentioned properties of explicit learning tasks are important modulators of sleep-dependent memory consolidation. Interestingly, in most of the studies participants are instructed to keep in memory the task items. However, a series of studies now shows that this instruction in itself is an important variable, gating if information is subject to sleep-dependent memory consolidation or not. These studies highlight the future-orientated and motivational character of sleep-dependent memory consolidation. The first study asked the participants to learn declarative (verbal and visual-spatial paired associates) tasks, and afterward participants slept or stayed awake (Wilhelm et al., 2011). Following the learning of the tasks the participants were or were not informed they would have to retrieve the information the next day. Sleep benefited the performance at delayed retrieval to a distinctly greater extent for the participants who were informed about the recall. In the wake condition expectancy had no effect on performance. Hence, access of a memory to sleep-dependent consolidation is mediated by the expectancy that this memory will be used again. Interestingly, in these experiments EEG slow oscillation activity during post-learning non-rapid-eye-movement sleep (NREM) was increased in the expectancy group and, accordingly, this increase correlated with the performance at delayed retrieval.

The strengthening of a memory trace through sleep consolidation may likewise be manipulated by reward expectancy (Fischer & Born, 2009). In this study participants learned one sequence (A or B) of a procedural finger-sequence tapping task, and immediately afterwards they learned the other. After learning they were instructed that good performance only for one of the sequences would be rewarded with money during a later recall test, whereas performance for the other sequence would not be rewarded. However, before recall they were informed that actually good performance on both sequences would be rewarded equally, thereby mitigating the immediate effects of reward expectancy. Nonetheless, participants who slept in the retention interval between learning and recall performed better on the finger sequence they had been expecting to be rewarded for during the retention interval. If the retention interval consisted of only wakefulness, there was no difference in performance. Importantly, in both these studies participants were informed about reward or retention only after learning had occurred, thus expectancy could not afflict encoding of the material and subsequent sleep-dependent processing of memories, rather than the expectancy itself, influenced properties of the consolidation process.

Another study shows that sleep also improves memory for the implementation of future plans (Diekelmann, Wilhelm, Wagner, & Born, 2011 submitted). The comparison of early and late sleep, consisting mainly of SWS or REM sleep, respectively, revealed that the implementation of plans relies on SWS. If the participant had the opportunity to fulfill the plan before sleep, sleep did not influence the plan. Executing the plan corrupted the enhancing effects of sleep for prospective memories and even of retrospective memories constituting the plan.

This series of experiments indicates that the consolidation of memory during sleep, though it is a process of the unconscious state, is motivationally driven and goal oriented, strengthening specifically those memories that are important for future actions and planned behavior. Importantly, this work underscores that the consolidation process is selective, not enhancing all memories, but preferentially those that are tagged during wakefulness as being relevant for the future. This tagging presumably occurs in the prefrontal cortex, which during anticipated retrieval of hippocampus-dependent declarative memories regulates the activation of memory traces (Cohen & O'Reilly, 1996; Hannula & Ranganath, 2009; Miller & Cohen, 2001; Polyn & Kahana, 2008). In fact, supporting this view, replay during subsequent SWS in neuron assemblies that were active during encoding is not only found in the hippocampus, but also in prefrontal areas (Euston, Tatsuno, & McNaughton, 2007; Peyrache, Khamassi, Benchenane, Wiener, & Battaglia, 2009). Replay of neuron assemblies is considered a key process mediating system consolidation during SWS. Hence, for admittance to sleep-dependent memory consolidation it may be essential that the memory encoded under explicit control of prefrontal-hippocampal networks be tagged by afferents from the prefrontal cortex. Interestingly, the prefrontal cortex is the region most involved in the generation of slow oscillations during SWS (Massimini, Huber, Ferrarelli, Hill, & Tononi, 2004; Murphy et al., 2009) tempting one to speculate that these slow oscillations preferentially originate from prefrontal circuitry contributing to memory representations tagged as being relevant for the future, and thus for sleep-dependent consolidation.

Consolidation during Sleep Causing Qualitative Changes of the Representation

An interesting necessity of memory redistribution at the neuronal level is a change in its qualitative aspects at the behavioral level. The gradual transfer of information reduces the dependence, possibly to zero, of memory traces on hippocampal structures (Frankland & Bontempi, 2005). Recent fMRI studies show that this transfer of information from hippocampus to neocortical networks is promoted by sleep (Gais et al., 2007; Takashima et al., 2006). These qualitative changes in neuronal representation of a certain memory are paralleled by obvious qualitative changes at a behavioral level. Adopting the behavioral view, studies indicate that, via these structural changes, sleep extracts repetitive and invariant features of a memory trace and, thereby, for example, renders formerly implicit aspects of a memory accessible to consciousness, making them explicit. Also, through this process sleep can decrease dependence on contextual cues during recall, thus favoring a "semantization" of representations (Cairney, Durrant, Musgrove, & Lewis, 2011).

Using a number reduction task the influence of sleep on the extraction of explicit knowledge was tested (Wagner, Gais, Haider, Verleger, & Born, 2004). This task consists of multiple strings of numbers and participants are asked to process the numbers to finally reach a solution for each string. The result can be calculated using a prior defined set of rules, by processing each of the digits sequentially. However, unbeknown to the participants the strings are constructed in such a way that it is sufficient to only process the first two digits in order to reach the correct answer, once the subject gains insight into this hidden structure. First, the participants performed the task on 90 digit strings, which was not sufficient to induce insight. After a period of either sleep or wakefulness the participants performed on further strings. At this retest, twice as many subjects gained insight into the hidden rule if they were allowed to sleep as compared to subjects who remained awake. Importantly, this effect was not evident if subjects did not practice the task before sleep, thus ruling out an effect of sleep on the general faculty of problem solving, e.g., through facilitation of creative thinking. In fact, the data indicate this to be a phenomenon of memory, such that a representation of the task is formed and the redistribution of respective representations during sleep leads to the extraction of insight through the accumulating effect of overlapping information. Split-night experiments show this beneficial effect of sleep can be attributed to SWS rather than REM sleep (Yordanova et al., 2008), and closer examination of these data revealed that slow spindle activity during SWS is one marker that favors the transformation of implicitly encoded information to an explicit representation over the course of sleep (Yordanova, Kolev, Wagner, Born, & Verleger, 2011).

Relational memory tasks (Ellenbogen, Hu, Payne, Titone, & Walker, 2007) and serial reaction time tasks (SRTT) (Fischer, Drosopoulos, Tsen, & Born, 2006) are other approaches to investigate the influence of sleep on the transformation from implicit to explicit representations of memory traces. For the SRTT, participants are trained under implicit conditions to press a repeatedly presented sequence of cued buttons as fast as possible, without knowledge of the underlying sequence of cue positions (Fig. 9.1). Whether training induced explicit knowledge of the cued sequence is then tested in a generation task, where the participant is asked to deliberately generate the sequence of cue positions. Sleep after training produced a significant gain in explicit sequence knowledge, whereas subjects tested after a corresponding wake retention interval still performed at chance level on the generation task. This sleep-dependent gain does not

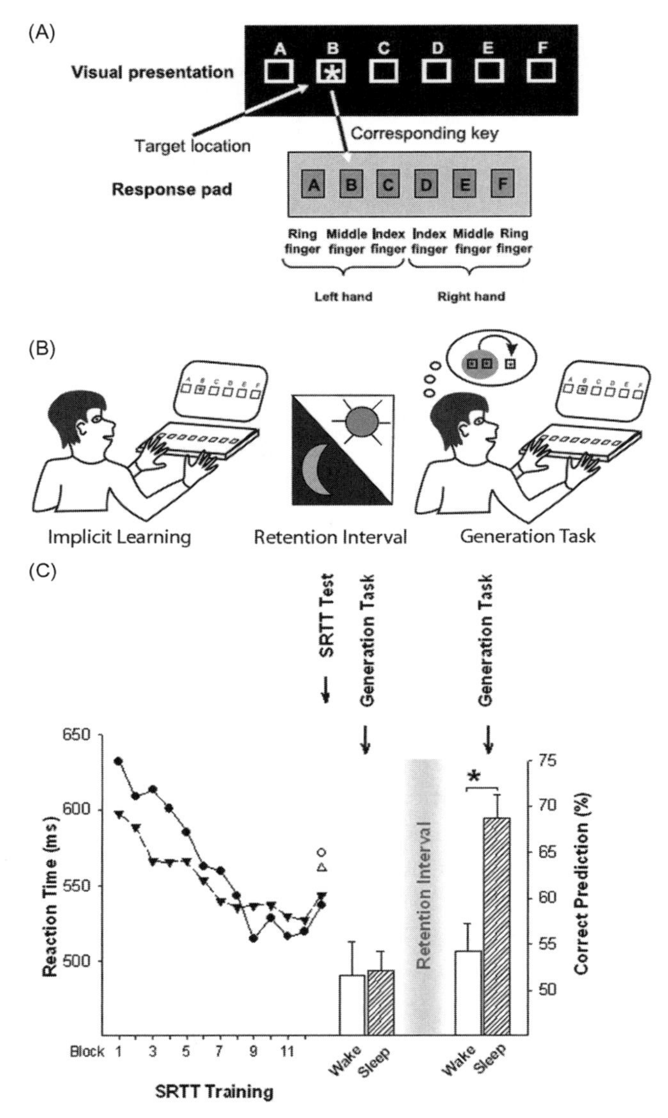

Figure 9.1 ***Effects of sleep on the generation of explicit sequence knowledge in a serial reaction time task (SRTT).*** (**A**) The SRTT: Subjects are presented six horizontally arranged target locations on a computer screen (white boxes). They are instructed to react as fast and as accurately as possible to the occurrence of a target stimulus (white star) at one of these locations by pressing a spatially corresponding key on a response pad. Upon responding, the target changes to another location. Unknown to the subject (i.e., implicit training conditions), the transitions of target locations follow a repeated sequence (grammar). (**B**) General experimental procedure: The participant is trained on the SRTT under implicit conditions (unaware of the regular repeating sequence underlying the target

depend on the participants knowing that there is an underlying grammar (Drosopoulos, Harrer, & Born, 2011). Preliminary findings point to an advantage in extraction of explicit sequence knowledge for children as compared to adults. This ability is connected to their ability to produce increased NREM slow wave activity (Wilhelm et al. 2012, submitted).

While these studies offer strong support for the extraction of explicit content from implicit training occurring during sleep and SWS, there may also be such reorganization of representations within a given memory system, i.e., for explicit memories that remain explicit and for implicit memories that remain implicit. Studies that use the false memories paradigm have provided such evidence, where sleep facilitates the extraction of explicit gist memory from a list of words, which was encoded explicitly before sleep (Diekelmann, Born, & Wagner, 2010; Payne et al., 2009). Applying procedural finger tapping tasks it is possible to show sleep benefits the transfer of skill from the hand used at training to the contralateral hand (D. A. Cohen, Pascual-Leone, Press, & Robertson, 2005; Witt, Margraf, Bieber, Born, & Deuschl, 2010). Sleep likewise benefited the implicit abstraction of statistical knowledge about tone strings and its generalization to novel strings, an effect associated with SWS (Durrant, Taylor, Cairney, & Lewis, 2011). Sleep even supports the tapping speed of a finger sequence if participants were only allowed to learn it by observation (van Der Werf, van Der Helm, Schoonheim, Ridderikhoff, & van Someren, 2009).

Figure 9.1 (*Continued*)
positions) before retention periods of daytime wakefulness or nocturnal sleep. At retrieval after the retention period, the participant performs on the "Generation task." For this task, the subject is informed that there was a repeating sequence in the SRTT he trained before the retention interval and, to test his explicit sequence knowledge, he is now asked to deliberately generate as accurately as possible this sequence. (**C**) Results from Fischer et al. (2006): Participants were trained (under implicit conditions) before the retention interval of either sleep (triangles) or wakefulness (circles) on 12 blocks of the SRTT (each block containing 196 target positions following a probabilistic 12-elements sequence). Mean (±SEM) reaction times (left ordinate) indicate a gradual decrease across blocks. Implicit sequence knowledge at the end of training is indicated by significantly faster reaction times to grammatically correct (filled symbols) as compared to random transitions of the target positions (empty symbols). Bar diagrams (left ordinate) indicate performance (number of correctly predicted target transitions) on the Generation task before and after retention intervals of sleep (hatched bar) and wakefulness (empty bar). Note, participants who slept showed a marked increase in explicit sequence knowledge, which was at chance level after the wake retention interval or before sleep. **$p < .005$. *Adapted from Fischer et al. (2006).*

Together these findings are indicative of a qualitative reorganization of implicitly encoded procedural and explicitly encoded declarative memories during sleep that also facilitates the extraction of explicit knowledge from implicit representations. The latter process of extracting explicit knowledge from implicit memories in the waking brain especially involves activity in the prefrontal cortical areas connected to the hippocampus and medial temporal lobe areas (Jung-Beeman et al., 2004; McIntosh, Rajah, & Lobaugh, 1999, 2003). An issue of future research is to what extent the reorganization and transformations of representations during slow wave sleep are subjected to a similar control by prefrontal-hippocampal circuitry.

MECHANISMS OF SLEEP-DEPENDENT MEMORY CONSOLIDATION

Active System Consolidation

As mentioned above, system consolidation in the two-stage model of memory formation heavily relies on the reactivation of memory traces in the temporary store to redistribute the representations to long-term storage sites. We here briefly review findings that speak for a causal role of reactivation for sleep-dependent memory consolidation.

Individual place cells in the rat brain fire at different locations during exploration of an arena and thereby may code spatial information. When investigating firing patterns of these individual neurons in the hippocampus, they show the same pattern of spatiotemporal firing during learning as during subsequent sleep (O'Neill, Pleydell-Bouverie, Dupret, & Csicsvari, 2010; Pavlides & Winson, 1989; Ribeiro et al., 2004; Sutherland & McNaughton, 2000; Wilson & McNaughton, 1994). This reactivation of neuron ensembles has been found robustly during SWS but only rarely during REM sleep. Additionally to the hippocampus, reactivation occurs in other regions, such as the striatum and neocortex (Euston et al., 2007; Lansink et al., 2008; 2009; Pennartz et al., 2004). Reactivations in these regions appear to be preceded by the reactivation in the hippocampus (Ji & Wilson, 2007; Lansink et al., 2009). By suppressing sharp-wave ripples that usually accompany neuronal reactivations in hippocampus during SWS, it is possible to impair the formation of hippocampus-dependent spatial memories in rats (Ego-Stengel & Wilson, 2010; Girardeau, Benchenane, Wiener, Buzsaki, & Zugaro, 2009).

Hints towards the reactivation of memory representations during sleep have been also provided by human imaging studies (Maquet et al., 2000; Peigneux et al., 2004). In fact, a study in humans was also the first to provide direct evidence for a causal role of reactivations to sleep-dependent memory consolidation as predicted by the two-stage model of memory formation (Rasch, Büchel, Gais, & Born, 2007). This study used odors that were associated to visuospatial memory (for card-pair locations) during a prior learning phase, allowing for cueing of these memories by presenting the odors during subsequent sleep (Fig. 9.2). Reexposure to the odor during post-learning SWS but not during periods of REM sleep significantly enhanced the memory for the card-pair locations learned prior to sleep. Odor stimuli have direct access to the hippocampus via the olfactory system. Consequently, in an imaging study the odor that was associated with the card-pair locations at learning led to an activation of the left hippocampus when reexposed during subsequent SWS. So the odor led to a reactivation of the card locations stored in the left hippocampus thereby strengthening their memory traces. The activations recorded during SWS were even larger than during a wake control condition, hinting at an enhanced responsiveness of the hippocampus to reactivating stimuli. While reactivation of memories through odor contexts is a most straightforward approach, as odor stimuli hardly influence sleep architecture (Badia, Wesensten, Lammers, Culpepper, & Harsh, 1990), this paradigm of cueing memories during sleep has also been shown to work with auditory stimuli (Rudoy, Voss, Westerberg, & Paller, 2009).

Reactivations also take place during wakefulness, when memories are retrieved or even when they are merely cued. However, during wakefulness, in contrast to sleep, the brain is constantly receiving interfering input from the sensory system. Importantly, it has been shown that after reactivation during wakefulness the memory trace is left in a destabilized state, which must be ameliorated by a subsequent phase of reconsolidation (Nader & Hardt, 2009; Sara, 2000). Adopting this reconsolidation framework to sleep-dependent memory consolidation, a study investigated if the reactivation of memories during SWS destabilizes these memories in the same way as during wakefulness (Diekelmann, Büchel, Born, & Rasch, 2010). The study used the same approach to reactivate visuospatial memories by odor cueing during sleep as in the above-mentioned study (Rasch et al., 2007; Fig. 9.2). As expected, in waking subjects reactivations induced by odor cueing destabilized the card-pair locations, because learning of an interference card-pair task immediately after the odor cueing distinctly impaired the memory for the originally learned card locations. Quite the opposite was true for the sleep condition. Here, the odor-induced reactivation, compared with a control without reactivation, produced immediate memory stabilization and, thus, enhanced recall

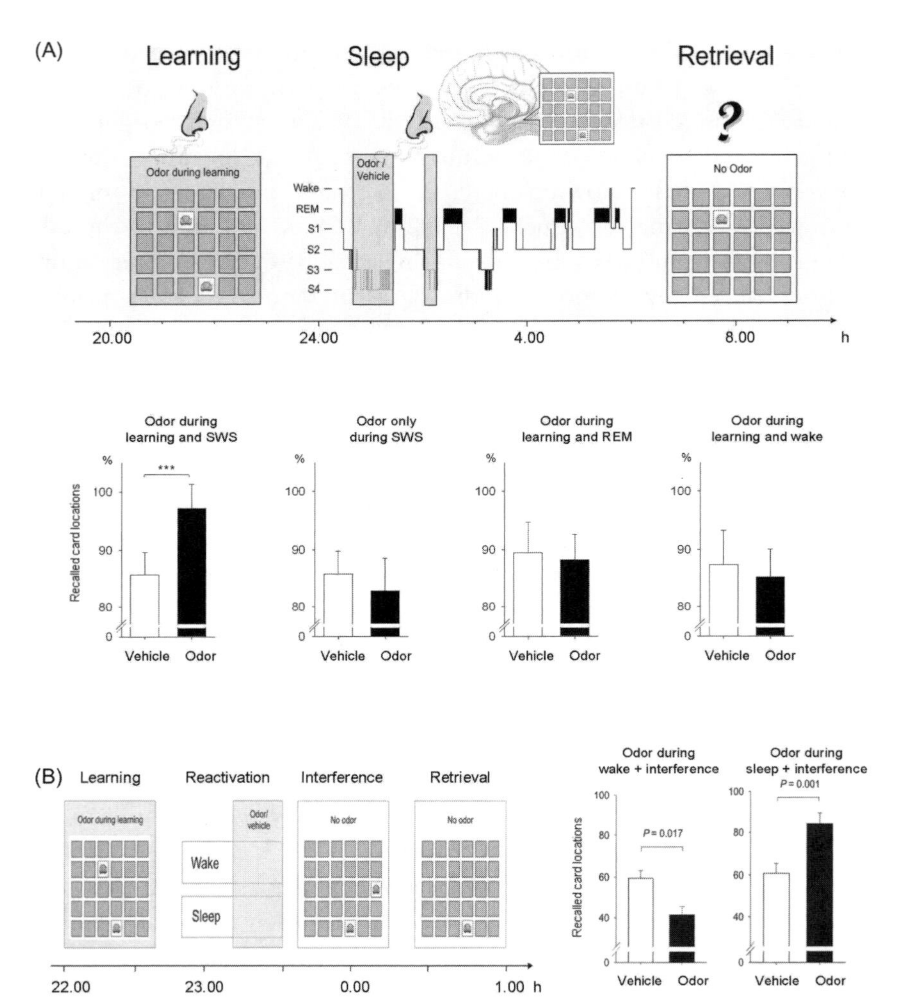

Figure 9.2 ***Reactivating memories by odor cues.*** (**A**) Procedure and task from Rasch et al. (2007): Participants learned a visuospatial 2D object-location task between 21.30 and 22.30 h while an odor was administered time-locked to the learning stimuli (Learning). During subsequent sleep (lights off at 23.00 h, awakening at 6.30 h), the same odor (versus vehicle) was delivered during the first two periods of slow wave sleep (SWS) in an alternating 30 s on/30 s off mode (to prevent habituation). Retrieval was tested between 7.00 and 7.30 h in the absence of odor. The 2D object-location task was similar to the game Concentration, consisting of 15 card pairs with each pair showing the same object (e.g., a red car) with all cards displayed upside down in the beginning. The participant is required to memorize the locations of the pairs of cards, i.e., at learning and for memory testing one card is displayed and he is asked to indicate the location of the corresponding second card of the pair. Underneath—Results: Participants performed significantly better on recall of the card-pair locations only if the odor was presented during learning and subsequent SWS, odor only during SWS,

for the originally learned card locations, despite the fact that the subjects had learned the interference card-pair task directly after odor cueing during SWS. Analyses of the fMRI data revealed that while reactivation during SWS took place mainly in the hippocampus and retrosplenial cortex, during wakefulness mainly right lateral prefrontal areas were involved. Hence, these findings indicate distinct functions of reactivation during wakefulness and sleep. It seems during wakefulness memories are destabilized after reactivation, possibly enabling an updating of the established memory network through lateral prefrontal networks, which introduces the need of subsequent reconsolidation, whereas reactivation during SWS leads to immediate stabilization of the memories, possibly through an enhanced hippocampus-mediated redistribution of the memory representations to neocortical networks. Through this redistribution memories become less dependent on traces deposited in the hippocampus, thereby becoming less vulnerable to retroactive interference, i.e., the overwriting of these still fresh memory traces through the encoding of new memory traces in the temporary hippocampal store (Kuhl, Shah, DuBrow, & Wagner, 2010).

Figure 9.2 *(Continued)*
during learning and REM sleep, or during learning and wakefulness had no effect. (**B**) Procedure from Diekelmann et al. (2011): Participants learned the object-location task in the evening in the presence of an odor. Half of the subjects stayed awake after learning; the other half slept for 40 min. To reactivate the newly encoded card-pair memories, the odor was again presented for about 20 min during SWS as well as during a corresponding time during wakefulness (reactivation). In a control session (no reactivation), odorless vehicle was presented. In order to probe the stability of the memory for the original object-location task, shortly after odor (or vehicle) reexposure—and after subjects in the sleep group were awakened—subjects learned an interference object-location task (without odor presentation) using the same card pairs as during learning, but with different locations (Interference). Retrieval of the original object-location task was tested 30 min after interference learning (without odor presentation). To the right—Results, left panel: Reactivating object-location memories by the reexposure of the odor during wakefulness, as expected, labilized these memories, as subsequent learning of the interference task significantly impaired recall of the originally learned task (black bar) as compared to recall performance without prior odor-cued reactivation (empty bar). Right panel: In contrast, odor induced reactivation during SWS significantly improved recall of the card-pair locations even if an interference task followed (right panel), indicating that reactivation of memory representations during SWS produces an immediate stabilization of the memory possibly because reactivations during SWS induce a redistribution of the memory representations to locations other than the hippocampus where they are at risk of becoming overwritten. *A Adapted from Rasch et al. (2007), B from Diekelmann et al. (2011).*

EEG Slow Oscillations for Memory Consolidation

Reactivation of memories during SWS has been proven to be a causal process implementing the redistribution of memories from temporary to long-term stores, i.e., in the declarative memory system from hippocampal to mainly neocortical storage sites. On the one hand, this information transfer requires a specific neurotransmitter milieu that appears to be established only during SWS. For example, only during SWS acetylcholinergic activity reaches an absolute minimum whereby the tonic presynaptic inhibition exerted via cholinergic afferents to the CA3 and CA1 regions is released (Hasselmo, 1999; Hasselmo & McGaughy, 2004). This disinhibition enables the occurrence of spontaneous neuronal (re-) activations in CA3 and greatly facilitates the outflow of reactivated memory information from hippocampus towards neocortical sites. In fact, experimentally increasing cholinergic activity during SWS-rich retention sleep completely blocked sleep-dependent consolidation of hippocampus-dependent word-pair memories (Gais & Born, 2004; Rasch, Born, & Gais, 2006).

On the other hand, the information transfer between hippocampus and neocortex is specifically regulated by EEG oscillatory events. There is increasing evidence that brain oscillations play a fundamental role in the coordination of information flow between distributed brain regions, and by synchronizing cycles of excitability in these regions determine spike-time dependent synaptic plasticity underlying the formation of distributed memory representations (Buzsaki, 2006; Varela, Lachaux, Rodriguez, & Martinerie, 2001). The EEG slow oscillations have been identified as a main rhythm supporting sleep-dependent memory consolidation in the declarative memory system. In the human EEG, the term slow oscillations refers to a rhythm <1 Hz with a spectral peak around ~0.75 Hz. The term "slow wave activity" instead refers to a wider frequency band (0.5–4 Hz) including besides the slow oscillations (0.5–1.0 Hz) also the delta frequencies (1.0–4.0 Hz). However, there is no clear evidence for functional differences between these rhythms. In fact, the slow oscillation rhythm, as measured in the scalp EEG, comprises also faster frequency components (especially in the falling and rising flanks of the oscillation) although its main constitutes are the lower frequencies.

Brain activity at the slow oscillation rhythm of <1 Hz was first described by Steriade's group in cats (Steriade et al., 1999a,b; Steriade, 2006) and is built up of rhythmically alternating up- and down-states that reflect the rhythmic changes in the membrane potential of neocortical neurons between depolarized and hyperpolarized levels. The slow oscillation rhythm synchronizes widespread neocortical networks including virtually all excitatory and inhibitory neurons, and also thalamic networks, which themselves generate and contribute to the EEG slow oscillation rhythm (Crunelli & Hughes, 2010). The generation of the depolarizing up-state of the slow oscillation involves a persistent Na^+ current and activation of Na^+-dependent K^+ currents whereas generation of the hyperpolarizing down-state has been linked to Ca^{2+}-dependent K^+ currents as well as synaptic depression and inhibitory neuronal activity (Bazhenov, Timofeev, Steriade, & Sejnowski, 2002; Crunelli & Hughes, 2010; Le Bon-Jego & Yuste, 2007). Initiation of depolarizing up-states in thalamocortical cells is linked to activation of T-type Ca^{2+} channels and accompanying burst activity (Destexhe, Hughes, Rudolph, & Crunelli, 2007). Also, miniature excitatory postsynaptic potentials, occurring during SWS as residual synaptic activity in neocortical networks used for information encoding during prior waking, have been proposed as a mechanism launching a depolarizing up-phase out of the global deactivation present during the down-state (Bazhenov et al., 2002). Within the neocortex, slow oscillations appear to originate preferentially from layer 5 neurons. However, human data suggest any layer could initiate firing at up-state onset (Csercsa, et al., 2010). Mostly, slow oscillations arise from anterior prefrontal cortex where they also show maximum amplitudes (Massimini et al., 2004; Murphy et al., 2009). It has been shown that the slow oscillation amplitude during SWS is higher the stronger the underlying cortical network was engaged in encoding during prior waking, indicating a direct dependence on prior learning (Huber, Ghilardi, Massimini, & Tononi, 2004; Mölle, Marshall, Gais, & Born, 2004).

Oscillating transcranial direct current stimulation (tDCS) allows for studying of the impact of EEG oscillations on sleep-dependent memory processing by manipulating the EEG. The approach allows one to stimulate the brain with a desired frequency and even to entrain neuronal firing to a certain firing frequency (Frohlich & McCormick, 2010; Ozen et al., 2010). Oscillating tDCS has been employed in humans to directly prove that slow oscillations are causal to the enhancing effect of sleep on memory consolidation. Two studies applied tDCS to the prefrontal cortex during the first 25–30 minutes of nocturnal NREM sleep. One use stimulation oscillating at the 0.75 Hz slow oscillation frequency; the other study used very slowly alternating 30 s on/30 s off stimulation (Marshall, Helgadottir, Mölle, & Born, 2006; Marshall, Mölle, Hallschmid, & Born, 2004). In both cases, stimulation produced an immediate increase of slow wave activity when compared to sham. The 0.75 Hz stimulation especially increased power in the <1 Hz slow oscillatory band, as measured during 1-minute breaks positioned between the 5-minute intervals of stimulation. tDCS-induced slow oscillation activity was associated with a significantly enhanced frontal slow spindle activity (10–12 Hz). Importantly, after sleep with tDCS, compared with sham stimulation, the participants showed an improved recall of declarative memories (lists of word pairs) they had learned before sleep. There was, however, no significant difference between tDCS and sham conditions in the sleep-associated improvement in procedural finger sequence tapping performance. If tDCS oscillating at 0.75 Hz was applied in the same way, but during the late half of nocturnal sleep, when REM sleep predominates, it did not produce any clear changes in slow wave activity and also did not improve memory performance.

It was further shown that the beneficial effect of tDCS during the SWS-rich beginning of nocturnal sleep critically depends on the stimulation frequency. tDCS oscillating at a faster rate of 5 Hz (in the theta range), instead of the 0.75 Hz rate, strongly suppressed endogenous slow oscillatory activity (Marshall, Kirov, Brade, Mölle, & Born, 2011). Slow frontal spindle activity was also suppressed, and the retention of learned word pairs was significantly impaired by this theta stimulation. Interestingly, applying tDCS oscillating at the 0.75-Hz slow oscillation frequency during wakefulness, when the brain is not prone to produce large amounts of endogenous slow wave activity, only induced very limited amounts of slow oscillations that were limited to areas close to the prefrontal sites of stimulation (Kirov, Weiss, Siebner, Born, & Marshall, 2009). Instead, that same stimulation that greatly increased slow oscillatory activity during NREM sleep boosted EEG theta activity (4–8 Hz) when applied during wakefulness. Consequently, the stimulation did not increase the consolidation of declarative memory (for word pairs), but actually improved encoding of such memories, as reflected by the improved immediate recall of word lists during acute stimulation. Overall these results suggest that the cortical networks that produce theta during explicit encoding of declarative memory in the wake period are functionally coupled to the slow oscillation producing networks that accomplish memory consolidation during subsequent sleep.

Overall, these data strongly suggest a causal role of the neocortical slow oscillation for the consolidation of hippocampus-dependent declarative memories. Of note, the estimated field potentials induced by the tDCS in these studies in the underlying cortical tissue are approximately the same size as those that occur naturally during SWS. Hence, these data suggest that the impact of slow oscillations on memory processing is not only mediated via the underlying synchronization of neuronal firing activity. Rather the potential fields accompanying the synchronized changes in membrane potential during natural slow oscillations in themselves contribute to the effect on memory processing (Frohlich & McCormick, 2010; Ozen et al., 2010).

Slow Oscillations Setting a Time Frame for Hippocampal-Neocortical Communication

Although tDCS studies revealed a causative contribution of slow oscillations to the consolidation of hippocampus-dependent memories during sleep, these studies provided little insight into the mechanisms that could mediate this consolidation process. Apart from the neocortical slow oscillation rhythm, two other rhythms, i.e., spindles activity arising from thalamic networks as well as sharp-wave ripples arising from hippocampal rhythms, have been implicated in memory processing. In this section, we will first summarize evidence that slow oscillations provide a *top-down* signal synchronizing the occurrence of spindles and ripples to the depolarizing up-phase of the slow oscillation. We will then discuss data suggesting that so called spindle-ripple events, as a result of the synchronizing influence of the slow oscillation, might be a mechanism that serves the effective *bottom-up* transfer of memory information reactivated during hippocampal ripples, from hippocampal to neocortical sites.

Slow oscillation-spindle interactions. Spindle activity in the human EEG is referred to as ongoing oscillatory activity in the ~12–15 Hz frequency band with a maximum over central and parietal cortical areas. In NREM sleep stage 2 this activity is expressed in discrete spindles with waxing and waning oscillations. SWS also contains this activity especially in the beginning of the night, however, activity is lower and forms less clearly discrete spindles (De Gennaro & Ferrara, 2003; Marshall, Mölle, & Born, 2003). Spindle activity originates in the thalamus from an interaction of pacemaking GABAergic neurons of the nucleus reticularis with glutamatergic thalamo-cortical projections, which mediate the synchronized and widespread propagation of spindles to cortical regions (Contreras, Destexhe, Sejnowski, & Steriade, 1996; De Gennaro & Ferrara, 2003; Marshall et al., 2003). Spindle activity within the neocortical networks is probably associated with strong calcium influx into neocortical pyramidal cells, which may facilitate plasticity through calcium-dependent signaling cascades (Contreras, Destexhe, & Steriade, 1997; Sejnowski & Destexhe, 2000). Spindles occur preferentially at synapses that were potentiated during encoding of information (Werk,

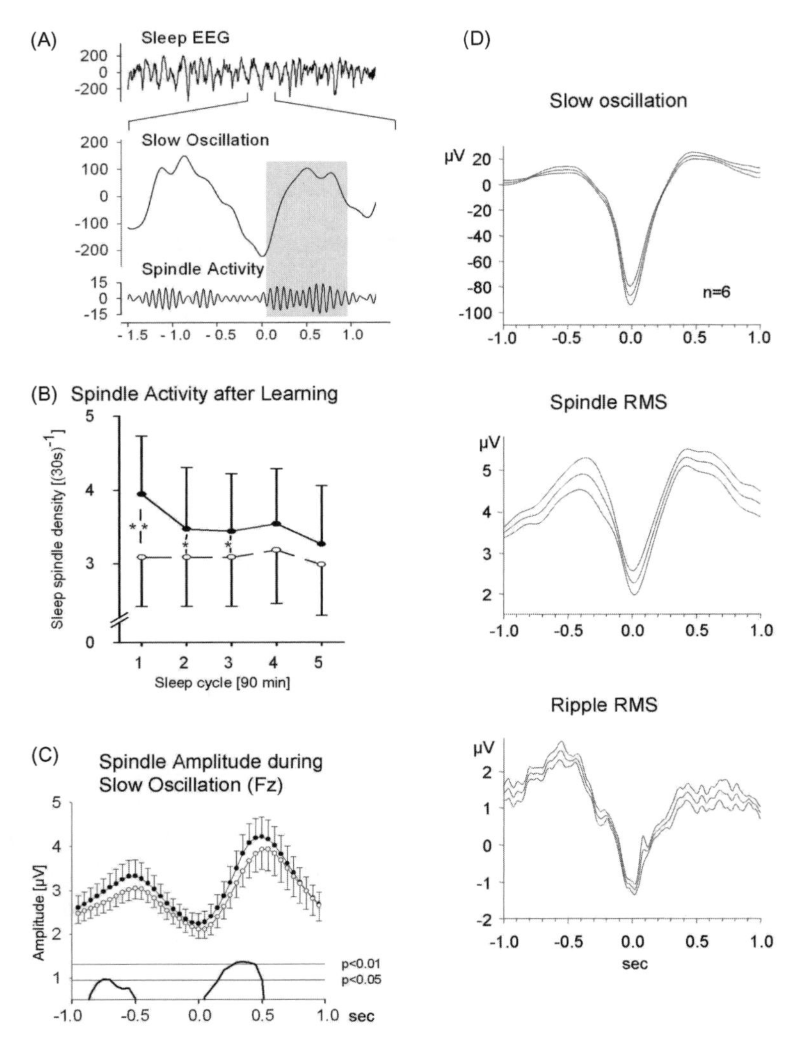

Figure 9.3 ***The slow oscillation up-phase driving spindle and ripple activity.*** (**A**) EEG recording during SWS (top panel), enlarged and filtered in the slow oscillation band (middle) and spindle band (bottom panel), respectively. Whereas spindle activity is suppressed during the hyperpolarizing slow oscillation down-phase, it is distinctly increased during the succeeding depolarizing up-state (gray box). (**B**) Spindle density during sleep after intense declarative learning (word pairs) and a nonlearning control condition. The mean (±SEM) number of sleep spindles (per 30-sec epoch of EEG-recording) during subsequent 90-min periods of a whole night of sleep is indicated. Note the distinct increase in spindle density during the first three 90-min periods in the learning condition. ** $p < .01$, * $p < .05$. (**C**) Amplitude of spindle activity during sleep slow oscillations after intense declarative learning (word pairs, filled circles) and a nonlearning control condition (empty circles). Averages (±SEM) time-locked to the negative half-wave of identified slow oscillations are indicated. Bottom lines indicate

Harbour, & Chapman, 2005). The participation of spindle activity in memory processing during sleep is indeed well established. Studies in both rats and humans have very consistently revealed increases in spindle density and activity during NREM sleep and SWS provoked by prior learning (e.g., Eschenko, Mölle, Born, & Sara, 2006; Fogel & Smith, 2006; Gais, Mölle, Helms, & Born, 2002; Morin et al., 2008; Schabus et al., 2004; Tamaki, Matsuoka, Nittono, & Hori, 2009). This increase, which in some cases is correlated to overnight gains in performance, is observed after both acquisition of declarative information and of procedural motor skills in humans (Clemens, Fabo, & Halasz, 2005, 2006; Nishida & Walker, 2007; Schabus et al., 2004; Tamaki et al., 2009). In some studies, the increase was found above cortical areas that were most strongly involved during encoding (Clemens et al., 2005, 2006; Nishida & Walker, 2007; Tamaki et al., 2009).

Steriade's group was not only the first to provide a thorough characterization of slow oscillations in cats but also the first to demonstrate that slow oscillations have a grouping influence on thalamocortical spindles as well as on neocortical fast oscillatory activity in the gamma frequency (20–60 Hz). During the depolarizing phase of the slow oscillations, the power of these rhythms is increased, whereas during the hyperpolarizing phase these rhythms are suppressed (Steriade, 1999, 2006; Steriade & Amzica, 1998). This synchronizing influence of slow oscillations on spindles was subsequently confirmed in the human EEG (Mölle, Bergmann, Marshall, & Born, 2011; Mölle, Eschenko, Gais, Sara, & Born, 2009; Mölle, Marshall, Gais, & Born, 2002; Mölle et al., 2004). The negative directed hyperpolarizing half-wave of the slow oscillations is associated with a distinct suppression of 12–15 Hz spindle activity, which is followed by rebound activity during the positive depolarizing half-wave (Fig. 9.3). This temporal relationship is also true for discrete spindles identified during NREM sleep stage 2 and SWS, as they

---

Figure 9.3 (*Continued*)
significance for pairwise comparisons of the learning and nonlearning conditions. Note the significant increase in spindle amplitude after learning that is restricted to the depolarizing up-phase of the slow oscillation. (**D**) EEG slow oscillation (upper panel) and associated spindle root mean square (RMS) and ripple RMS activity (middle and bottom panel, respectively). Averages (±SEM) are indicated, time-locked to the negative half-wave of the slow oscillation (0 sec). Slow oscillations and spindles were recorded from frontal electrode sites, ripples from intracranial parahippocampal sites. Recordings are from epileptic patients. Note that ripple RMS activity in parallel with spindle activity is decreased during the hyperpolarizing down-state of the slow oscillation and rebounds during the successive depolarizing up-state. *Data for A–C are from Gais et al. (2002). Data for D are from Clemens et al. (2007).*

preferentially occur during the depolarizing up-phase of the slow oscillations. A parallel but weaker modulation can be observed in the faster beta band and gamma rhythms (Compte et al., 2008; Csercsa et al., 2010).

In light of the strong grouping effect slow oscillations exert on spindle activity, the question arises whether increase in spindle activity that is induced by prior learning can be found, especially in the up-phase of slow oscillations. Indeed, not only is the learning-related increase in spindle activity concentrated on the depolarizing up-phase, but also learning induces an increase in the slow oscillation amplitude itself. In a comparative study in humans and rats Mölle et al. (2009) investigated the effects of learning of hippocampus-dependent materials (word pairs, odor-reward associations) on subsequent sleep slow oscillations and associated spindle activity. Compared with a nonlearning control condition, learning induced a distinct increase in 12–15 Hz spindle activity, and, as predicted, this increase in activity was clearly focused on the depolarizing up-phase of the slow oscillation. No significant changes occurred during the hyperpolarizing down-state. The effect did not occur after learning of odor-reward associations in rats, which probably reflects that this task does not essentially rely on the recruitment of thalamocortical circuitry. Prior learning also altered the slow oscillation in humans such that the up-state was more pronounced and the negative half-wave was sharpened. Consistent with these findings, another study revealed that intense prior learning of word pairs produced a distinct increase in EEG coherence between different cortical regions during the slow oscillation up-states of subsequent SWS in both the slow spindle frequency and the slow oscillation bands (Mölle et al., 2004). A joint increase in slow oscillation amplitude and local fast spindle activity was similarly observed during sleep after subjects had trained on a visuomotor rotation adaptation task (Huber et al., 2004). The increase in slow oscillation activity was most pronounced for the neocortical area, which was most strongly involved in training the task and was correlated with the skill performance after sleep. Overall, these findings support the notion that encoding of information during learning has an increasing effect on sleep spindles and slow oscillations and that this happens in a temporally coordinated fashion, such that the slow oscillation exerts a top-down control bundling learning-dependent increases of spindle activity into their depolarizing up-states.

Importantly, evidence is growing for the existence of two kinds of spindles that may differ in function with regard to memory processing, i.e., the classical fast spindles that show a peak frequency of 12–15 Hz and are more widely distributed with maximum amplitudes over central and parietal cortical areas, and the slow spindles that are focused topographically over the frontal cortex areas and peak at about 8–12 Hz (Anderer et al., 2001; De Gennaro & Ferrara, 2003; Terrier & Gottesmann, 1978). Slow spindles are predominantly found during SWS, setting them apart from fast spindles that are usually present during NREM sleep stage 2, on average even more prominently so than during SWS. Fine-grained analyses of temporal relationships revealed a strikingly different occurrence of fast and slow spindles during the slow oscillation cycle: whereas fast spindles are closely associated with the depolarizing down-to-up transition, slow spindles occur mainly in association to the fading depolarization, i.e., at the transition to the down-state (Mölle et al., 2011).

This differential coupling to slow oscillations indicates that the widely held view of emergent slow oscillation depolarization driving the thalamic generation of spindles (Contreras & Steriade, 1995; Destexhe, 1999; Steriade, 2006; Timofeev & Bazhenov, 2005) holds true for the classical fast spindles only, but not for slow frontal spindles occurring when depolarization is already waning. Although thalamic contributions cannot be excluded (Schabus et al., 2007), slow spindles, unlike fast spindles, may be of mainly cortical origin, as a product of preceding peak network depolarization that was reached during the slow oscillation up-state. Fitting this view, blocking of Na^+ channels by administration of carbamazepine enhanced slow frontal spindle activity conjointly with slow oscillation power (Ayoub, Hörschelmann, Born, & Marshall, 2012 submitted). Contrary to slow spindles, fast centroparietal spindles were decreased by carbamazepine. It has been proposed in this context that slow frontal spindles might functionally be associated with corticocortical coupling whereas fast centroparietal spindles reflect activity in thalamocortical loops (Doran, 2003).

A look at trains of several succeeding slow oscillations admitted deeper insights into the finely tuned relationships between slow oscillations and spindles (Mölle et al., 2011). Indeed, slow spindles tend to follow fast spindles with a delay of approximately 500 ms in the depolarizing up-phase. Moreover, these analyses suggested that the fast centroparietal spindles are not only driven by the depolarizing slow oscillation up-phase but in fact exert themselves an increasing influence on succeeding slow oscillations and thus the likelihood of frontal spindles. Importantly, this dynamic was markedly increased if the participants had learned before sleep a lengthy list of word pairs. In detail, prior learning in this study not only increased fast spindles together with the occurrence of trains of slow oscillations, but in particular increased the fast spindle activity during the initial slow oscillation of such a train. Slow spindle activity on the other hand was increased less clearly and such increases were found mainly at the end of the trains.

Thus the fast spindle seems to receive a pivotal role in launching and maintaining sleep-dependent memory processing, as prior learning increased slow oscillation train length in combination with a spindle increase during the first up-phase of this train. There may be a loop-like process at work, where the fast spindle, e.g., by way of promoting Ca^{2+}-influx into cortical pyramidal cells (Sejnowski & Destexhe, 2000), enhances the likelihood and amplitude of the next slow oscillation and in turn the development of slow frontal spindles during the waning depolarizing phase of these slow oscillations. The next emergent slow oscillation up-state exerts an even stronger drive on fast spindle generation. However, refractoriness induces a gradual decrease of fast spindle amplitude along the train of slow oscillations (Luthi & McCormick, 1998). The driving force of fast spindles on the generation of the slow oscillation-spindle cycles is increased by prior learning. This view that fast spindles serve a priming role in the process of sleep-dependent memory processing remains in line with the numerous findings that learning induces a robust increase of fast spindles also during NREM sleep stage 2 (e.g., Clemens et al., 2005; Gais et al., 2002; B. Rasch, Pommer, Diekelmann, & Born, 2009; Schabus et al., 2004), in which by definition slow wave activity is distinctly diminished, if one assumes that, on a larger time scale, the generally high amounts of spindles during NREM support the generation of slow oscillations during subsequent periods of SWS.

Slow oscillation-ripple interaction. As detailed above, memory processing during sleep is believed to rely on the basic mechanism of reactivation of neuronal firing patterns, which are associated with encoding during prior wakefulness. Reactivations occur mostly in association with sharp wave-ripples in hippocampal networks (Diba & Buzsaki, 2007; O'Neill et al., 2010; Wilson & McNaughton, 1994). Ripples are bursts of high frequency oscillations around 180 Hz that originate in the CA1 region of the hippocampus (Csicsvari, Hirase, Czurko, Mamiya, & Buzsaki, 1999) and occur in association with sharp waves emerging from the CA3 region (Buzsaki, 1986; Buzsaki, Horvath, Urioste, Hetke, & Wise, 1992). Compared with rats, human hippocampal ripples are of somewhat lower frequency (80–140 Hz). Importantly, there is evidence that both reactivated neuronal firing patterns in the hippocampal CA1 region as well as ripples are entrained to the slow oscillation up-states (Battaglia, Sutherland, & McNaughton, 2004; Ji & Wilson, 2007; Mölle, Yeshenko, Marshall, Sara, & Born, 2006; Peyrache, et al.

2009; Sirota, Csicsvari, Buhl, & Buzsaki, 2003), although the exact timing of ripples during the slow oscillation cycle depends also on the hippocampal region examined as well as on the method of recording (O'Neill et al., 2010; Sirota & Buzsaki, 2005). The hippocampus itself does not generate slow oscillations but is reached by neocortical slow oscillations with a slight delay of about 50 ms (Isomura et al., 2006; Sirota et al., 2003; Wolansky, Clement, Peters, Palczak, & Dickson, 2006). Peyrache et al. (2009) reported an entrainment to the slow oscillation up-state of hippocampal sharp wave-ripples that occurred in conjunction with assembly reactivations in medial prefrontal cortex during SWS after learning a rule shifting task. Importantly, the synchronizing influence of the slow oscillation on hippocampal ripple activity was confirmed also in human studies with epileptic patients, where in intracranial recordings from parahippocampal electrodes, sharp wave-ripple activity was clearly time locked to the depolarizing up-phase of the neocortical slow oscillation (Clemens et al., 2007; Fig. 9.3D).

Recent studies have compellingly demonstrated the involvement of hippocampal ripples in memory processing. In rats, training of an odor-reward association task was followed by a robust increase of ripple number during the first hour of subsiding SWS, and during the first 2 hours of SWS after learning ripple magnitude remained increased (Eschenko, Ramadan, Mölle, Born, & Sara, 2008). Similarly, in epileptic patients the number of ripples detected in the rhinal cortex was correlated with the consolidation of picture stimuli acquired before a nap (Axmacher, Elger, & Fell, 2008). Two studies in rats pointed towards a causal role of hippocampal ripples for sleep-dependent memory consolidation (Ego-Stengel & Wilson, 2010; Girardeau et al., 2009). They showed that selective electrical disruption of local hippocampal ripples during the rest periods after learning corrupted the later performance on the spatial learning task. Whether such increases in the expression of ripples during SWS after learning per se concentrate on the intervals of the slow oscillation up-state periods has not been thoroughly studied so far. Such modulation of ripples might occur only in tasks heavily involving prefrontal-hippocampal circuitry during encoding (Benchenane et al., 2010; Mölle et al., 2009; Peyrache et al., 2009).

Spindle-Ripple Events Supporting Bottom-up Transfer of Memory Information

The joint synchronization of hippocampal sharp wave-ripple generation and the generation of spindles in the thalamus by slow oscillations has been proposed to support the formation of so called spindle-ripple events where ripples and associated reactivated hippocampal memory information becomes fed into the excitatory cycles of the spindle oscillation (Clemens et al., 2011; Mölle & Born, 2009; Siapas & Wilson, 1998; Sirota et al., 2003). Spindle-ripple events may thus represent a mechanism that effectively mediates the transfer of memory information from the hippocampus to the neocortex in a temporally fine-tuned manner, and may also prime the integration of this information within neocortical networks. In fact, there is an early demonstration of a weak albeit highly significant temporal association between hippocampal ripples and thalamocortical spindles in rats that was also reflected in the correlated activity of single neurons in these networks (Siapas & Wilson, 1998). In mice a robust temporal correlation between spindles recorded at somatosensory cortical sites and hippocampal ripples is evident (Sirota et al., 2003). These studies showed that the synchronous cortical unit discharge associated with spindles can increase firing in hippocampal neurons within ~50 ms. Because this increase in hippocampal activity was often related to the occurrence of a sharp wave-ripple, the authors concluded that spindle-associated discharges can promote hippocampal ripples and thus reactivation of memories (Sirota & Buzsaki, 2005; Sirota et al., 2003).

There is, however, also the possibility of a reverse effect in spindle-ripple events. In CA1 neurons and downstream in subicular and entorhinal neurons ripple-associated synchronous discharges provide a most effective output to the neocortex (Chrobak & Buzsaki, 1994). There exists a nonlinear relationship between the magnitude of hippocampal ripple bursts and spindle-like prefrontal cortical neuronal responses (Wierzynski, Lubenov, Gu, & Siapas, 2009). Smaller ripple bursts only lead to a single peaked, short latency cortical firing response, but larger bursts could issue an additional peak occurring 100 ms later, thus mimicking a spindle cycle. This observation suggests a promoting influence of ripple discharge on activity in the spindle range, if corticothalamic networks are sufficiently excited. A bottom-up influence of sharp wave-ripples on thalamic generation of spindle activity was likewise revealed by event-correlation histograms of data from intracranial recordings in epileptic patients, indicating that neocortical spindles are preceded by an increase in ripple activity recorded from parahippocampal sites (Clemens et al., 2007).

Again learning prior to sleep seems to be a relevant factor in the temporal relationship between spindles and hippocampal ripples during SWS. Like in epileptic patients, studies in rats on the cooccurrence of spindle and ripple activity during time intervals of several seconds revealed that spindle activity was distinctly increased in the presence of hippocampal ripples, with these increases starting on average about 200 ms before the onset of a ripple (Mölle et al., 2009). Of note, such increases in spindle activity persisted for up to 1 second, i.e., long after the end of the ripple, and this enhancement was further prolonged to up to 2 seconds, if the rats had learned an odor-reward association task before sleep. Taken together, these data suggest a loop-like scenario, where emergent thalamocortical spindle activity and associated neocortical firing drives ripple activity and subsequent neuronal discharge in the CA1 output region of the hippocampus, which themselves feedback positively onto the generating mechanism of spindle activity. This feedback mechanism of spindles is possibly facilitated through generation of ripples in select hippocampal circuitry that was potentiated during prior learning.

Ripples in this model provide the memory reactivations and feed exactly into the excitatory phases of the spindle cycle to be transferred to neocortical sites. This view is substantially backed by two recent studies in humans (Bergmann, Mölle, Diedrichs, Born, & Siebner, 2011 in revision; Clemens et al., 2011). Using functional magnetic resonance imaging, Bergmann et al. provided evidence that spindles cooccur with reactivation of memory representations in hippocampal networks and, conjointly, also in cortical networks. Clemens, Mölle, et al. (2011) recorded spindles from neocortical sites (frontal and parietal) together with activity from intracranial parahippocampal sites in epileptic patients. In fact, spindles observed at parahippocampal sites were closely synchronized with parietal spindles. Averaging of ripple activity time locked to the maximum (trough) of a spindle indeed revealed that this activity becomes nested into the single spindle troughs. This holds for the hippocampal spindles and to a lesser degree for the parietal spindles (Fig. 9.4A). In both spindle types the peak of ripple activity was reached ~10 ms before the troughs (Fig. 9.4B) and was very consistent within the patient population and further is remarkably similar to the lag found in rodents (Sirota et al., 2003). There was no such effect for the slower frontal spindles, which underlines their distinct functionality.

A recent magnetencephalographic study showed that there is a phase-amplitude coupling not only between spindles and ripples, but also between (fast) spindles and neocortical gamma band activity between 30–80 Hz (Ayoub, Mölle, Preissl, & Born, 2011). The phase-locked generation of local gamma oscillations is presumably mediated by spindle associated thalamocortical inputs to inhibitory fast spiking neurons known to exert a common regulatory influence on both rhythms (Bartos, Vida, & Jonas, 2007; Gibson, Beierlein, & Connors, 1999; Peyrache, Battaglia, & Destexhe, 2011; Puig, Ushimaru, & Kawaguchi, 2008). Gamma band activity is a marker of ongoing

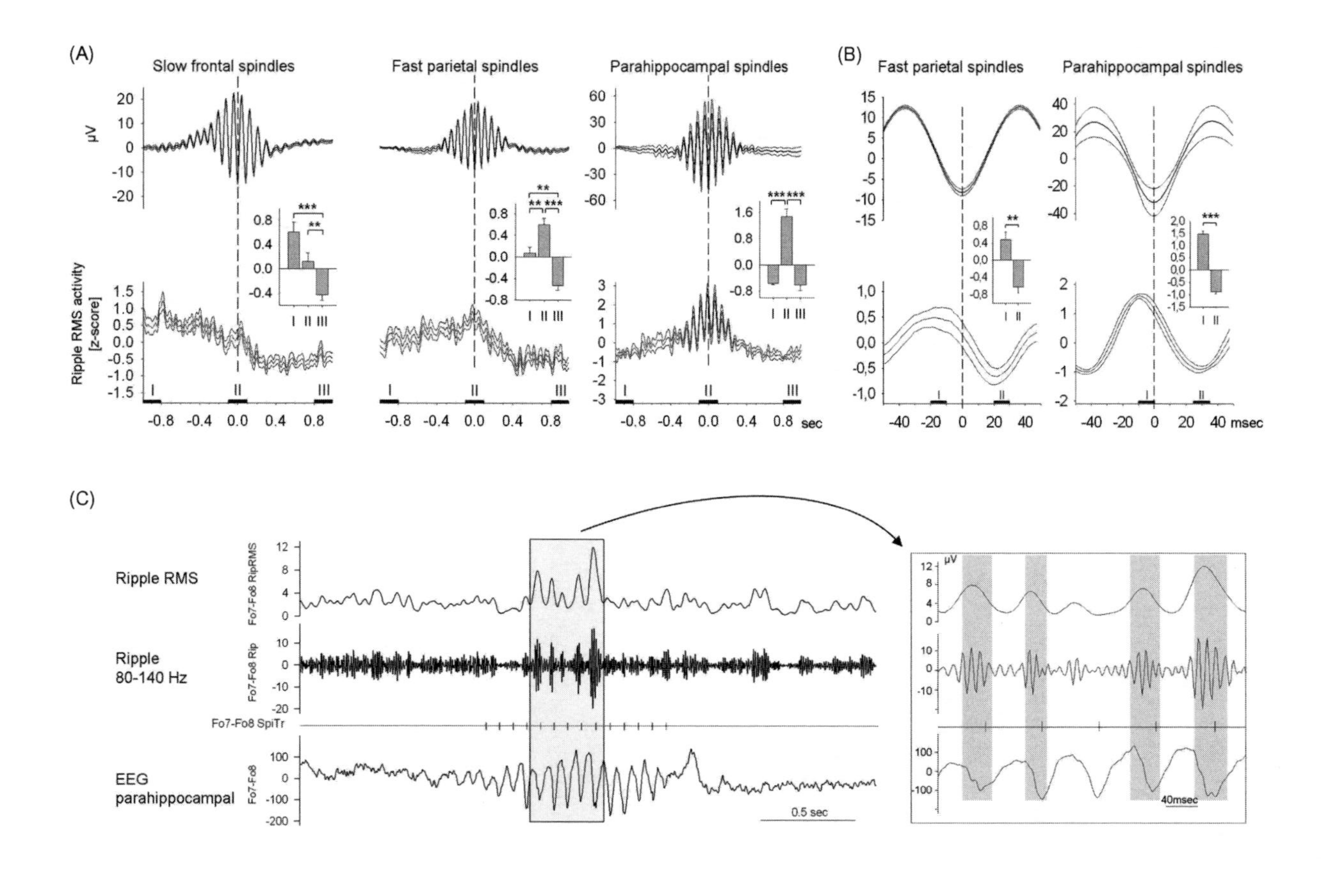
(A)
Slow frontal spindles
Fast parietal spindles
Parahippocampal spindles
µV
Ripple RMS activity [z-score]
sec
(B)
Fast parietal spindles
Parahippocampal spindles
msec
(C)
Ripple RMS
Ripple 80-140 Hz
EEG parahippocampal
Fo7-Fo8 RipRMS
Fo7-Fo8 Rip
Fo7-Fo8 SpiTr
Fo7-Fo8
0.5 sec
µV
40msec

coherent information processing in local neocortical networks, and is associated also with network synaptic potentiation (Bikbaev & Manahan-Vaughan, 2008; Fries, Nikolic, & Singer, 2007). Thus, gamma power fluctuating in the rhythm of spindles nicely fits the view that spindles provide a fine-tuned time frame for the integration of hippocampal memory information within neocortical networks. Spindles arriving at neocortical sites may act by priming the respective networks, for example by inducing Ca^{2+}-influx, which is an important agent for subsequent plastic processes. This in turn would render ripples in hippocampal networks and gamma activity in local neocortical networks carrying the incoming information in the troughs of spindles to be most efficient in producing adaptive plastic changes of memories in the long-term store of the neocortical network. Consistent with this concept, spindle-associated spike discharge has been demonstrated to efficiently trigger long-term potentiation in neocortical networks (Rosanova & Ulrich, 2005; Timofeev et al., 2002). Inasmuch as spindles are able to synchronize gamma band activity between distant cortical locations, they may also contribute to linking quite distributed cortical aspects of the cortical representation.

To sum up, the close temporal relationship between spindles and ripples that provoke the term spindle-ripple event could well account for a basic mechanism of bottom-up information transfer between the hippocampus

---

◀ **Figure 9.4** ***Relationship between spindles and ripples in recordings from epileptic patients.*** (**A**) Upper panels show spindle recorded over frontal (left) and parietal (middle) cortical sites, and from intracranial parahippocampal sites (right), averaged time-locked to the maximum trough of a spindle. Lower panels show the associated root mean square (RMS) ripple activity recorded from parahippocampal sites. Means (±SEM) are indicated. Inserted bar plots indicate means of ripple RMS activity during the three marked 200 ms epochs at the beginning, the middle, and the end of the ±1 sec interval surrounding the spindle maximum, which were used for inference statistics of ripple RMS activity. ** $p < .01$, *** $p < .001$. Note the increase in ripple activity around the maximum of parietal and parahippocampal spindles, but not around the maximum of frontal spindles. (**B**) Finer grained analysis of temporal relationship between parahippocampal ripple RMS activity (lower panel) and parietal (left) and parahippocampal (right) spindle oscillations, both averaged time-locked to the troughs of a spindle (0 sec). Means (±SEM) for z-transformed values are indicated. Bar plots provide means of 10 ms intervals used for inference statistics. ** $p < .01$, *** $p < .001$. Note the strongly increased ripple activity shortly before the peak of a spindle trough indicating that ripples become nested into spindle troughs thus forming spindle-ripple events. (**C**) Example of recording (from parahippocampal sites) used for analysis of spindle-ripple events (box—enlarged to the right), ripples are nested in the troughs of the parahippocampal spindles. Upper trace—ripple RMS recording site, middle—EEG filtered between 80–140 Hz, bottom—original EEG recording. *Adapted from Clemens et al. (2010).*

and the neocortex (Marshall & Born, 2007; Fig. 9.5). So, to the degree that the hippocampal output during ripples originates in reactivation of memory, the nesting of hippocampal ripples in succeeding troughs of a spindle, together with a similar phase-amplitude coupling of neocortical

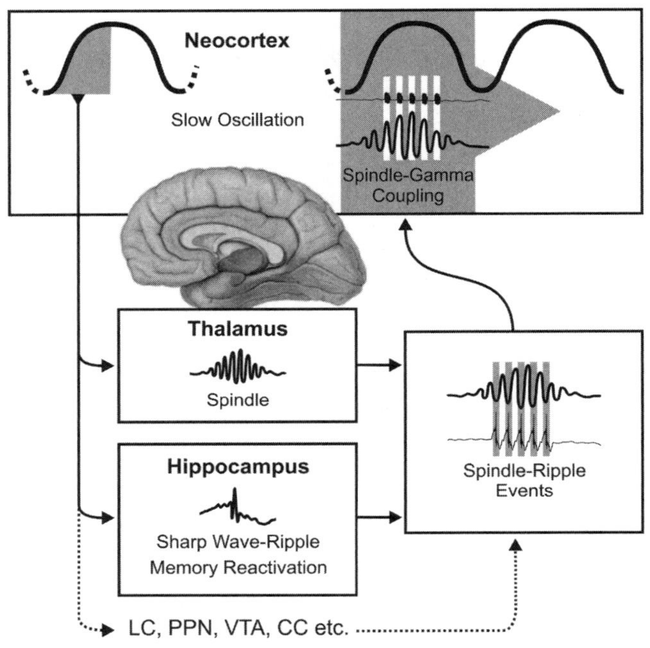

Figure 9.5 ***EEG rhythms involved in system consolidation during slow wave sleep (SWS).*** Slow oscillations (<1 Hz) mainly originating in the prefrontal cortex and spreading through the whole neocortex and brain issue a top-down control with the depolarizing up-phases of these oscillations driving fast (12–15 Hz) spindles originating in the thalamus and sharp wave-ripples that carry reactivated memory information from the hippocampus, as well as activity of other subcortical regions possibly involved in this process, e.g., locus coeruleus (LC), pedunculopontine nucleus (PPN), ventral tegmental area (VTA), and cerebellar cortex (CC). The concurrent drive enables the formation of spindle-ripple events where ripples and associated memory information are nested in the succeeding troughs of a spindle. Spindle-ripple events serve the bottom-up transport and integration of reactivated memory information into neocortical networks. They reach the cortex in a time-locked manner still during the depolarizing up-state of the slow oscillation when neocortical excitability is enhanced and induction of plastic synaptic processes may be facilitated. In fact, not only hippocampal ripple activity but also neocortical gamma band power shows a phase-coupling to spindle oscillations, suggesting that spindles exert a coherent regulatory influence on the processing of memory information allowing—during the phase of enhanced excitability of the spindle cycle—for the binding of information between distributed neocortical networks. In addition, spindle activity in a feed forward manner favors the occurrence of another slow oscillation cycle and thereby contributes to the perpetuation of memory processing.

gamma band activity to spindles, may also serve the integration of the reactivated memory information in distributed neocortical representations. Importantly, such spindle-ripple events are formed under top-down control of the neocortical slow oscillation. Thus, the driving force of the depolarizing up-state of these slow oscillations enables that the reactivated memory representation, enwrapped in these spindle-ripple events, arrives at the neocortical networks still during the slow oscillation up-state, i.e., a period of facilitated neocortical excitability.

Synaptic Homeostasis

A different theory does not consider memory consolidation during sleep an active process, but focuses on the homeostatic regulation of synaptic strength (Tononi & Cirelli, 2003, 2006). According to this theory the uptake of information and potentiation of involved synapses during the daytime wake period leads to an increase of net strength of synaptic connections in neocortical networks. Processes during sleep allow for global downscaling of synaptic strength to a sustainable level, in terms of energy consumption, tissue volume demands, and future capacity of potentiation for encoding (Cirelli & Tononi, 2000; Vyazovskiy, Cirelli, Pfister-Genskow, Faraguna, & Tononi, 2008). It is assumed that slow oscillations play a critical role in downscaling. They show their maximum amplitudes at the beginning of the night, when synapses are most potentiated, due to encoding of information during prior wakefulness, and reduce their amplitude as a function of synaptic downscaling in subsequent sleep cycles. Memories are thought to be strengthened because they benefit from an improved signal-to-noise-ratio, as downscaling is thought to be linear and affect weak synapses to the same degree as strong synapses, thereby eliminating the weak ones and associated noise, and ameliorating information processing in networks with strong synapses (Tononi & Cirelli, 2006).

Although there is compelling evidence for synaptic downscaling to occur during sleep at a global scale, a mechanism of how slow oscillations induce synaptic downscaling remains elusive. The sequence of depolarization and hyperpolarization of slow oscillations may promote synaptic depression (N. Kemp & Bashir, 2001). While long-term depression is in fact favored over long-term potentiation by slow oscillations and associated activity of T-type Ca^{2+}-channels (Czarnecki, Birtoli, & Ulrich, 2007), thalamocortical spindles and hippocampal ripples that are time-locked to slow oscillations are associated to long-term potentiation (Buzsaki, Haas, & Anderson, 1987; King, Henze, Leinekugel, & Buzsaki, 1999; Rosanova & Ulrich, 2005). Moreover, synaptic long-term depression, though mediating a depotentiation, may be more linked to specific learning rather than to unspecific desaturating processes globally affecting the entire neocortical network (Kemp & Manahan-Vaughan, 2007; Tsanov & Manahan-Vaughan, 2008). Also, the assumption that memory enhancement during sleep occurs on the basis of a linear influence of downscaling is difficult to reconcile with findings indicating that the slow oscillations mediating downscaling are not only highest in amplitude over the regions most closely involved in learning of a skill but also that this amplitude increase is correlated with the memory enhancement across sleep (Huber et al., 2004). Also, against the conclusion that synaptic downscaling benefits strong memory representations over weak ones, empirically, sleep benefits either both weak and strong traces to the same degree or even, under certain conditions, favors weak memory traces (Drosopoulos et al., 2007; Kuriyama et al., 2004).

Nevertheless, a desaturating of the effect of sleep slow oscillations, freeing the capacity of synaptic networks for encoding of new information, has also been demonstrated in human subjects. Inducing slow oscillations during a nap by transcranial direct current stimulating (oscillating at 0.75 Hz) in healthy young volunteers significantly improved subsequent encoding of a large amount of word pairs and picture stimuli (landscapes and houses) in these subjects, in comparison with a sham control condition, and this effect was not due to unspecific improvements in vigilance at post-sleep encoding (Antonenko, Diekelmann, Olsen, Born, & Mölle, 2011 submitted). Results of this study complement those of a previous study indicating that the explicit (i.e., hippocampus-dependent) encoding of similar materials was deteriorated after slow wave activity had been suppressed by auditory stimulation during a prior nap (Van Der Werf et al., 2009). Interestingly, in this study impaired encoding after suppression of slow wave activity was associated with reduced activation specifically in the hippocampus, i.e., a region that is not capable for generating slow oscillations itself (Isomura et al., 2006), rather than in the neocortex.

Given the positive evidence for synaptic downscaling the theory might be simply combined with that of active system consolidation, with the former acting nonspecifically at the global level to desaturate networks but sparing local networks in which select memories are actively consolidated. Accordingly, although markers for synaptic potentiation are globally reduced after a period of sleep, they are increased locally, especially if sleep was preceded by learning (Ribeiro et al., 2004; Ribeiro, Goyal, Mello, & Pavlides, 1999; Ribeiro et al., 2007), which indicates synaptic potentiation is possible during sleep. Further evidence for the cooperation of both processes during sleep is provided by neuroimaging studies that show reduced task-related activity in cortical regions after sleep is accompanied by increased activity in other regions (Fischer, Nitschke, Melchert, Erdmann, & Born, 2005; Gais et al., 2007; Orban et al., 2006; Takashima et al., 2006). Such findings argue in favor of complimentary roles of downscaling for synaptic homeostasis in the memory system and active system consolidation for integration of freshly encoded traces into the memory system. To what extent these processes rely on directly interdependent mechanisms during sleep, particularly during the slow oscillations of SWS, is an issue of future research.

CONCLUDING REMARKS

Intense research in the last 20 years has confirmed that sleep plays a pivotal role for memory consolidation. However, we are only starting to understand the processes that are involved. Behavioral data indicate there are multiple variables that determine if a given memory gains access to sleep's strengthening power. Electrophysiological research both in humans and in animals points towards SWS as the key player in this process, which has been identified as a system consolidation process. Slow oscillations, spindles, and sharp wave-ripples coordinate reactivation and transfer of memory information from temporary to long-term stores that is thought to underlie the redistribution and qualitative reorganization of representations during SWS. Although this concept provides a promising approach to enlighten sleep's contribution to memory consolidation, a number of issues at the core of this concept remain to be solved. Thus, it will be interesting to understand how the brain translates information about relevance into an electrical code and how it is able to integrate old and new information into a unified representation and eventually into adaptive behavior.

Throughout this endeavor a role for REM sleep needs to be specified. Early research in this field indeed focused on REM sleep as the driving force in memory consolidation. However, recent studies indicated that sleep can stabilize and enhance memory regardless of whether REM sleep occurs or not (Diekelmann, Büchel, Born, & Rasch, 2011; Rasch et al., 2009). A "dual process" account on memory consolidation during sleep exists, which assumes that REM sleep profits procedural and emotional aspects of memory not essentially depending on hippocampal function whereas SWS enhances hippocampus-dependent declarative memories (Born, Rasch, & Gais, 2006; Plihal & Born, 1997). However, although this view may hold for emotional memory (Walker & van der Helm, 2009), its validity with regard to procedural memories must be questioned in light of accumulating evidence also indicating the superior importance of NREM sleep for skill memories (e.g., Aeschbach, Cutler, & Ronda, 2008; Huber et al., 2004; B. Rasch et al., 2009). Considering these shortcomings of the dual process view, we have recently proposed a sequential hypothesis, following an idea by Antonio Giuditta (Diekelmann & Born, 2010; Giuditta, 1985). Based on the fact that in normal sleep REM sleep always follows SWS, the sequential model basically assumes that memory processing during sleep likewise follows a sequence in which REM sleep serves functions complementing those of SWS. Specifically, we proposed that during SWS memories are redistributed within the memory network through system consolidation and during REM sleep synaptic consolidation of the achieved new connections is accomplished (Diekelmann & Born, 2010; Giuditta, 1985). However, research elaborating on this hypothesis is still in its infancy (e.g., Ribeiro et al., 2007).

ACKNOWLEDGEMENTS

We thank Sabine Groch for critical comments on a previous version of the manuscript. This work has been supported by grants from the DFG (SFB 654) and BMBF (Bernstein Focus on Learning—F 280125).

REFERENCES

Aeschbach, D., Cutler, A. J., & Ronda, J. M. (2008). A role for non-rapid-eye-movement sleep homeostasis in perceptual learning. *The Journal of Neuroscience*, *28*(11), 2766–2772.

Albouy, G., Sterpenich, V., Balteau, E., Vandewalle, G., Desseilles, M., Dang-Vu, T., et al. (2008). Both the hippocampus and striatum are involved in consolidation of motor sequence memory. *Neuron*, *58*(2), 261–272.

Anderer, P., Klosch, G., Gruber, G., Trenker, E., Pascual-Marqui, R. D., Zeitlhofer, J., et al. (2001). Low-resolution brain electromagnetic tomography revealed simultaneously active frontal and parietal sleep spindle sources in the human cortex. *Neuroscience*, *103*(3), 581–592.

Antonenko, D., Diekelmann, S., Olsen, c., Born, J., & Mölle, M. (2012 submitted). Take a deep nap: Slow oscillating brain stimulation during an afternoon nap enhances subsequent declarative encoding performance.

Axmacher, N., Elger, C. E., & Fell, J. (2008). Ripples in the medial temporal lobe are relevant for human memory consolidation. *Brain*, *131*(Pt 7), 1806–1817.

Ayoub, A., Hörschelmann, A., Born, J., & Marshall, L. (2012 submitted). Blocking voltage-dependent Na^{+} channels and $Ca2^{+}$ channels in humans reveals differential effects on fast and slow spindles and associated sleep slow oscillations.

Ayoub, A., Mölle, M., Preissl, H., & Born, J. (2011). Grouping of MEG gamma oscillations by EEG sleep spindles. *Neuroimage*, *59*(2), 1491–1500.

Badia, P., Wesensten, N., Lammers, W., Culpepper, J., & Harsh, J. (1990). Responsiveness to olfactory stimuli presented in sleep. *Physiology & Behavior*, *48*(1), 87–90.

Bartos, M., Vida, I., & Jonas, P. (2007). Synaptic mechanisms of synchronized gamma oscillations in inhibitory interneuron networks. *Nature Reviews Neuroscience*, *8*(1), 45–56.

Battaglia, F. P., Sutherland, G. R., & McNaughton, B. L. (2004). Hippocampal sharp wave bursts coincide with neocortical "up-state" transitions. *Learning & Memory, 11*(6), 697–704.

Bazhenov, M., Timofeev, I., Steriade, M., & Sejnowski, T. J. (2002). Model of thalamocortical slow-wave sleep oscillations and transitions to activated States. *The Journal of Neuroscience, 22*(19), 8691–8704.

Benchenane, K., Peyrache, A., Khamassi, M., Tierney, P. L., Gioanni, Y., Battaglia, F. P., et al. (2010). Coherent theta oscillations and reorganization of spike timing in the hippocampal- prefrontal network upon learning. *Neuron, 66*(6), 921–936.

Bergmann, T. O., Mölle, M., Diedrichs, J., Born, J., & Siebner, H. R. (2011). Sleep spindle-related reactivation of category-specific cortical regions after learning face-scene associations. *Neuroimage, 59*(3), 2733–2742.

Bikbaev, A., & Manahan-Vaughan, D. (2008). Relationship of hippocampal theta and gamma oscillations to potentiation of synaptic transmission. *Frontiers in Neuroscience, 2*(1), 56–63.

Born, J., Rasch, B., & Gais, S. (2006). Sleep to remember. *Neuroscientist, 12*(5), 410–424.

Buzsaki, G. (1986). Hippocampal sharp waves: Their origin and significance. *Brain Research, 398*(2), 242–252.

Buzsaki, G. (2006). *Rhythms of the brain*. New York: Oxford University Press.

Buzsaki, G., Haas, H. L., & Anderson, E. G. (1987). Long-term potentiation induced by physiologically relevant stimulus patterns. *Brain Research, 435*(1–2), 331–333.

Buzsaki, G., Horvath, Z., Urioste, R., Hetke, J., & Wise, K. (1992). High-frequency network oscillation in the hippocampus. *Science, 256*(5059), 1025–1027.

Cairney, S. A., Durrant, S. J., Musgrove, H., & Lewis, P. A. (2011). Sleep and environmental context: Interactive effects for memory. *Experimental Brain Research, 214*(1), 83–92.

Carpenter, G. A., & Grossberg, S. (1988). The Art of Adaptive Pattern-Recognition by a Self-Organizing Neural Network. *Computer, 21*(3), 77–88.

Chrobak, J. J., & Buzsaki, G. (1994). Selective activation of deep layer (V-VI) retrohippocampal cortical neurons during hippocampal sharp waves in the behaving rat. *The Journal of Neuroscience, 14*(10), 6160–6170.

Cirelli, C., & Tononi, G. (2000). Differential expression of plasticity-related genes in waking and sleep and their regulation by the noradrenergic system. *The Journal of Neuroscience, 20*(24), 9187–9194.

Clemens, Z., Fabo, D., & Halasz, P. (2005). Overnight verbal memory retention correlates with the number of sleep spindles. *Neuroscience, 132*(2), 529–535.

Clemens, Z., Fabo, D., & Halasz, P. (2006). Twenty-four hours retention of visuospatial memory correlates with the number of parietal sleep spindles. *Neuroscience Letters, 403*(1–2), 52–56.

Clemens, Z., Mölle, M., Eross, L., Barsi, P., Halasz, P., & Born, J. (2007). Temporal coupling of parahippocampal ripples, sleep spindles and slow oscillations in humans. *Brain, 130*(Pt 11), 2868–2878.

Clemens, Z., Mölle, M., Eross, L., Jakus, R., Rasonyi, G., Halasz, P., et al. (2011). Fine-tuned coupling between human parahippocampal ripples and sleep spindles. *The European Journal of Neuroscience, 33*(3), 511–520.

Cohen, D. A., Pascual-Leone, A., Press, D. Z., & Robertson, E. M. (2005). Off-line learning of motor skill memory: A double dissociation of goal and movement. *Proceedings of the National Academy of Sciences of the United States of America, 102*(50), 18237–18241.

Cohen, J., & O'Reilly, R. (1996). A preliminary theory of the interactions between prefrontal cortex and hippocampus that contribute to planning and prospective memory. In M. Brandimonte, G. Einstein, & M. McDaniel (Eds.), *Prospective memory: Theory and applications* (pp. 267–295). Mahwah, NJ: Erlbaum.

Compte, A., Reig, R., Descalzo, V. F., Harvey, M. A., Puccini, G. D., & Sanchez-Vives, M. V. (2008). Spontaneous high-frequency (10–80 Hz) oscillations during up states in the cerebral cortex in vitro. *The Journal of Neuroscience, 28*(51), 13828–13844.

Contreras, D., Destexhe, A., Sejnowski, T. J., & Steriade, M. (1996). Control of spatiotemporal coherence of a thalamic oscillation by corticothalamic feedback. *Science*, *274*(5288), 771–774.

Contreras, D., Destexhe, A., & Steriade, M. (1997). Intracellular and computational characterization of the intracortical inhibitory control of synchronized thalamic inputs in vivo. *Journal of Neurophysiology*, *78*(1), 335–350.

Crunelli, V., & Hughes, S. W. (2010). The slow (<1 Hz) rhythm of non-REM sleep: A dialogue between three cardinal oscillators. *Nature Neuroscience*, *13*(*1*), 9–17.

Csercsa, R., Dombovari, B., Fabo, D., Wittner, L., Eross, L., Entz, L., et al. (2010). Laminar analysis of slow wave activity in humans. *Brain*, *133*(9), 2814–2829.

Csicsvari, J., Hirase, H., Czurko, A., Mamiya, A., & Buzsaki, G. (1999). Fast network oscillations in the hippocampal CA1 region of the behaving rat. *The Journal of Neuroscience*, *19*(16), RC20.

Czarnecki, A., Birtoli, B., & Ulrich, D. (2007). Cellular mechanisms of burst firing-mediated long-term depression in rat neocortical pyramidal cells. *The Journal of Physiology*, *578*(Pt 2), 471–479.

De Gennaro, L., & Ferrara, M. (2003). Sleep spindles: An overview. *Sleep Medicine Reviews*, 7(5), 423–440.

Destexhe, A., Hughes, S. W., Rudolph, M., & Crunelli, V. (2007). Are corticothalamic 'up' states fragments of wakefulness? *Trends in Neurosciences*, *30*(7), 334–342.

Diba, K., & Buzsaki, G. (2007). Forward and reverse hippocampal place-cell sequences during ripples. *Nature Neuroscience*, *10*(10), 1241–1242.

Diekelmann, S., & Born, J. (2010). The memory function of sleep. *Nature Reviews Neuroscience*, *11*(2), 114–126.

Diekelmann, S., Born, J., & Wagner, U. (2010). Sleep enhances false memories depending on general memory performance. *Behavioural Brain Research*, *208*(2), 425–429.

Diekelmann, S., Büchel, C., Born, J., & Rasch, B. (2011). Labile or stable: Opposing consequences for memory when reactivated during waking and sleep. *Nature Neuroscience*, *14*(3), 381–386.

Diekelmann, S., Wilhelm, I., Wagner, U., & Born, J. (2012 submitted). Sleep to implement intentions.

Doran, S. M. (2003). The dynamic topography of individual sleep spindles. *Sleep Research Online*, *5*, 133–135.

Drosopoulos, S., Harrer, D., & Born, J. (2012). Sleep and awareness about presence of regularity speed the transition from implicit to explicit knowledge. *Biological Psychology*, *86*(3), 168–173.

Drosopoulos, S., Schulze, C., Fischer, S., & Born, J. (2007). Sleep's function in the spontaneous recovery and consolidation of memories. *Journal of Experimental Psychology. General*, *136*(2), 169–183.

Dudai, Y. (2004). The neurobiology of consolidations, or, how stable is the engram? *Annual Review of Psychology*, *55*, 51–86.

Durrant, S. J., Taylor, C., Cairney, S., & Lewis, P. A. (2011). Sleep-dependent consolidation of statistical learning. *Neuropsychologia*, *49*(5), 1322–1331.

Ego-Stengel, V., & Wilson, M. A. (2010). Disruption of ripple-associated hippocampal activity during rest impairs spatial learning in the rat. *Hippocampus*, *20*(1), 1–10.

Ellenbogen, J. M., Hu, P. T., Payne, J. D., Titone, D., & Walker, M. P. (2007). Human relational memory requires time and sleep. *Proceedings of the National Academy of Sciences of the United States of America*, *104*(18), 7723–7728.

Eschenko, O., Mölle, M., Born, J., & Sara, S. J. (2006). Elevated sleep spindle density after learning or after retrieval in rats. *The Journal of Neuroscience*, *26*(50), 12914–12920.

Eschenko, O., Ramadan, W., Mölle, M., Born, J., & Sara, S. J. (2008). Sustained increase in hippocampal sharp-wave ripple activity during slow-wave sleep after learning. *Learning & Memory*, *15*(4), 222–228.

Euston, D. R., Tatsuno, M., & McNaughton, B. L. (2007). Fast-forward playback of recent memory sequences in prefrontal cortex during sleep. *Science*, *318*(5853), 1147–1150.

Fischer, S., & Born, J. (2009). Anticipated reward enhances offline learning during sleep. *Journal of Experimental Psychology. Learning, Memory, and Cognition*, *35*(6), 1586–1593.

Fischer, S., Drosopoulos, S., Tsen, J., & Born, J. (2006). Implicit learning – explicit knowing: A role for sleep in memory system interaction. *Journal of Cognitive Neuroscience*, *18*(3), 311–319.

Fischer, S., Nitschke, M. F., Melchert, U. H., Erdmann, C., & Born, J. (2005). Motor memory consolidation in sleep shapes more effective neuronal representations. *The Journal of Neuroscience*, *25*(49), 11248–11255.

Fogel, S. M., & Smith, C. T. (2006). Learning-dependent changes in sleep spindles and Stage 2 sleep. *Journal of Sleep Research*, *15*(3), 250–255.

Frankland, P. W., & Bontempi, B. (2005). The organization of recent and remote memories. *Nature Reviews Neuroscience*, *6*(2), 119–130.

Fries, P., Nikolic, D., & Singer, W. (2007). The gamma cycle. *Trends in Neurosciences*, *30*(7), 309–316.

Frohlich, F., & McCormick, D. A. (2010). Endogenous electric fields may guide neocortical network activity. *Neuron*, *67*(1), 129–143.

Gais, S., Albouy, G., Boly, M., Dang-Vu, T. T., Darsaud, A., Desseilles, M., et al. (2007). Sleep transforms the cerebral trace of declarative memories. *Proceedings of the National Academy of Sciences of the United States of America*, *104*(47), 18778–18783.

Gais, S., & Born, J. (2004). Low acetylcholine during slow-wave sleep is critical for declarative memory consolidation. *Proceedings of the National Academy of Sciences of the United States of America*, *101*(7), 2140–2144.

Gais, S., Mölle, M., Helms, K., & Born, J. (2002). Learning-dependent increases in sleep spindle density. *The Journal of Neuroscience*, *22*(15), 6830–6834.

Gibson, J. R., Beierlein, M., & Connors, B. W. (1999). Two networks of electrically coupled inhibitory neurons in neocortex. *Nature*, *402*(6757), 75–79.

Girardeau, G., Benchenane, K., Wiener, S. I., Buzsaki, G., & Zugaro, M. B. (2009). Selective suppression of hippocampal ripples impairs spatial memory. *Nature Neuroscience*, *12*(10), 1222–1223.

Giuditta, A. (1985). A sequential hypothesis for the function of sleep. In W. P. Koella, E. Ruther, & H. Schulz (Eds.), *Sleep '84* (pp. 222–224). Stuttgart: Fischer.

Hannula, D. E., & Ranganath, C. (2009). The eyes have it: Hippocampal activity predicts expression of memory in eye movements. *Neuron*, *63*(5), 592–599.

Hasselmo, M. E. (1999). Neuromodulation: Acetylcholine and memory consolidation. *Trends in Cognitive Sciences*, *3*(9), 351–359.

Hasselmo, M. E., & McGaughy, J. (2004). High acetylcholine levels set circuit dynamics for attention and encoding and low acetylcholine levels set dynamics for consolidation. *Progress in Brain Research*, *145*, 207–231.

Huber, R., Ghilardi, M. F., Massimini, M., & Tononi, G. (2004). Local sleep and learning. *Nature*, *430*(6995), 78–81.

Isomura, Y., Sirota, A., Ozen, S., Montgomery, S., Mizuseki, K., Henze, D. A., et al. (2006). Integration and segregation of activity in entorhinal-hippocampal subregions by neocortical slow oscillations. *Neuron*, *52*(5), 871–882.

Ji, D., & Wilson, M. A. (2007). Coordinated memory replay in the visual cortex and hippocampus during sleep. *Nature Neuroscience*, *10*(1), 100–107.

Jung-Beeman, M., Bowden, E. M., Haberman, J., Frymiare, J. L., Arambel-Liu, S., & Greenblatt, R., et al. (2004). Neural activity when people solve verbal problems with insight. *PLoS Biology*, *2*(4), E97.

Kemp, A., & Manahan-Vaughan, D. (2007). Hippocampal long-term depression: Master or minion in declarative memory processes? *Trends in Neurosciences*, *30*(3), 111–118.

Kemp, N., & Bashir, Z. I. (2001). Long-term depression: A cascade of induction and expression mechanisms. *Progress in Neurobiology*, *65*(4), 339–365.

King, C., Henze, D. A., Leinekugel, X., & Buzsaki, G. (1999). Hebbian modification of a hippocampal population pattern in the rat. *The Journal of Physiology*, *521*(Pt 1), 159–167.

Kirov, R., Weiss, C., Siebner, H. R., Born, J., & Marshall, L. (2009). Slow oscillation electrical brain stimulation during waking promotes EEG theta activity and memory encoding. *Proceedings of the National Academy of Sciences of the United States of America*, *106*(36), 15460–15465.

Kuhl, B. A., Shah, A. T., DuBrow, S., & Wagner, A. D. (2010). Resistance to forgetting associated with hippocampus-mediated reactivation during new learning. *Nature Neuroscience*, *13*(4), 501–506.

Kuriyama, K., Stickgold, R., & Walker, M. P. (2004). Sleep-dependent learning and motor-skill complexity. *Learning & Memory*, *11*(6), 705–713.

Lansink, C. S., Goltstein, P. M., Lankelma, J. V., Joosten, R. N., McNaughton, B. L., & Pennartz, C. M. (2008). Preferential reactivation of motivationally relevant information in the ventral striatum. *The Journal of Neuroscience*, *28*(25), 6372–6382.

Lansink, C. S., Goltstein, P. M., Lankelma, J. V., McNaughton, B. L., & Pennartz, C. M. (2009). Hippocampus leads ventral striatum in replay of place-reward information. *PLoS Biology*, 7(8), e1000173.

Le Bon-Jego, M., & Yuste, R. (2007). Persistently active, pacemaker-like neurons in neocortex. *Frontiers in Neuroscience*, *1*(1), 123–129.

Lewis, P. A., & Durrant, S. J. (2011). Overlapping memory replay during sleep builds cognitive schemata. *Trends in Cognitive Sciences*, *15*(8), 343–351.

Luthi, A., & McCormick, D. A. (1998). Periodicity of thalamic synchronized oscillations: The role of Ca2+-mediated upregulation of Ih. *Neuron*, *20*(3), 553–563.

Maquet, P., Laureys, S., Peigneux, P., Fuchs, S., Petiau, C., Phillips, C., et al. (2000). Experience-dependent changes in cerebral activation during human REM sleep. *Nature Neuroscience*, *3*(8), 831–836.

Marr, D. (1971). Simple memory: A theory for archicortex. *Philosophical Transactions of the Royal Society of London. Series B, Biological Sciences*, *262*(841), 23–81.

Marshall, L., & Born, J. (2007). The contribution of sleep to hippocampus-dependent memory consolidation. *Trends in Cognitive Sciences*, *11*(10), 442–450.

Marshall, L., Helgadottir, H., Mölle, M., & Born, J. (2006). Boosting slow oscillations during sleep potentiates memory. *Nature*, *444*(7119), 610–613.

Marshall, L., Kirov, R., Brade, J., Mölle, M., & Born, J. (2011). Transcranial electrical currents to probe EEG brain rhythms and memory consolidation during sleep in humans. *PLoS One*, *6*(2), e16905.

Marshall, L., Mölle, M., & Born, J. (2003). Spindle and slow wave rhythms at slow wave sleep transitions are linked to strong shifts in the cortical direct current potential. *Neuroscience*, *121*(4), 1047–1053.

Marshall, L., Mölle, M., Hallschmid, M., & Born, J. (2004). Transcranial direct current stimulation during sleep improves declarative memory. *The Journal of Neuroscience*, *24*(44), 9985–9992.

Massimini, M., Huber, R., Ferrarelli, F., Hill, S., & Tononi, G. (2004). The sleep slow oscillation as a traveling wave. *The Journal of Neuroscience*, *24*(31), 6862–6870.

McClelland, J. L., McNaughton, B. L., & O'Reilly, R. C. (1995). Why there are complementary learning systems in the hippocampus and neocortex: Insights from the successes and failures of connectionist models of learning and memory. *Psychological Review*, *102*(3), 419–457.

McIntosh, A. R., Rajah, M. N., & Lobaugh, N. J. (1999). Interactions of prefrontal cortex in relation to awareness in sensory learning. *Science*, *284*(5419), 1531–1533.

McIntosh, A. R., Rajah, M. N., & Lobaugh, N. J. (2003). Functional connectivity of the medial temporal lobe relates to learning and awareness. *The Journal of Neuroscience, 23*(16), 6520–6528.

Miller, E. K., & Cohen, J. D. (2001). An integrative theory of prefrontal cortex function. *Annual Review of Neuroscience, 24*, 167–202.

Mölle, M., Bergmann, T. O., Marshall, L., & Born, J. (2011). Fast and Slow Spindles during the Sleep Slow Oscillation: Disparate Coalescence and Engagement in Memory Processing. *Sleep, 34*(10), 1411–1421.

Mölle, M., & Born, J. (2009). Hippocampus whispering in deep sleep to prefrontal cortex–for good memories? *Neuron, 61*(4), 496–498.

Mölle, M., Eschenko, O., Gais, S., Sara, S. J., & Born, J. (2009). The influence of learning on sleep slow oscillations and associated spindles and ripples in humans and rats. *The European Journal of Neuroscience, 29*(5), 1071–1081.

Mölle, M., Marshall, L., Gais, S., & Born, J. (2002). Grouping of spindle activity during slow oscillations in human non-rapid eye movement sleep. *The Journal of Neuroscience, 22*(24), 10941–10947.

Mölle, M., Marshall, L., Gais, S., & Born, J. (2004). Learning increases human electroencephalographic coherence during subsequent slow sleep oscillations. *Proceedings of the National Academy of Sciences of the United States of America, 101*(38), 13963–13968.

Mölle, M., Yeshenko, O., Marshall, L., Sara, S. J., & Born, J. (2006). Hippocampal sharp wave-ripples linked to slow oscillations in rat slow-wave sleep. *Journal of Neurophysiology, 96*(1), 62–70.

Morin, A., Doyon, J., Dostie, V., Barakat, M., Hadj Tahar, A., Korman, M., et al. (2008). Motor sequence learning increases sleep spindles and fast frequencies in post-training sleep. *Sleep, 31*(8), 1149–1156.

Murphy, M., Riedner, B. A., Huber, R., Massimini, M., Ferrarelli, F., & Tononi, G. (2009). Source modeling sleep slow waves. *Proceedings of the National Academy of Sciences of the United States of America, 106*(5), 1608–1613.

Nadel, L., Samsonovich, A., Ryan, L., & Moscovitch, M. (2000). Multiple trace theory of human memory: Computational, neuroimaging, and neuropsychological results. *Hippocampus, 10*(4), 352–368.

Nader, K., & Hardt, O. (2009). A single standard for memory: The case for reconsolidation. *Nature Reviews Neuroscience, 10*(3), 224–234.

Nishida, M., & Walker, M. P. (2007). Daytime naps, motor memory consolidation and regionally specific sleep spindles. *PLoS One, 2*(4), e341.

O'Neill, J., Pleydell-Bouverie, B., Dupret, D., & Csicsvari, J. (2010). Play it again: Reactivation of waking experience and memory. *Trends in Neurosciences, 33*(5), 220–229.

Orban, P., Rauchs, G., Balteau, E., Degueldre, C., Luxen, A., Maquet, P., et al. (2006). Sleep after spatial learning promotes covert reorganization of brain activity. *Proceedings of the National Academy of Sciences of the United States of America, 103*(18), 7124–7129.

Ozen, S., Sirota, A., Belluscio, M. A., Anastassiou, C. A., Stark, E., Koch, C., et al. (2010). Transcranial electric stimulation entrains cortical neuronal populations in rats. *The Journal of Neuroscience, 30*(34), 11476–11485.

Packard, M. G., & McGaugh, J. L. (1996). Inactivation of hippocampus or caudate nucleus with lidocaine differentially affects expression of place and response learning. *Neurobiology of Learning and Memory, 65*(1), 65–72.

Pavlides, C., & Winson, J. (1989). Influences of hippocampal place cell firing in the awake state on the activity of these cells during subsequent sleep episodes. *The Journal of Neuroscience, 9*(8), 2907–2918.

Payne, J. D., Schacter, D. L., Propper, R. E., Huang, L. W., Wamsley, E. J., Tucker, M. A., et al. (2009). The role of sleep in false memory formation. *Neurobiology of Learning and Memory, 92*(3), 327–334.

Peigneux, P., Laureys, S., Fuchs, S., Collette, F., Perrin, F., Reggers, J., et al. (2004). Are spatial memories strengthened in the human hippocampus during slow wave sleep? *Neuron*, *44*(3), 535–545.

Pennartz, C. M., Lee, E., Verheul, J., Lipa, P., Barnes, C. A., & McNaughton, B. L. (2004). The ventral striatum in off-line processing: Ensemble reactivation during sleep and modulation by hippocampal ripples. *The Journal of Neuroscience*, *24*(29), 6446–6456.

Peyrache, A., Battaglia, F. P., & Destexhe, A. (2011). Inhibition recruitment in prefrontal cortex during sleep spindles and gating of hippocampal inputs. *Proceedings of the National Academy of Sciences of the United States of America*, *108*(41), 17207–17212.

Peyrache, A., Khamassi, M., Benchenane, K., Wiener, S. I., & Battaglia, F. P. (2009). Replay of rule-learning related neural patterns in the prefrontal cortex during sleep. *Nature Neuroscience*, *12*(7), 919–926.

Plihal, W., & Born, J. (1997). Effects of early and late nocturnal sleep on declarative and procedural memory. *Journal of Cognitive Neuroscience*, *9*(4), 534–547.

Polyn, S. M., & Kahana, M. J. (2008). Memory search and the neural representation of context. *Trends in Cognitive Sciences*, *12*(1), 24–30.

Puig, M. V., Ushimaru, M., & Kawaguchi, Y. (2008). Two distinct activity patterns of fast-spiking interneurons during neocortical UP states. *Proceedings of the National Academy of Sciences of the United States of America*, *105*(24), 8428–8433.

Rasch, B., Büchel, C., Gais, S., & Born, J. (2007). Odor cues during slow-wave sleep prompt declarative memory consolidation. *Science*, *315*(5817), 1426–1429.

Rasch, B., Pommer, J., Diekelmann, S., & Born, J. (2009). Pharmacological REM sleep suppression paradoxically improves rather than impairs skill memory. *Nature Neuroscience*, *12*(4), 396–397.

Rasch, B. H., Born, J., & Gais, S. (2006). Combined blockade of cholinergic receptors shifts the brain from stimulus encoding to memory consolidation. *Journal of Cognitive Neuroscience*, *18*(5), 793–802.

Ribeiro, S., Gervasoni, D., Soares, E. S., Zhou, Y., Lin, S. C., Pantoja, J., et al. (2004). Long-lasting novelty-induced neuronal reverberation during slow-wave sleep in multiple forebrain areas. *PLoS Biology*, *2*(1), E24.

Ribeiro, S., Goyal, V., Mello, C. V., & Pavlides, C. (1999). Brain gene expression during REM sleep depends on prior waking experience. *Learning & Memory*, *6*(5), 500–508.

Ribeiro, S., Shi, X., Engelhard, M., Zhou, Y., Zhang, H., Gervasoni, D., et al. (2007). Novel experience induces persistent sleep-dependent plasticity in the cortex but not in the hippocampus. *Frontiers in Neuroscience*, *1*(1), 43–55.

Robertson, E. M., Pascual-Leone, A., & Press, D. Z. (2004). Awareness modifies the skill-learning benefits of sleep. *Current Biology*, *14*(3), 208–212.

Rosanova, M., & Ulrich, D. (2005). Pattern-specific associative long-term potentiation induced by a sleep spindle-related spike train. *The Journal of Neuroscience*, *25*(41), 9398–9405.

Rudoy, J. D., Voss, J. L., Westerberg, C. E., & Paller, K. A. (2009). Strengthening individual memories by reactivating them during sleep. *Science*, *326*(5956), 1079.

Sara, S. J. (2000). Retrieval and reconsolidation: Toward a neurobiology of remembering. *Learning & Memory*, 7(2), 73–84.

Schabus, M., Dang-Vu, T. T., Albouy, G., Balteau, E., Boly, M., Carrier, J., et al. (2007). Hemodynamic cerebral correlates of sleep spindles during human non-rapid eye movement sleep. *Proceedings of the National Academy of Sciences of the United States of America*, *104*(32), 13164–13169.

Schabus, M., Gruber, G., Parapatics, S., Sauter, C., Klosch, G., Anderer, P., et al. (2004). Sleep spindles and their significance for declarative memory consolidation. *Sleep*, *27*(8), 1479–1485.

Schendan, H. E., Searl, M. M., Melrose, R. J., & Stern, C. E. (2003). An FMRI study of the role of the medial temporal lobe in implicit and explicit sequence learning. *Neuron*, *37*(6), 1013–1025.

Sejnowski, T. J., & Destexhe, A. (2000). Why do we sleep? *Brain Research, 886*(1–2), 208–223.

Siapas, A. G., & Wilson, M. A. (1998). Coordinated interactions between hippocampal ripples and cortical spindles during slow-wave sleep. *Neuron, 21*(5), 1123–1128.

Sirota, A., & Buzsaki, G. (2005). Interaction between neocortical and hippocampal networks via slow oscillations. *Thalamus & Related Systems, 3*(4), 245–259.

Sirota, A., Csicsvari, J., Buhl, D., & Buzsaki, G. (2003). Communication between neocortex and hippocampus during sleep in rodents. *Proceedings of the National Academy of Sciences of the United States of America, 100*(4), 2065–2069.

Squire, L. R. (1992). Memory and the hippocampus: A synthesis from findings with rats, monkeys, and humans. *Psychological Review, 99*(2), 195–231.

Steriade, M. (1999). Coherent oscillations and short-term plasticity in corticothalamic networks. *Trends in Neurosciences, 22*(8), 337–345.

Steriade, M. (2006). Grouping of brain rhythms in corticothalamic systems. *Neuroscience, 137*(4), 1087–1106.

Steriade, M., & Amzica, F. (1998). Coalescence of sleep rhythms and their chronology in corticothalamic networks. *Sleep Research Online, 1*(1), 1–10.

Sutherland, G. R., & McNaughton, B. (2000). Memory trace reactivation in hippocampal and neocortical neuronal ensembles. *Current Opinion in Neurobiology, 10*(2), 180–186.

Takashima, A., Petersson, K. M., Rutters, F., Tendolkar, I., Jensen, O., Zwarts, M. J., et al. (2006). Declarative memory consolidation in humans: A prospective functional magnetic resonance imaging study. *Proceedings of the National Academy of Sciences of the United States of America, 103*(3), 756–761.

Tamaki, M., Matsuoka, T., Nittono, H., & Hori, T. (2009). Activation of fast sleep spindles at the premotor cortex and parietal areas contributes to motor learning: A study using sLORETA. *Clinical Neurophysiology, 120*(5), 878–886.

Terrier, G., & Gottesmann, C. L. (1978). Study of cortical spindles during sleep in the rat. *Brain Research Bulletin, 3*(6), 701–706.

Timofeev, I., Grenier, F., Bazhenov, M., Houweling, A. R., Sejnowski, T. J., & Steriade, M. (2002). Short- and medium-term plasticity associated with augmenting responses in cortical slabs and spindles in intact cortex of cats in vivo. *The Journal of Physiology, 542*(Pt 2), 583–598.

Tononi, G., & Cirelli, C. (2003). Sleep and synaptic homeostasis: A hypothesis. *Brain Research Bulletin, 62*(2), 143–150.

Tononi, G., & Cirelli, C. (2006). Sleep function and synaptic homeostasis. *Sleep Medicine Reviews, 10*(1), 49–62.

Tsanov, M., & Manahan-Vaughan, D. (2008). Synaptic plasticity from visual cortex to hippocampus: Systems integration in spatial information processing. *Neuroscientist, 14*(6), 584–597.

Van Der Werf, Y. D., Van Der Helm, E., Schoonheim, M. M., Ridderikhoff, A., & Van Someren, E. J. (2009). Learning by observation requires an early sleep window. *Proceedings of the National Academy of Sciences of the United States of America, 106*(45), 18926–18930.

Varela, F., Lachaux, J. P., Rodriguez, E., & Martinerie, J. (2001). The brainweb: Phase synchronization and large-scale integration. *Nature Reviews Neuroscience, 2*(4), 229–239.

Vyazovskiy, V. V., Cirelli, C., Pfister-Genskow, M., Faraguna, U., & Tononi, G. (2008). Molecular and electrophysiological evidence for net synaptic potentiation in wake and depression in sleep. *Nature Neuroscience, 11*(2), 200–208.

Wagner, U., Gais, S., Haider, H., Verleger, R., & Born, J. (2004). Sleep inspires insight. *Nature, 427*(6972), 352–355.

Wagner, U., Hallschmid, M., Rasch, B., & Born, J. (2006). Brief sleep after learning keeps emotional memories alive for years. *Biological Psychiatry, 60*(7), 788–790.

Walker, M. P., & van der Helm, E. (2009). Overnight therapy? The role of sleep in emotional brain processing. *Psychological Bulletin, 135*(5), 731–748.

Werk, C. M., Harbour, V. L., & Chapman, C. A. (2005). Induction of long-term potentiation leads to increased reliability of evoked neocortical spindles in vivo. *Neuroscience, 131*(4), 793–800.

Wierzynski, C. M., Lubenov, E.V., Gu, M., & Siapas, A. G. (2009). State-dependent spike-timing relationships between hippocampal and prefrontal circuits during sleep. *Neuron, 61*(4), 587–596.

Wilhelm, I., Diekelmann, S., Molzow, I., Ayoub, A., Mölle, M., & Born, J. (2011). Sleep selectively enhances memory expected to be of future relevance. *The Journal of Neuroscience, 31*(5), 1563–1569.

Wilson, M.A., & McNaughton, B. L. (1994). Reactivation of hippocampal ensemble memories during sleep. *Science, 265*(5172), 676–679.

Winocur, G., & Moscovitch, M. (2011). Memory transformation and systems consolidation. *Journal of the International Neuropsychological Society, 17*(5), 766–780.

Winocur, G., Moscovitch, M., & Bontempi, B. (2010). Memory formation and long-term retention in humans and animals: Convergence towards a transformation account of hippocampal-neocortical interactions. *Neuropsychologia, 48*(8), 2339–2356.

Witt, K., Margraf, N., Bieber, C., Born, J., & Deuschl, G. (2010). Sleep consolidates the effector-independent representation of a motor skill. *Neuroscience, 171*(1), 227–234.

Wolansky, T., Clement, E. A., Peters, S. R., Palczak, M. A., & Dickson, C. T. (2006). Hippocampal slow oscillation: A novel EEG state and its coordination with ongoing neocortical activity. *The Journal of Neuroscience, 26*(23), 6213–6229.

Yordanova, J., Kolev, V., Verleger, R., Bataghva, Z., Born, J., & Wagner, U. (2008). Shifting from implicit to explicit knowledge: Different roles of early- and late-night sleep. *Learning & Memory, 15*(7), 508–515.

Yordanova, J., Kolev, V., Wagner, U., Born, J., & Verleger, R. (2011). Increased alpha (8–12 Hz) activity during slow wave sleep as a marker for the transition from implicit knowledge to explicit insight. *Journal of Cognitive Neuroscience, 24*(1), 119–132.

Zola-Morgan, S., & Squire, L. R. (1990). The neuropsychology of memory. Parallel findings in humans and nonhuman primates. *Annals of the New York Academy of Sciences, 608*, 434–450. (discussion 450–436).

CHAPTER 10

Sleep Slow Oscillations and Cortical Maturation

Salomé Kurth[1,2] and Reto Huber[1,3]
[1]Child Development Center, University Children's Hospital Zurich, [2]University of Colorado at Boulder, Department of Integrative Physiology, [3]Zurich Center for Integrative Human Physiology, University of Zurich

SLOW WAVES AND SLEEP HOMEOSTASIS

Waves of high amplitude (>75 μV) and low frequency (<4.5 Hz) are a dominant characteristic of the non-rapid-eye-movement (NREM) sleep electroencephalogram (EEG). Such slow waves are most prevalent during deep sleep. Already in the 1930s slow waves were shown to parallel sleep depth (Blake & Gerard, 1937). Today, it is commonly accepted that sleep depth/pressure is homeostatically regulated—it increases in proportion to the time spent awake and decreases during sleep (Borbély & Achermann, 2005). Slow wave activity (SWA, EEG spectral power between 1 and 4.5 Hz) represents a well-established marker for the homeostatic decline of sleep depth during sleep. Recent evidence suggests that slow waves are closely related to synaptic changes (Cirelli, 2009; Diekelmann & Born, 2010; Tononi & Cirelli, 2006). This possible functional relationship of slow waves might be of particular relevance in brain development during childhood and adolescence, when brain structures experience major transformations (Giedd et al., 1999; Huttenlocher, 1979). Thus, it will be fundamental to understand whether slow waves even actively impact synaptic changes during development.

Sleep Slow Oscillations

At the single neuron level, intracellular recordings show that during deep sleep, membrane potentials of cortical neurons alternate between a depolarized up-state and a hyperpolarized down-state with a frequency of about 1 Hz (Steriade, Contreras, Curro Dossi, & Nunez, 1993; Timofeev, Grenier, & Steriade, 2001). Thus, during such slow oscillations, cortical neurons are bistable, oscillating between two distinct states, each lasting a few hundred milliseconds. When large networks of neurons undergo such oscillations in high synchrony, the scalp surface EEG measures high-amplitude voltage changes (Vyazovskiy et al., 2011; Vyazovskiy et al., 2009). Only a

Sleep and Brain Activity
DOI: http://dx.doi.org/10.1016/B978-0-12-384995-3.00010-1

few human studies have investigated slow oscillation on the surface EEG (Achermann & Borbely, 1997; Bersagliere & Achermann, 2010). The authors found such slow oscillations among slow waves (<4.5 Hz). The question of the origin of slow waves with higher frequencies (1–4.5 Hz) is unresolved. However, since slow oscillations seem to travel across the cortex with a distinct origin and propagation path (Massimini, Huber, Ferrarelli, Hill, & Tononi, 2004), it might be possible that such waves originate from the collision of slow oscillations starting at a similar time, however with different origin. Such collisions of traveling slow oscillations may produce the higher frequency waves (Riedner et al., 2007). Traditionally, on the surface EEG, sleep slow waves are quantified by slow wave activity during NREM sleep (Borbély & Achermann, 2005).

Regulation of Sleep

We have a good understanding of the regulation of sleep. The *Two-Process Model* of sleep regulation describes two main processes involved in the regulation of sleep in humans and in a variety of other species (Borbély & Achermann, 2005; Tobler, 2000). The model proposes that sleep pressure is determined by the interaction of a circadian process C and a homeostatic process S. Process C represents the circadian component that maintains an approximate 24-hour rhythm. The homeostatic process S increases during waking and decreases during sleep. The immediate history of sleep and waking thus determines the level of process S. The probability to fall asleep is explained by the interplay of processes S and C. Understanding the physiological mechanisms of both processes and examining the dynamic response of the model to challenge (e.g., sleep deprivation) has been a major focus of sleep research during the past 30 years (Borbély & Achermann, 2005). It is known that the circadian rhythm (process C) is generated by an intrinsic pacemaker located in the suprachiasmatic nucleus (SCN). During the last decade the molecular machinery responsible for the clock-like output of SCN cells has been uncovered in great detail (Aston-Jones, 2005; Mistlberger, 2005; Saper, Lu, Chou, & Gooley, 2005; Zee & Manthena, 2007). In contrast, the knowledge about the mechanisms underlying process S is limited. Fortunately, SWA was uncovered as a well-established electrophysiological marker of process S. Numerous studies have confirmed that SWA during NREM sleep is homeostatically regulated—it increases as a function of time awake and decreases during sleep (Borbély & Achermann, 2005). One assumption of the *Two-Process Model* is that the loss of NREM sleep can be recovered by an intensification of

NREM sleep, measured as a SWA increase, and not necessarily by an increase in sleep duration. This assumption strengthens the important role of SWA in sleep regulation. A second important concept of the *Two-Process Model* is that the homeostatic and the circadian processes operate independently. This has been confirmed by sleep deprivation studies in SCN-lesioned rats. These animals no longer exhibit circadian modulation of sleep and wakefulness. Nevertheless sleep deprivation still results in an increase of SWA (Mistlberger, Bergmann, Waldenar, & Rechtschaffen, 1983; Tobler, Borbely, & Groos, 1983; Trachsel, Edgar, Seidel, Heller, & Dement, 1992).

Local Use-Dependent Sleep

In recent years it has become more and more evident that sleep is regulated locally. Unihemispheric sleep is known in birds and cetaceans (Siegel, 2009; Tobler, 2000), whereby selective deprivation gives rise to a unihemispheric slow-wave sleep rebound (Oleksenko, Mukhametov, Polyakova, Supin, & Kovalzon, 1992). Although not in a unihemispherical fashion, an increasing number of studies provide evidence for local and use-dependent sleep in humans (e.g., Huber et al., 2006; Huber, Ghilardi, Massimini, & Tononi, 2004; Kattler, Dijk, & Borbely, 1994; Landsness et al., 2009; Maatta et al., 2010). For example, in young adults extensive sensory stimulation of the hand lead to an increase of EEG power in the SWA frequency range during subsequent sleep (Kattler et al., 1994). The question was raised whether local sleep is adaptive or maladaptive (Vyazovskiy et al., 2011). In rats, it was shown that restricted sleep over days reveals sleep EEG characteristics during the behavioral state of waking (Leemburg et al., 2010). During subsequent sleep, the frontal regions exhibited the largest SWA rebound. Another study identified sleep-like features in the local waking EEG, associated with worsened performance, after a long period of waking (Vyazovskiy et al., 2011). Rats sleep approximately 12–14 hours per day with sleep cycles of approximately 10–20 minutes (Cirelli & Tononi, 2008). When these cycles are interrupted, rats are able to quickly adapt to shorter sleep periods and to maintain their overall daily time of NREM sleep (Tartar et al., 2006). These two examples describe an adaptive role of local sleep. In the animal kingdom, other forms of adaptive inactivity are also observed: hibernation, torpor, periods of migration in birds, mating season (Siegel, 2009). Therefore, while unihemispheric sleep seems to be adaptive, behaviorally dissociated states as sleep walking, REM sleep behavior disorder, and other parasominas are clearly maladaptive as Vyazovskiy et al. pointed out (Mahowald & Schenck, 2005; Terzaghi et al., 2009; Vyazovskiy et al., 2011).

Topographical assessments of the sleep EEG support a region-specific need for sleep (Huber et al., 2004; Landsness et al., 2009; Maatta et al., 2010). This region specific sleep SWA is possibly use-dependent, and could represent neuronal tiredness such as synaptic overload as introduced in Tononi & Cirelli (2006). This might be particularly important for the developing brain undergoing a period of major synaptic change.

At the cellular level two recent studies provide evidence for local differences in sleep measures: Nir and coauthors showed by means of intracerebral EEG in humans that sleep slow waves appear as both global or local waves (Nir et al., 2011). They showed that slow waves became more local in late sleep as compared to early sleep. In addition, local waves were predominantly of low amplitude, and low amplitude waves typically occur in late sleep, when sleep pressure has largely dissipated (Nir et al., 2011; Riedner et al., 2007). The largest discrepancy of activity states within the brain is found when some brain regions are asleep, while others are awake. Long periods of waking in rats were associated with EEG and behavior clearly recognized as waking, however, individual local neuronal populations exhibited aspects of sleep at the same time. In addition, this state was associated with impaired performance (Vyazovskiy et al., 2011). In adult epileptic patients, a similar dissociation of vigilance states was observed (Nobili et al., 2011). Nobili et al. found coexisting wake-like and sleep-like patterns simultaneously in different cortical areas. This regional occurrence of sleep corroborates the local regulation of sleep slow waves.

Learning and Slow Wave Activity

Numerous studies propose a close relationship between waking activity and subsequent EEG activity during sleep (Diekelmann & Born, 2010). For example, visuomotor learning was associated with a region-specific local increase of sleep SWA (Huber et al., 2004). In contrast, a local reduction of SWA was found as a consequence of reduced activity in sensorimotor areas as a result of arm immobilization (Huber et al., 2006). Several studies show a direct link between changes in SWA and sleep-dependent performance improvements. For example, the local increase of SWA after visuomotor learning was positively correlated with the performance improvement the next morning (Huber et al., 2004). A causal relationship between sleep SWA and performance improvements was shown in protocols artificially manipulating the level of SWA. Slow wave deprivation by acoustic stimuli hindered sleep-dependent performance gains of visuomotor learning (Landsness et al., 2009). In contrast, artificial boosting of slow waves, by means of transcranially applied oscillating currents, was associated with performance improvement in the declarative memory system (Marshall, Helgadottir, Molle, & Born, 2006).

The above findings provide good evidence for an involvement of sleep slow waves in plastic processes related to learning. Long-term potentiation is the cellular basis of learning and involves activity-induced molecular changes at the synapse (Bliss & Lomo, 1973; Kandel, 2001; Whitlock, Heynen, Shuler, & Bear, 2006). How sleep SWA and such plastic processes may be related is illustrated in the hypothesis about a function of sleep below.

SLOW WAVE SLEEP IS THE PRICE PAID FOR PLASTICITY IN A COMPLEX NEURAL SYSTEM

The Synaptic Homeostasis Hypothesis

The *Synaptic Homeostasis Hypothesis* proposed by Tononi and Cirelli provides an interesting explanation for the close relationship between neuronal activity during waking and sleep (Tononi & Cirelli, 2006). The hypothesis links the homeostatic regulation of sleep to the homeostatic regulation of synaptic strength. Changes in synaptic strength by means of long term potentiation processes are thought to be the underlying mechanism of learning and memory (Bliss & Lomo, 1973; Whitlock et al., 2006). We learn throughout our life. To prevent a saturation of overall synaptic strength due to such learning and to ensure a rather stable learning capacity across decades during adulthood synaptic strength needs to be balanced. Thus, recent work proposes the idea that average synaptic strength is homeostatically regulated to promote stability of the system (Turrigiano, 1999). The Synaptic Homeostasis Hypothesis now proposes that the rebalancing of synaptic strength takes place during sleep. More specifically, according to the hypothesis the homeostatic SWA decrease during sleep reflects the decrease of synaptic strength. The hypothesis further proposes that SWA is not only reflecting synaptic strength but is also responsible for its reduction. Thus, according to the Synaptic Homeostasis Hypothesis SWA actively triggers synaptic downscaling, i.e., renormalizes synaptic strength to a baseline level. The hypothesis assumes that such global "downscaling" of cortical connections does not happen in a selective manner, but depends on the strength of synapses in cortical regions: strong synaptic connections are downscaled into weaker connections but maintained, whereas weak connections may be entirely deleted.

Recent *in vitro* and *in vivo* animal experiments have found evidence for a net prevalence of synaptic potentiation mechanisms during wakefulness and a net reduction of synaptic strength during sleep. Thus, prolonged wakefulness was found to increase the frequency and amplitude of miniature excitatory postsynaptic currents (EPSCs) in cortical slices—a key *in vitro* measure of synaptic strength (Liu, Faraguna, Cirelli, Tononi, & Gao, 2010). In *Drosophila melanogaster* prolonged wakefulness was associated with an increase in the protein levels of key components of central synapses (Gilestro, Tononi, & Cirelli, 2009), as well as an increase in the number and size of synapses (Bushey, Tononi, & Cirelli, 2011). In rats, wakefulness increased the rate and synchrony of firing of cortical neurons (Vyazovskiy et al., 2009) and the slope of the local field potentials (LFPs) evoked by electrical cortical stimulation (Vyazovskiy, Cirelli, Pfister-Genskow, Faraguna, & Tononi, 2008), a classic measure of synaptic strength *in vivo*. As SWA decreases in the course of sleep these measures of synaptic strength decrease as well, i.e., the responsiveness of cortical neurons and the protein levels of key components, as well as the size and number of synapses and the frequency and amplitude of miniature EPSCs (Bushey et al., 2011; Gilestro et al., 2009; Liu et al., 2010; Vyazovskiy et al., 2008). These observations suggest that the homeostatic regulation of sleep, as reflected by SWA, parallels the regulation of measures of synaptic strength. Using a combination of transcranial magnetic stimulation (TMS) and high-density EEG (hdEEG), a recent study observed that human cortical excitability, measured as the immediate EEG reaction to TMS, also progressively increases with time awake, from morning to evening and after one night of total sleep deprivation, and that it decreases after recovery sleep (Huber et al., 2012).

Synchronization of Neural Activity during Sleep

The strength of population excitatory postsynaptic currents is best reflected by the slope of LFPs evoked by electrical stimuli (Rall, 1967). Vyazovskiy et al. used this technique in the rat to show that the slope of LFPs increases as a function of the time spent awake and decreases during sleep (Vyazovskiy et al., 2008). Furthermore, the slope of LFPs was positively correlated with SWA during the first hour of NREM sleep (Vyazovskiy et al., 2008). Multiunit recordings in the rat allow a simultaneous assessment of the firing patterns across numerous neurons. Thus, at the beginning of the night, when synaptic strength is high, most individual neurons start and stop firing in near synchrony with the rest of the population (Vyazovskiy et al., 2009). Moreover, synchronous transitions at the unit level were associated with steep slopes of slow waves during early sleep and less synchronous transitions with reduced slopes at the end of the night. This observation proposes the slope of slow waves as a direct measure of synaptic strength. In humans (Riedner et al., 2007), rats (Vyazovskiy, Riedner, Cirelli, & Tononi, 2007), and *in computo* (Esser, Hill, & Tononi, 2007) the slope of slow waves was shown to decrease from the beginning to the end of a sleep period. This slope decrease of slow waves during sleep was explained as homeostatic reduction of synaptic strength.

These observations show that changes in synaptic strength affect the characteristics of sleep slow waves via the changes in synchronization. Several cortical network properties affect neuronal synchronization and therefore are directly linked to the generation of sleep slow waves. For example, the capacity of white matter fiber tracts impacts the speed of signal propagation, and gray matter volume/thickness, which is proportional to synapse number/density, in turn impacts the efficiency of signal proliferation from the pre- to the postsynapse. In neuronal systems with reduced connectivity, a lower synchronization between the elements (e.g., neurons or neuronal populations) is expected. The occurrence of a down-state in a neuron is always maintained by its input, i.e., the number of and connectivity to neighbored neurons: the greater the degree of presynaptic activity (number of neurons/synapses) impinging onto a target neuron, the higher the probability that an action potential is elicited in this neuron. Network modeling, *in vivo*, and *in vitro* recordings agree that the role of short-range connections is important in maintaining and synchronizing the slow oscillation (Hill & Tononi, 2005; Sanchez-Vives & McCormick, 2000; Timofeev, Grenier, Bazhenov, Sejnowski, & Steriade, 2000). Computer simulations of the activity of the thalamocortical system reveal that synchronization and amplitude of the slow oscillation are dependent on the strength of both intra- and interareal corticocortical connections (Hill & Tononi, 2005), again consistent with *in vivo* (Amzica & Steriade, 1995) and *in vitro* (Sanchez-Vives & McCormick, 2000) results.

It is hence conceivable that the maturation of anatomical connectivity impacts the EEG in both wake and sleep. The first two decades of life are the only time where such dramatic changes in our brain's architecture occur under healthy conditions. Accordingly, sleep should also change during development. How sleep changes and how those changes parallel anatomical changes is discussed in the next sections.

Maturation of Sleep Homeostasis

The homeostatic regulation of sleep and its electrophysiological marker SWA are not present right from the start. First signs for a homeostatic regulation of sleep can be observed in the first months of life (Bes, Schulz, Navelet, & Salzarulo, 1991; Jenni & Carskadon, 2004). Similar results come from animal studies, which also allow challenging sleep homeostasis. They show that while sleep deprivation does not impact SWA in 12-day-old rats, a clear increase in SWA follows sleep deprivation in 24-day-old rats (Frank, Morrissette, & Heller, 1998). Instead, the 12-day-old rats compensate for the sleep loss with increased sleep time and continuity. Similar results are found in human neonates, where sleep restriction leads to compensatory increases in NREM sleep duration only (Anders & Roffwarg, 1973; Thomas et al., 1996). In humans, the exact timing of when sleep deprivation starts to produce an increase in SWA is unknown. Yet, a SWA decline in the course of the night is first visible during the second postnatal month (Bes et al., 1991). An early indication for homeostatic regulation of sleep in infants might be seen in the decline of theta power in the course of a sleep episode that occurs between 6 and 9 months of age (Jenni, Borbely, & Achermann, 2004). Later in life, the homeostatic process continues to show maturational changes: a slowing in the buildup of homeostatic sleep pressure during wakefulness was proposed to occur from childhood to adolescence (Jenni, Achermann, & Carskadon, 2005). In contrast, the homeostatic decline during sleep remains stable (Rusterholz & Achermann, 2011; Jenni, et al., 2005).

Clear maturational changes of the homeostatic regulation of sleep are present during early development. The question arises whether and how these changes in sleep homeostasis are related to structural and functional brain maturation. The first months of human life are characterized by a huge outgrowth of synaptic connections (Huttenlocher, 1979). Also, several neurotransmitter systems are not yet fully developed (Lidow, Goldman-Rakic, & Rakic, 1991). Interestingly, in rats it was shown that the start of the homeostatic regulation of sleep coincides with the occurrence of brain-derived neurotrophic factor (BDNF) (Hairston et al., 2004), a key marker of synaptic plasticity (Alonso et al., 2005; Cancedda et al., 2004; Castillo, Figueroa-Guzman, & Escobar, 2006; Chakravarthy et al., 2006; Genoud, Knott, Sakata, Lu, & Welker, 2004; Kleim et al., 2006). On the one hand, BDNF is well known as a major mediator of synaptic plasticity (Jiang, Akaneya, Hata, & Tsumoto, 2003). On the other hand, in adult rats, BDNF was closely linked to the regulation of SWA: BDNF increases after sleep deprivation (Hairston et al., 2004) and it is more expressed in rats showing more SWA after exploring enriched environments (Huber, Tononi, & Cirelli, 2007). A causal relationship between BDNF and sleep SWA was shown in a more recent study (Faraguna, Vyazovskiy, Nelson, Tononi, & Cirelli, 2008): blockage of

BDNF by direct infusion of anti-BDNF resulted in a subsequent reduction of SWA. In contrast, local infusion of BDNF resulted in a corresponding increase of SWA. Taken together it seems that the evolvement of synaptic plasticity is closely linked to the homeostatic regulation of sleep during development. Interestingly, in humans, the increase of BDNF mRNA levels in the dorsolateral prefrontal cortex during adolescence coincides with the time when the frontal cortex matures both structurally and functionally (Webster, Weickert, Herman, & Kleinman, 2002).

Even after the homeostatic regulation of sleep is fully functional the developmental changes of this system continue.

THE DEVELOPMENT OF SLOW WAVES PARALLELS CORTICAL MATURATION

Slow Waves and the Maturation of Gray Matter

A major developmental change during the first two decades of life concerns the number/density of synapses (Huttenlocher, 1979). It is clear that the formation and elimination of synapses is intensified during the developmental period (Goldman & Nauta, 1977; Huttenlocher, 1979; Majewska & Sur, 2003; Rakic, Bourgeois, Eckenhoff, Zecevic, & Goldman-Rakic, 1986). In early childhood, neurons explore much wider areas than their final targets (Gao, Yue, Cerretti, Dreyfus, & Zhou, 1999) and the number of synapses exceeds adult levels by far (Huttenlocher, 1979). Then, in the course of adolescence, more connections are eliminated than formed (Zuo, Lin, Chang, & Gan, 2005). Postmortem studies in the macaque reveal that most pronounced synaptic overproduction occurs in supragranular layers, suggesting that synaptic pruning and stabilization are more important for the maturation of corticocortical circuits (Petanjek, Judas, Kostovic, & Uylings, 2008; Woo, Pucak, Kye, Matus, & Lewis, 1997). The elimination processes occur in an activity-dependent fashion (Hua & Smith, 2004). Evidence is increasing that in a system of concurrent synapse formation and elimination, activity modulates development, by controlling the formation of new and the maintenance of existing connections (Hua & Smith, 2004). Synaptic plasticity in adult neural circuits may involve the strengthening or weakening of existing synapses as well as structural plasticity, including synapse formation and elimination (Holtmaat & Svoboda, 2009).

The pattern of an initial overproduction followed by elimination and refinement associated with increased specificity is also found in other biological systems. This mechanism of refinement is in the neuronal context termed pruning (Huttenlocher, 1990), and seems to follow the principle of *Trial and Error* (Holmes, 1907; Jennings, 1904). In terms of neuronal plasticity, the synaptic activity can be considered as stimulus (trial), while inactivity leads to elimination of connections (error). The advantage of systems with higher complexity is the ability to flexibly respond. But flexibility carries also costs: although synaptic plasticity enables high-capacity learning, it also introduces increased energy consumption and a reduction in efficiency. Interestingly, across development, brain energy consumption follows a similar time course as synapse density (brain oxygen consumption (Kety, 1956; Kety & Schmidt, 1948) or glucose utilization (Chugani, 1998)). The maintenance of synapses is one of the most energy-demanding processes (see Tononi & Cirelli, 2006). A system with high synaptic connectivity may also be less efficient, compared to a system with lower synaptic connectivity but instead more long-range fibers. In the high synaptic density system, signals need to be transmitted via numerous synaptic connections, which is time consuming (millisecond range (Borst & Sakmann, 1996)). The transmission speed of the action potential across the synaptic cleft is estimated to be about 0.2 mm/s, i.e. >1000 times slower than when the electrical impulse travels down the nerve axon (>25 m/s) (Wheatley, 1998). Of course, myelination, which significantly changes during development (see Deoni et al. (2011) for quantitative analyses or Paus et al. (2001) for a review of magnetic resonance imaging (MRI) studies), also plays an important role in signal transduction. Oligodendrocytes wrap the axon with myelin, consisting of multiple layers of closely opposed glial membranes. Myelin acts as insulator of the axonal membrane, improving the passive flow of electrical current. Myelin speeds up action potential conduction, e.g., from 0.5–10 m/s in unmyelinated axons, to 150 m/s in myelinated axons.

In summary, we have seen that synapse density and energy consumption follow a similar time course across development. Interestingly, Feinberg already in the 1980s found that SWA parallels both the time course of synapse density and energy consumption (Feinberg, 1982), suggesting a possible relationship (Fig. 10.1). At the age when synaptic growth is maximal the amplitude of slow waves also reaches a peak. Thus SWA reaches a ceiling level at the time of maximal synaptic density during childhood. Adolescence is then accompanied by a sharp decrease in the amount of deep sleep (Feinberg, 1982), synaptic density (Huttenlocher, 1979) and in the rate of brain metabolism (Chugani, 1998).

A possible explanation for the close relationship between SWA and synaptic density is provided by the observation that the amplitude of slow

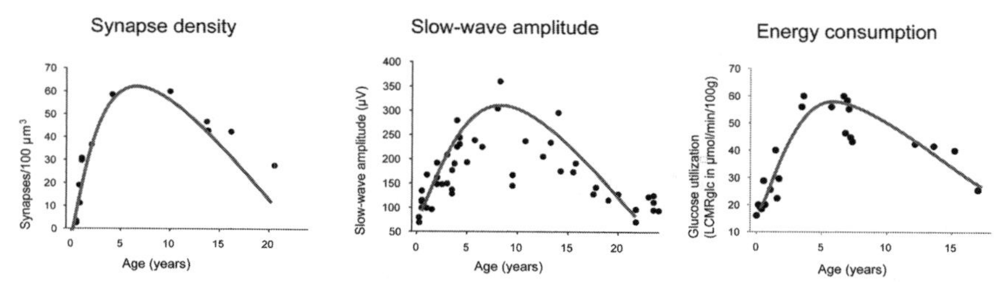

Figure 10.1 Development of synapse density (left, postmortem data, reconstructed from Huttenlocher & Dabholkar (1997)), slow wave amplitude (middle, reconstructed from Feinberg (1982)), and energy consumption (right, glucose utilization (reconstructed from Chugani, (1998)). The three measures follow a similar time course across the first two decades of life.

waves, the major contributing factor to spectral power in the slow wave frequency range, depends on the ability of a cortical network to synchronize its activity (Vyazovskiy et al., 2009). How fast a cortical network can synchronize its activity depends on the strength and density of its connections (Esser et al., 2007; Vyazovskiy et al., 2009). Thus, maximal synaptic density reached during childhood results in maximal synchronization of cortical activity during sleep, which in turn gives rise to maximal amplitude slow waves. If this mechanism holds true it would also explain why the maturational changes of the EEG are not restricted to sleep, but also seen in waking (Gasser, Verleger, Bacher, & Sroka, 1988).

In the sleep EEG, the analysis of the slope of slow waves provides a useful tool to study the level of synchronization and associated change in synaptic strength (Kurth, Jenni et al., 2010; Riedner et al., 2007; Vyazovskiy et al., 2007). A computer model of the thalamocortical system reproduced in detail the cortical slow oscillations underlying EEG slow waves during sleep (Esser et al., 2007). The model showed that the slope of sleep slow waves is the best indicator for changes in synaptic strength (Esser et al., 2007). Moreover, local field potential recordings in the rat also showed a decline of the slope of slow waves across a sleep period, again indicating that the slow-wave slope may reflect the level of synchronization (Vyazovskiy et al., 2007), This finding was also observed in adult humans, where the slope of slow waves decreased across the night (Riedner et al., 2007). We used this measure to compare cortical synchronization in prepubertal children and mature adolescents (Kurth, Jenni et al., 2010). We found that the slow-wave slope was steeper in prepubertal children compared to mature adolescents, even when controlling for amplitude. The steeper slope in premature children may indicate greater synaptic strength in prepubertal children compared to mature adolescents.

Gray matter, however, is not the only part maturing during childhood and adolescence. Thus, the next section illustrates the major maturational changes of white matter and how these changes may be related to slow waves.

Slow Waves and the Maturation of White Matter

Myelination

In general, and in contrast to gray matter maturation, white matter maturation follows a linear increase. The efficiency of long-range cortical connections in the vertebrate cerebral cortex is influenced by myelin sheets surrounding axons (Deoni et al., 2011; Paus, Keshavan, & Giedd, 2008). Myelination of axons accelerates the propagation of action potentials, through a mechanism that does not require large amounts of additional space or energy (Frankenhaeuser, 1952; Halter & Clark, 1991; Huxley & Stampfli, 1949). During development myelination increases (Lebel & Beaulieu, 2011; Paus et al., 2001), which results in increased efficiency of action potential propagation. Well-established MRI measures offer insights into structural maturation, via white matter volumetry or diffusion tensor imaging (DTI) (Gogtay et al., 2004; Paus et al., 2008). During cortical maturation, the relaxation times, upon which such analyses are based, are influenced by different facets of development, such as changes in axonal density, size, membrane structure, and the content of lipids, macromolecules, and proteins, as well as water compartmentalization (Deoni et al., 2011). State-of-the-art imaging technologies can provide a quantification of myelination processes (Deoni et al., 2011). Also, electrophysiological measures offer the possibility to assess white matter efficiency. For example, coherence measures assess the connectivity between distant cortical regions (Achermann & Borbely, 1998). Interestingly, the increasing myelination across the developmental period is accompanied by an increase in EEG coherence (Tarokh, Carskadon, & Achermann, 2010). In a dataset of 59 subjects, we confirm these findings for the age range 2–26 years (Fig. 10.2). This linear increase in coherence, as exemplified for a left central region in the SWA frequency band, is in line with the increase of white matter until the 30s (Lebel & Beaulieu, 2011; Paus et al., 2001).

Another measure used in sleep EEG recordings that might be related to white matter maturation is the characteristics of sleep slow waves to behave as traveling waves. In the NREM sleep EEG, Massimini et al. showed that slow oscillations sweep across the cortex as waves (Massimini et al., 2004). Thereby, each wave maintains a definite site of onset, spatial propagation, and speed of traveling. Source localization of spontaneous and evoked slow

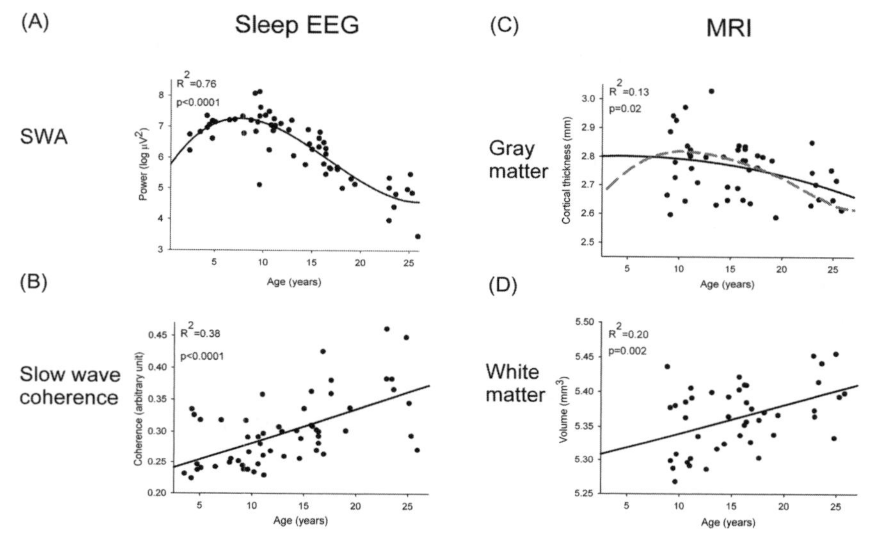

Figure 10.2 ***Comparison of the maturation of the sleep EEG and MRI-derived anatomical markers of cortical maturation.*** (**A**) Slow wave activity (SWA) across age for the derivation C4A1 (n = 62, 2.4–25.9y, $R^2 = 0.76$, $p < 0.0001$). (**B**) EEG coherence in the slow-wave frequency range (1–4.5 Hz) for the derivations F3C3-C3P3 (n = 59, 2.4–25.9y, $R^2 = 0.38$, $p < 0.0001$). Coherence was calculated using Welch's averaged periodogram method (see Achermann & Borbely, 1998) and represents a measure for the similarity of the EEG activity at different derivations. (**C**) MRI-derived gray matter thickness across age, shown for the left hemisphere (n = 45, 8.8–25.8y, $R^2 = 0.13$, $p = 0.02$). The dotted gray line indicates the expected developmental time course during childhood, as known from previous studies (Giedd et al., 1999; Shaw et al., 2008). (**D**) MRI-derived white matter volume of the left hemisphere is shown across age (n = 45, 8.8–25.8y, $R^2 = 0.20$, $p = 0.002$). For a detailed description of the MR data analysis see Buchmann et al. (2011) and Buchmann et al. (2010).

waves in adults identified frequent traveling activity in the anteroposterior direction (Murphy et al. 2009), possibly associated with the superior longitudinal fascicle. Thus, traveling waves might serve as an indicator for fiber maturation during brain development.

Interaction of Gray and White Matter Maturation

Local Progression of the Cortical Developmental Changes

Already the studies by Huttenlocher have shown that the time course of synapse density is not uniform across the cortex (Huttenlocher, 1979; Huttenlocher & Dabholkar, 1997). More specifically, depending on the cortical area peak synaptic density was reached at different ages: occipital cortex reached peak synaptic density first and frontal cortex last. MRI-derived gray

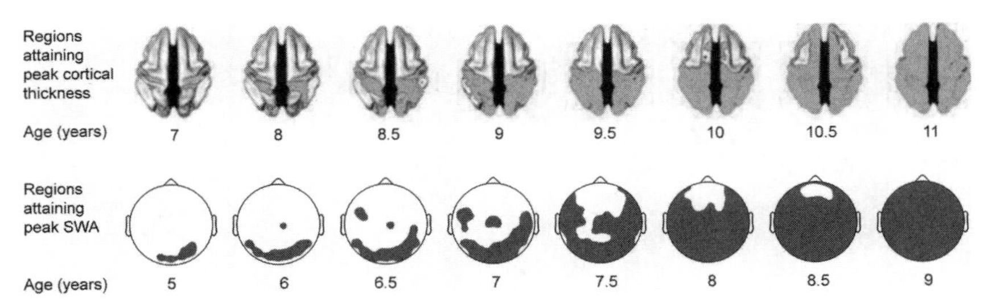

Figure 10.3 ***Posterior-to-anterior maturation of the cortex.*** *Upper panel*: Age of attaining peak cortical thickness across the cerebral cortex (adapted after Shaw et al., 2008). *Lower panel*: Topographical distribution of sleep slow wave activity (SWA) maturation. Electrodes that have reached peak SWA at the age given at the bottom are shaded. SWA was normalized for common regional differences across age (e.g., minimal SWA over temporal regions) by normalizing SWA power for each electrode and individual to the average power across all ages.

matter volume or thickness show a very similar spatial trajectory (Giedd, 2004; Shaw et al., 2008; Sowell et al., 2004). In general, these studies show a posterior-to-anterior maturational pattern, where occipital regions exhibit an early peak, and frontal regions do so late (Shaw et al., 2008). Interestingly, the close relationship between gray matter maturation and changes in sleep SWA across development (see Fig. 10.2) can be extended to the topography. hdEEG recordings in children and adolescents showed that maximal SWA is located over posterior regions in childhood, and shifts over parietal/central regions during adolescence (Fig. 10.3), resulting in the well-known frontal maximum in adults (Finelli, Borbely, & Achermann, 2001; Kurth, Ringli et al., 2010). In analogy to the analysis of cortical thickness by Shaw et al. (2008), we identified the age of peak of SWA for every electrode. For each age range (see Fig. 10.3) electrodes were colored when peak SWA had been reached. Our main observation of this analysis was the posterior-to-anterior maturation of SWA. Other sleep EEG studies across development support this finding of a posterior-to-anterior maturation of power in the low-frequency range (Campbell & Feinberg, 2009). A central feature of this comparison is the close temporal relationship of the electrophysiological and anatomical development (Fig. 10.3). It appears that the maturation of SWA precedes the maturation of gray matter thickness by approximately 2 years, e.g., at the age of 11 years the entire cerebral cortex has attained peak thickness, whereas SWA peaks are entirely attained with the age of 9 years.

The pattern of posterior-to-anterior maturation can also be observed when looking at the maturation of brain functions. Thus, the visual cortex, located in posterior areas, is known to undergo early maturation that

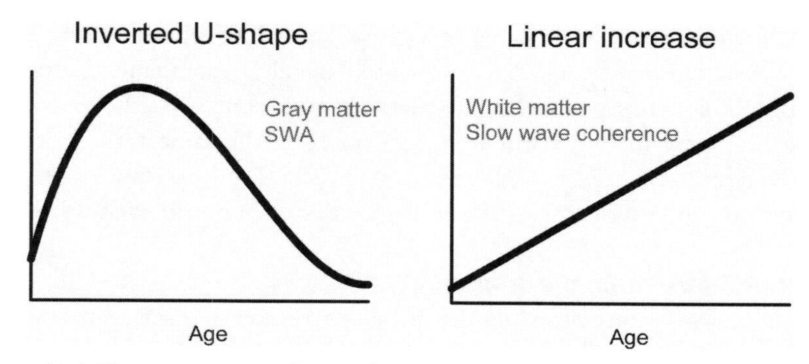

Figure 10.4 The two patterns of cortical maturation during the first two decades of life.

can be assessed for example by visual acuity (Teller, 1981). Motor abilities, for example assessed by the Zurich Neuromotor Assessment (ZNA; Gasser, Rousson, Caflisch, & Jenni, 2010; Largo et al., 2001), mature in adolescence, and cognitive abilities, for example needed for executive functions, continue to mature into young adulthood (Luna & Sweeney, 2004). Although these observations on EEG, MRI, and behavior describe the maturation of the brain from a different perspective, they show the same posterior-to-anterior spatial trajectory along the cortex. We have explored the relationship between the maturation of brain functions and changes in SWA topography in a cross-sectional sample of developing subjects (Kurth et al., 2012). In our analyses, SWA topography preceded the maturation of behavioral skills by approximately 3 years. Moreover, when gray matter thinning was included, it was the last to reach maturity.

It is clear though that the maturation of brain functions not only relies on the maturation of gray matter but is also dependent on white matter maturation. In particular, higher cognitive skills (like executive functions) located in the prefrontal cortex (PFC) profit from the white matter connectivity changes, because such functions depend on the interplay of different brain regions. Only efficient long range connections allow the timely integration of complex information originating from different modalities. Thus, the PFC receives projections from virtually all cortical sensory systems, motor systems, and many subcortical structures allowing it to coordinate a wide range of neuronal processes (Miller & Cohen, 2001).

In summary, there exist two particular maturational patterns (Fig. 10.4): (1) the inverted U-shape time course of synaptic density/gray matter, which shows an approximate posterior-to-anterior propagation, and (2) the maturation of white matter/myelin, which increases linearly until the third decade, also following a particular spatial trajectory (Paus, 2005; Tarokh et al., 2010). We have shown that sleep SWA very closely parallels gray matter maturation and EEG coherence parallels white matter maturation. Since these processes (white and gray matter maturation) take place at the same time, it is clear that the development of brain functions reflects both of these processes.

Gray and white matter maturation may also depend on each other.

A Competition within the Brain?

One limitation occurring while the number of synapses grows is the restriction of space. It is well known that the largest growth of the human brain volume takes place before adolescence (Courchesne et al., 2000). During this time, MRI reveals a growth of gray matter together with an increase of white matter. At the peak of this process, before puberty, also slow waves are most pronounced. Sleep staging such EEG recordings is impressive, since slow waves during deep sleep can easily reach 1000 microvolts in amplitude and continue to occur for minutes. Then, during adolescence things change. Now gray matter volume decreases while the volume of white matter keeps increasing. As pointed out before, during adolescence, SWA decreases dramatically, while at the same time EEG coherence keeps increasing. The decrease in gray matter volume most likely reflects the documented decrease of synapse density (Huttenlocher, 1979; Huttenlocher & Dabholkar, 1997). Thus, the huge decrease in synapse density during adolescence is accompanied by continuous growing of long range fibers (Lebel & Beaulieu, 2011). Particularly, the myelination of long-range fibers accounts for the linear white matter volume increase during the first two decades (Lebel & Beaulieu, 2011; Paus et al., 2001). Recent DTI analyses delineate the microstructural maturation of brain white matter tracts and reveal a regional succession of fibers: commissural and projection fibers mature earliest, while association fibers and frontal-temporal connections demonstrate the most prolonged development (Lebel & Beaulieu, 2011; Lebel, Walker, Leemans, Phillips, & Beaulieu, 2008). Space needs, together with the demand for efficient functioning, thus requires a reduction in energy consumption while the brain system is growing.

The increase in myelination may result in more efficient signal transfer. The gray matter reduction, on the other hand, is associated with fewer synapses, which results in space gains and a reduction of energy consumption (Chugani, Phelps, & Mazziotta, 1987). It is noteworthy that age-related changes in synaptic strength or density not only should result in changes of the sleep EEG in the SWA frequency range, but might also explain the age-related amplitude changes of oscillations in other frequency ranges and other vigilance states observed (Feinberg et al., 1990; Gasser et al., 1988; Gaudreau, Carrier, & Montplaisir, 2001).

The question arises whether space constraints are a limiting factor during cortical maturation and are overcome by pruning processes. Thus, do space constraints initiate the "reversal point," where synaptic density stops growing and starts decreasing? It is likely that space restriction is associated with competition over supplies. Regarding neurons it would mean a competition over access to the postsynaptic cells. When a new synapse is formed the presynaptic axon influences the postsynapse to outgrow cell plasma (Yuste & Bonhoeffer, 2004). Moreover, there are indications that used synapses restrain unused synapses in their neighborhood (Balice-Gordon & Lichtman, 1994). An extreme example of maturational synaptic alteration is found in the neural maturation of *Caenorhabditis elegans*, where existing synapses in DD motoneurons are completely eliminated and new synapses are formed without changing cell morphology (Park et al., 2011). The current literature is, however, very limited and does not allow a conclusive understanding of these processes. Important question remain, like, which genes are involved? How does this switch of the changes in synaptic density occur without interference of behavioral output? Clearly these processes need to be studied in more detail. Our data show that the sleep EEG might be a useful tool to explore these processes (even more so due to the following observations).

Quantifying Cortical Maturation during Development

As introduced above, three main methods have been used to quantify developmental changes in the human cerebral cortex: structural MRI, EEG, and postmortem synapse counting using electron microscopy or phosphotungstic acid staining. Although different methods were used, the observed developmental changes were similar: an increase during childhood, a decrease during adolescence, and a stabilization in adulthood. MRI allows for the estimation of cortical thickness. Studies show that cortical thickness is reduced with a rate of about 0.1 mm/year during adolescence with local maxima in reduction of up to 0.3 mm/year (Table 10.1). However, whether these maturational changes in cortical thickness are related to changes in neuropil, neuronal size, or dendritic arborization can only be answered with postmortem or animal studies.

To assess neuronal connectivity, counting and estimating synapse number is restricted to animal studies or the investigation of postmortem human brains. Synaptic densities obtained by phosphotungstic acid staining have been similar to those obtained by conventional electron microscopy

Table 10.1 Cortical changes during Adolescence

	Annual Change	%	Reference	Method
Number of synapses	$5.7 \star 10^7$ synapses/mm^3 of cortical tissue/year (middle frontal gyrus) $4.8 \star 10^{13}$ synapses/year (entire cortex)	55%	(Huttenlocher, 1979); (Good et al., 2001) for cortical volume estimation (young adults: 0.85 liter)	Postmortem, middle frontal gyrus layer 3, phosphotunstic acid method (see Bloom and Aghajanian, 1968), synaptic decrease quantified (slope of decline)
	$3.15 \star 10^{12}$ synapses/year (entire cortex)	50%	(Bourgeois & Rakic, 1993; Rakic, et al., 2009)	Electron microscopy, extrapolation from primary visual cortex (sections of calcarine fissure) in macaque to human cortex, decrease in synapse number quantified during puberty, estimation for entire cortex
Sleep slow waves	155 μV^2/year		(Campbell et al., 2011)	Slow wave activity (SWA) in sleep EEG decrease quantified according to slope of SWA decline in electrode C3 (C4) referenced to mastoid
	215 μV^2/year (Fz) 265 μV^2/year (Cz) 120 μV^2/year (O1)	170% (Fz) 320% (Cz) 340% (O1)	(Feinberg, de Bie, Davis, & Campbell, 2011)	SWA decrease in sleep EEG quantified according to slope of SWA decline in electrodes Fz, Cz, C3, C4, and O1 referenced to mastoid
Cortical thickness	About 0.1 mm cortical thickness reduction per year	40%	(Shaw et al., 2008)	MRI, superior frontal gyri
	Up to 0.3 mm cortical thickness reduction per year		(Sowell et al., 2004)	MRI, topographical annual change (mm), in kids 5–11y; changes that happen after 11y are not considered

Findings from literature are summarized based on three different methods: postmortem synapse number in humans or macaques, slow waves in the sleep EEG, and cortical thickness deriving from MRI. Calculations of annual changes are based on the phase of steepest decline of the particular measure (slope). Percent change was calculated as the difference between maximal values in late childhood and stabilized levels in young adulthood.

(Aghajanian & Bloom, 1967; Armstrong-James & Johnson, 1970). However, the former method might underestimate immature synapses, requires extrapolation from single sections, and synapses close to the synaptic cleft may not be recognized. Bourgeois & Rakic show in the macaque that the mean maximum density of 90 synapses/100 μm^3 of neuropil is reached by the third postnatal month (corresponding to childhood in humans), and decreases during puberty to reach adult levels of 40–50 synapses/100 μm^3 (Bourgeois & Rakic, 1993). This corresponds to a reduction rate of $3.2 * 10^{12}$ synapses/year for the entire cortex during this period (see Table 10.1, and Bourgeois & Rakic, 1993; Rakic, Arellano, & Breunig, 2009). Huttenlocher's postmortem quantification of the reduction of synapses in humans (Huttenlocher, 1979) reveals a reduction rate of $5.7 * 10^7$ synapses/mm^3 of cortical tissue per year. This corresponds to an approximate reduction of $4.8 * 10^{13}$ synapses/year for the entire cortex (Table 10.1). Comparing this human synaptic loss with the extrapolated data from macaque reveals only a small discrepancy of 6.5%.

In electron microscopy analyses, the cortex of old animals is not noticeably thinner than that of adolescent animals, despite a decreasing percentage of neuropil (Bourgeois & Rakic, 1993). This discrepancy might at least in part be due to the continuous process of myelination (Deoni et al., 2011; Paus et al., 2008). In addition, the magnitude of pruning in the basal dendritic trees of pyramidal cells differs dramatically between cortical regions and layers (Elston, Oga, & Fujita, 2009; Rabinowicz, Petetot, Khoury, & de Courten-Myers, 2009). Thus, such region- and layer-specific changes of cortical structure might also impact the subdivision into gray and white matter in MRI analyses. Estimation of cortical thickness and volume may be affected by these difficulties.

The sleep slow waves of the scalp EEG preferentially originate from extragranular layers of the cortex (Rappelsberger, Pockberger, & Petsche, 1982), which are sites of increased plasticity not only in the adult but also in the developing brain (Heynen & Bear, 2001; Trachtenberg, Trepel, & Stryker, 2000). Local field potential recordings in rats show that the rhythmic polarization changes of the membrane potential during deep sleep (i.e., slow oscillation) is highly synchronized between neighbored neuronal units (Vyazovskiy et al., 2009). A firing burst (ON state) occurs simultaneously with the positive deflection of the surface EEG signal, whereas neuronal silence during the oscillation (OFF state) corresponds to the negative deflection of the slow wave in the surface EEG (Vyazovskiy et al., 2009). These studies show that neuronal activity is directly reflected in the SWA EEG signal. On this background, maturational changes of cortical connectivity might specifically be reflected in the EEG. Thereby, the sleep

EEG represents a tool with minimized external disturbance. In the past, the major restriction of the EEG has been its spatial resolution. Today, electrode nets containing up to 256 electrodes enable higher spatial resolution.

A longitudinal approach to monitor sleep SWA across development revealed a yearly decline of about $155\,\mu V^2$ during adolescence (Table 10.1). The reduction of SWA depends on the cortical region, ranging from $120\,\mu V^2$ over central regions to $265\,\mu V^2$ over occipital regions (Table 10.1). These major regional differences in SWA reduction point out the high sensitivity of this method for developmental changes in cortical activity and connectivity. In addition, comparing the magnitude of change across the different measuring methods, it becomes evident that changes in the sleep EEG are most extreme (up to 340%, Table 10.1). Hence, the EEG provides a powerful tool to identify maturational changes due to its sensitivity and specificity in detecting regional differences (see Kurth, Ringli, et al., 2010).

As noted above, these measures all undergo an inverted U-shape maturational time course across development. However, the maturational peak appears at different time points. Synapse density reaches a maximum at 2.5–5 years (Huttenlocher & Dabholkar, 1997), while gray matter volume and thickness attained maxima at 7–11 years (Giedd, 2008; Shaw et al., 2008). SWA reached a maximum at 5–9 years (Figs. 10.1 and 10.3). The discrepancy between synapse numbers and gray matter changes might arise from the different methods. However, a similar spatial maturation across the cortex has been found for gray matter cortical thickness and SWA. In a recent analysis we have measured the maturation of SWA topography, gray matter thickness, and behavioral skills in the same subjects (Kurth et al., 2012). Exemplified for the motor cortex, we found that SWA matured first, then motor skills, and finally gray matter thinning. Again, this shows that SWA closely reflects cortical maturation. Moreover, SWA labels the early processes of cortical functional maturation, and might thus even be involved in the functional refinement of cortical connectivity

ACTIVE ROLE OF SLOW WAVES?

The Synaptic Homeostasis Hypothesis provides a possible mechanism to describe how sleep slow waves may actively contribute to cortical maturation (Tononi & Cirelli, 2006). According to the hypothesis sleep slow waves are responsible for the renormalization of overall synaptic strength. This process is believed to be fundamental for optimal learning throughout life. Several studies have shown that the manipulation of slow waves causally impacts sleep-dependent performance improvements (Aeschbach, Cutler, & Ronda, 2008; Landsness et al., 2009; Marshall et al., 2006). Both the "wiring" (connecting) and "reshaping" (pruning) of cortical neurons during childhood and adolescence depend on use (Chklovskii, Mel, & Svoboda, 2004; Hensch, 2004, 2005; Knudsen, 2004; Mataga, Mizuguchi, & Hensch, 2004; Sur & Rubenstein, 2005), i.e., through learning processes. Such learning processes, according to the Synaptic Homeostasis Hypothesis, are shaped by synaptic renormalization during sleep.

Childhood and adolescence is characterized by massive learning. Moreover, it might not be by chance that sleep also plays a more central role during that time; for example, sleep takes up to 50% of the time during early childhood and sleep depth, assessed by sleep SWA (see Fig. 10.2), peaks in late childhood. As introduced above, one of the most prominent changes in sleep is the previously introduced change in SWA. This inverted U-shape time course of SWA led already early on to the proposal that sleep slow waves may be important for cortical maturation. Before going into a model explaining how the Synaptic Homeostasis Hypothesis may interact and shape maturational processes, the next paragraph highlights existing experimental evidence supporting an active role of sleep during cortical maturation.

Sleep Slow Waves Seem to Be Important for the Developing Cortex

Evidence for an important role of sleep in maturational processes stems from experiments investigating one of the best studied developmental processes, the ocular dominance plasticity (White & Fitzpatrick, 2007). Using this model Frank et al. showed that sleep enhanced the synaptic changes induced by a preceding period of monocular deprivation, while wakefulness in complete darkness did not (Frank, Issa, & Stryker, 2001). In contrast, the reversible silencing of visual cortices during sleep led to reduced ocular dominance plasticity (Jha et al., 2005). Thus, plastic changes not only require specific cortical stimulation (activity) during waking, but also activity during sleep. The close relationship between cortical stimulation and sleep SWA is illustrated by the following experiment: cats and mice reared in darkness showed a reduction of sleep SWA, which was restricted to visual cortex and reversible (Miyamoto, Katagiri, & Hensch, 2003). This effect was impaired by a reduction of NMDA receptor function, revealing that both ocular dominance plasticity as well as SWA are sensitive to NMDA receptor function (Miyamoto et al., 2003).

These experiments show that indeed sleep, specifically slow waves during sleep, may be related to cortical maturation. Still the question remains as to how cortical activity during sleep may interact with maturational processes. In the framework of the Synaptic Homeostasis Hypothesis we propose a working hypothesis providing a possible mechanism how sleep impacts on cortical maturation.

A Model for Sleep-Dependent Synaptic Changes during Maturation

Analyzing regional SWA distribution across age uncovered that the topographical distribution of SWA underwent maturational changes (Kurth, Ringli, et al., 2010) that closely resembled the spatial maturation of changes in cortical thickness (Shaw, et al., 2008) (see also Fig. 10.3), and the maturation of functions attributable to the underlying cortical regions (Luna & Sweeney, 2004). Our current analyses show that the maturation of EEG SWA topography correlates with the maturation of skills, as measured with behavioral tasks (Kurth et al., 2012). Moreover, in this cross-sectional sample, the maturation of SWA preceded the maturation of skills by ~3 years. Thus SWA may actively impact cortical structure during this sensitive period.

One possible mechanism for how slow waves may actively contribute to cortical maturation is proposed in a model by Ringli and Huber (2011), which is based on the Synaptic Homeostasis Hypothesis by Tononi and Cirelli (2006). The *Synaptic Homeostasis Hypothesis* assumes that synapses are potentiated during waking and downscaled during slow wave sleep ensuring an overall balance of synaptic strength across 24 hours (Tononi & Cirelli, 2006). Numerous studies show that synaptic potentiation may also lead to growth of synapses, whereas a reduction of synaptic strength causes synapses and spines to retract or shrink (Holtmaat & Svoboda, 2009). Hence, the balancing an overall synaptic strength may be closely related to the balancing of synaptic density. Thus, according to the model by Ringli and Huber, in adults, the sum of synapse formation is in equilibrium with the sum of synapse elimination. However, during development, this balance might be tilted (Fig. 10.5). Ringli and Huber suggest that synapse formation excels synapse elimination during childhood, leading to a net increase in synaptic strength/synapse number (Ringli & Huber, 2011). This is exemplified in Figure 10.5 (first phase), where synaptic potentiation during wakefulness is greater than synaptic downscaling during sleep, which results in a net increase of synaptic strength by 20 arbitrary units. This increase might be very small, hardly measurable on a daily basis, but persisting over several years leading to a

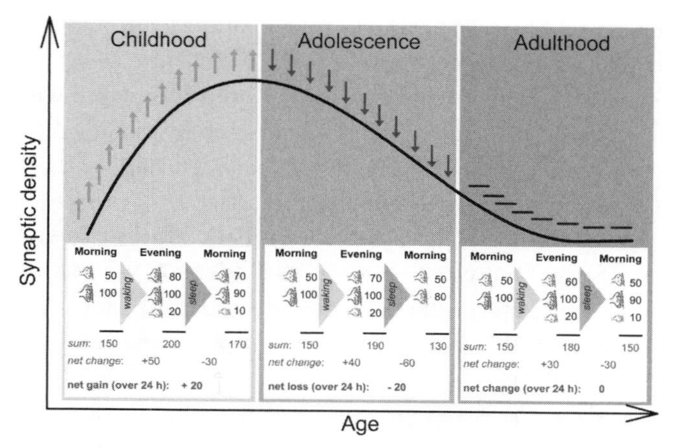

Figure 10.5 ***Model explaining a possible active role of sleep in cortical maturation.*** This model is based on the assumption that synaptic strength increases during waking (synaptic potentiation) and decreases during sleep (synaptic downscaling, Tononi & Cirelli, 2006). Three distinct phases are delineated: childhood, adolescence, and adulthood. The model by Ringli & Huber suggests alterations in the balance of synaptic strength/density during the three phases (Ringli & Huber, 2011). Synapse potentiation/formation (taking place during waking) exceeds synapse elimination and attenuation (taking place during sleep) during childhood, leading to a net increase in synapses. The change in synaptic strength across 24 hours (from the morning to the next morning) for a few exemplary synapses is illustrated as arbitrary values. In addition, the synaptic strength is summed up for each time point (sum), the resulting net changes of synaptic strength across waking and sleep are presented (net change), and finally the change across 24 hours is calculated. Elimination exceeds formation during adolescence, associated with a net decrease in synaptic strength. An equilibrium between synapse elimination and formation is found in adolescence, with no net change in synaptic strength. As presented in the model, these three phases account for the inverted U-shape time course of synaptic density across development.

pronounced overall increase of synaptic strength/density. In the model, the contrary is assumed during adolescence, during which synapse elimination excels synapse formation. Finally, synapse formation and elimination levels off in an equilibrium during adulthood. These three phases together account for the inverted U-shape time course of synaptic density across development, as known from human studies in postmortem or macaques (Bourgeois & Rakic, 1993; Huttenlocher, 1979).

A recent study using two-photon imaging in mice supports at least part of this model. This study showed that while waking is associated with spinogenesis, sleep seems to facilitate spine elimination during adolescence (Maret, Faraguna, Nelson, Cirelli, & Tononi, 2011). In contrast in adult mice, spine turnover was not affected by sleep and waking. These findings reveal that sleep is critically involved in synaptic turnover during a sensitive period like adolescence (Maret et al., 2011). How sleep would result in a net decrease in synaptic density/strength remains unknown. However, potential mechanisms favoring synaptic depression/elimination during deep sleep may require the repeated alternation between depolarization (associated with synchronous firing) and hyperpolarization (with complete neuronal silence) of slow oscillations (of about 1 Hz) (Birtoli & Ulrich, 2004; Lante, Toledo-Salas, Ondrejcak, Rowan, & Ulrich, 2011). In addition low levels of neuromodulators such as noradrenaline and of plasticity-related molecules such as BDNF may also favor synaptic depression during sleep (Tononi & Cirelli, 2006).

Indirect evidence for a change in the balance of synaptic strength comes from the observation of changes in the dynamics of the homeostatic regulation of sleep during childhood and adolescence. If indeed our marker of sleep homeostasis—SWA—reflects synaptic homeostasis, then quantifying the dynamics of sleep homeostasis may elude developmental changes in synaptic plasticity. Accordingly, the faster increase in the homeostatic process during childhood (Jenni et al., 2005) might possibly also indicate a faster increase and a sooner saturation of synaptic strength during childhood compared to adolescence. Why would synaptic saturation be reached sooner in children than adolescents and adults?

The highly connected cortical network during childhood results in higher synchronized neuronal network activity on a larger scale as compared to more focused activity in adults. This assumption is in line with functional MRI studies revealing activity patterns of greater magnitude in children compared to adults (Casey, Giedd, & Thomas, 2000). Over time, increased and more widespread (and thus less specific) cortical activation results in a faster saturation of synaptic strength during childhood. As a result sleep pressure saturates faster. Only the pruning of synapses during adolescence and the gain in specificity would bring about a slowing of the buildup of sleep pressure during the day. In contrast to the age-dependent changes in the homeostatic increase of sleep pressure, the dynamics of the homeostatic decrease seems to be stable across age (Jenni et al., 2005; Rusterholz & Achermann, 2011). These studies show that the balance between the buildup and dissipation of sleep pressure changes across development. However, how the dynamics of sleep homeostasis relate to changes in, for example, synaptic strength on the neuronal level needs to be determined in future studies.

Moreover, as mentioned earlier the maturation of short-range connectivity (number and strength of synapses) and the maturation of long-range connectivity (degree of fiber myelination) follow a different time course

(Paus et al., 2001; Shaw et al., 2008). Thus, the changes in white matter may also contribute to the observed changes in the dynamics of sleep homeostasis during childhood and adolescence.

Since adolescence in humans is a period of increasing incidence of psychiatric and mood disorders, schizophrenia, anxiety, substance abuse, and personality disorders, it appears that this period of synapse elimination is possibly sensitive for disturbances (Paus et al., 2008).

TRANSLATION INTO CLINICS

Pruning and Psychiatric Disorders

Pruning during adolescence may unmask preexisting synaptic deficits (Hoffman & McGlashan, 1993). In 1982, based on sleep slow-wave amplitude measures, Feinberg hypothesized a faulty pruning during adolescence in schizophrenia patients (Feinberg, 1982), a phenotype that suffers from disturbances in basic cognitive functions (Lewis & Lieberman, 2000). During adolescence, when schizophrenia symptoms often start (Keshavan, Anderson, & Pettegrew, 1994), "overpruning" could explain the reduced expression of synaptic proteins and the decreased volume of neuropil in prefrontal circuits observed in schizophrenic patients (Woo & Crowell, 2005). Indeed later studies in childhood-onset schizophrenia patients showed an altered time course of cortical development, as measured in grey matter loss during adolescence (Gogtay et al., 2008; Rapoport & Gogtay, 2008).

Changes in Downscaling and Sleep Topography

Two studies further broach the issue that pathological phenotypes are probably reflected in the sleep EEG. First, Bölsterli et al. show that in patients with continuous spike wave epilepsy during slow wave sleep, the slope of slow waves does not decrease across the night (Bolsterli et al., 2011). The authors interpret this finding as a deficient overnight synaptic downscaling process in these patients, resulting in the progressive deterioration of cerebral functioning, as is the case in these patients (Tassinari, Rubboli, Volpi, Billard, & Bureau, 2005). Since spike wave epilepsy predominantly occurs in childhood, the faulty downscaling process, mediated by SWA, might impact particular brain regions while they undergo a sensitive period for development (Bolsterli et al., 2011). This might account for the major cognitive dysfunctions in these patients.

Second, a study in progress investigates the sleep EEG in children and adolescents suffering from attention deficit hyperactivity disorder (ADHD) (Ringli et al., 2011). Sleep EEG topography uncovers local deviations of SWA in these patients when contrasted with age-matched healthy controls. Interestingly, the topographical pattern of SWA seen in children with ADHD is typical for healthy children of younger age (Ringli et al., 2011). Thus, these data point to a maturational delay in sleep EEG topography of patients with ADHD. Imaging findings confirm this implication of a maturational delay in terms of gray matter thickness (Shaw, Gogtay, & Rapoport, 2010). Such findings highlight the clinical relevance of establishing hdEEG during sleep as diagnostic tool.

The Sleep SWA Topography as a Diagnostic/Prognostic Marker

The use of SWA topography as an imaging tool to diagnostically or even prognostically mark cortical maturation provides several advantages: First, SWA is a direct reflection of underlying spontaneous neuronal activity (Steriade, Timofeev, & Grenier, 2001; Timofeev et al., 2001; Vyazovskiy, et al., 2009). SWA topography accordingly represents the spatial variation in neuronal activity. Such cortical activity may be a more straightforward predictor of functional discrepancies than brain anatomy. Second, during sleep possible confounding factors related to waking activities can be excluded, such as attention, motivation, and distractibility. This might be particularly relevant for studies in children and patients with impairments of cognition or performance. The fact that SWA topography is highly reproducible across nights (Finelli, Achermann, & Borbely, 2001; Huber et al., 2004) illustrates the consistency of this measure. Thus, the measurement is objective, reliable, and robust. These factors are also of interest in study conditions, since many functional neuroimaging studies face similar problems as diagnostic settings. Third, sleep SWA is a tool that is easy to apply and not susceptible to moving artifacts, which is important for the investigation of children and patients. In addition, sleep recordings allow the data collection of hours of brain activity representing a stable assessment. Fourth, it is cost-efficient and can be repeatedly applied without any ethical concerns. Finally, brain activity, as measured by SWA, may be more predictive for functional outcome than is gray matter volume. In sum, sleep may be a rich source of information to gain insight into the healthy and disturbed processes of brain function and maturation.

REFERENCES

Achermann, P., & Borbely, A. A. (1997). Low-frequency (<1 Hz) oscillations in the human sleep electroencephalogram. *Neuroscience*, *81*(1), 213–222. doi: S0306-4522(97)00186-3.

Achermann, P., & Borbely, A. A. (1998). Coherence analysis of the human sleep electroencephalogram. *Neuroscience*, *85*(4), 1195–1208. doi: S0306-4522(97)00692-1.

Aeschbach, D., Cutler, A. J., & Ronda, J. M. (2008). A role for non-rapid-eye-movement sleep homeostasis in perceptual learning. *The Journal of Neuroscience, 28*(11), 2766–2772. doi: 28/11/2766 10.1523/JNEUROSCI.5548-07.2008.

Aghajanian, G. K., & Bloom, F. E. (1967). The formation of synaptic junctions in developing rat brain: A quantitative electron microscopic study. *Brain Research, 6*(4), 716–727. doi: 0006-8993(67)90128-X.

Alonso, M., Bekinschtein, P., Cammarota, M., Vianna, M. R., Izquierdo, I., & Medina, J. H. (2005). Endogenous BDNF is required for long-term memory formation in the rat parietal cortex. *Learning & Memory, 12*(5), 504–510. doi: 12/5/504 10.1101/lm.27305.

Amzica, F., & Steriade, M. (1995). Disconnection of intracortical synaptic linkages disrupts synchronization of a slow oscillation. *The Journal of Neuroscience, 15*(6), 4658–4677.

Anders, T. F., & Roffwarg, H. P. (1973). The effects of selective interruption and deprivation of sleep in the human newborn. *Developmental Psychobiology, 6*(1), 77–89. doi:10.1002/dev.420060110.

Armstrong-James, M., & Johnson, R. (1970). Quantitative studies of postnatal changes in synapses in rat superficial motor cerebral cortex. An electron microscopical study. *Zeitschrift für Zellforschung und Mikroskopische Anatomie, 110*(4), 559–568.

Aston-Jones, G. (2005). Brain structures and receptors involved in alertness. *Sleep Medicine, 6*(Suppl. 1), S3–7. doi: S1389-9457(05)80002-4.

Balice-Gordon, R. J., & Lichtman, J. W. (1994). Long-term synapse loss induced by focal blockade of postsynaptic receptors. *Nature, 372*(6506), 519–524. doi:10.1038/372519a0.

Bersagliere, A., & Achermann, P. (2010). Slow oscillations in human non-rapid eye movement sleep electroencephalogram: Effects of increased sleep pressure. *Journal of Sleep Research, 19*(1 Pt 2), 228–237. doi: JSR775 10.1111/j.1365-2869.2009.00775.x.

Bes, F., Schulz, H., Navelet, Y., & Salzarulo, P. (1991). The distribution of slow-wave sleep across the night: A comparison for infants, children, and adults. *Sleep, 14*(1), 5–12.

Birtoli, B., & Ulrich, D. (2004). Firing mode-dependent synaptic plasticity in rat neocortical pyramidal neurons. *The Journal of Neuroscience, 24*(21), 4935–4940. doi:10.1523/JNEUROSCI.0795-04.2004. (24/21/4935)

Blake, H., & Gerard, R. W. (1937). Brain potentials during sleep. *The American Journal of Physiology, 119*, 692–703.

Bliss, T. V., & Lomo, T. (1973). Long-lasting potentiation of synaptic transmission in the dentate area of the anaesthetized rabbit following stimulation of the perforant path. *The Journal of Physiology, 232*(2), 331–356.

Bolsterli, B. K., Schmitt, B., Bast, T., Critelli, H., Heinzle, J., Jenni, O. G., et al. (2011). Impaired slow wave sleep downscaling in encephalopathy with status epilepticus during sleep (ESES). *Clinical Neurophysiology* doi: S1388-2457(11)00173-8 10.1016/j.clinph.2011.01.053.

Borbély, A. A., & Achermann, P. (2005). Sleep homeostasis and models of sleep regulation. In M. H. Kryger, T. Roth, & W. C. Dement (Eds.), *Principles and practice of sleep medicine* (pp. 405–417). Philadelphia: Elsevier Saunders.

Borst, J. G., & Sakmann, B. (1996). Calcium influx and transmitter release in a fast CNS synapse. *Nature, 383*(6599), 431–434. doi:10.1038/383431a0.

Bourgeois, J. P., & Rakic, P. (1993). Changes of synaptic density in the primary visual cortex of the macaque monkey from fetal to adult stage. *The Journal of Neuroscience, 13*(7), 2801–2820.

Buchmann, A., Kurth, S., Ringli, M., Geiger, A., Jenni, O. G., & Huber, R. (2011). Anatomical markers of sleep slow wave activity derived from structural magnetic resonance images. *Journal of Sleep Research*. doi:10.1111/j.1365-2869.2011.00916.x.

Buchmann, A., Ringli, M., Kurth, S., Schaerer, M., Geiger, A., Jenni, O. G., et al. (2010). EEG Sleep Slow-Wave Activity as a Mirror of Cortical Maturation. *Cerebral Cortex* doi: bhq129 10.1093/cercor/bhq129.

Bushey, D., Tononi, G., & Cirelli, C. (2011). Sleep and synaptic homeostasis: Structural evidence in Drosophila. *Science, 332*(6037), 1576–1581. doi: 332/6037/1576 10.1126/science.1202839.

Campbell, I. G., Darchia, N., Higgins, L. M., Dykan, I. V., Davis, N. M., de Bie, E., et al. (2011). Adolescent changes in homeostatic regulation of EEG activity in the delta and theta frequency bands during NREM sleep. *Sleep, 34*(1), 83–91.

Campbell, I. G., & Feinberg, I. (2009). Longitudinal trajectories of non-rapid eye movement delta and theta EEG as indicators of adolescent brain maturation. *Proceedings of the National Academy of Sciences of the United States of America, 106*(13), 5177–5180.

Cancedda, L., Putignano, E., Sale, A., Viegi, A., Berardi, N., & Maffei, L. (2004). Acceleration of visual system development by environmental enrichment. *The Journal of Neuroscience, 24*(20), 4840–4848. doi:10.1523/JNEUROSCI.0845-04.2004. (24/20/4840)

Casey, B. J., Giedd, J. N., & Thomas, K. M. (2000). Structural and functional brain development and its relation to cognitive development. *Biological Psychology, 54*(1–3), 241–257. doi: S0301051100000582.

Castillo, D. V., Figueroa-Guzman, Y., & Escobar, M. L. (2006). Brain-derived neurotrophic factor enhances conditioned taste aversion retention. *Brain Research, 1067*(1), 250–255. doi: S0006-8993(05)01548-9 10.1016/j.brainres.2005.10.085.

Chakravarthy, S., Saiepour, M. H., Bence, M., Perry, S., Hartman, R., Couey, J. J., et al. (2006). Postsynaptic TrkB signaling has distinct roles in spine maintenance in adult visual cortex and hippocampus. *Proceedings of the National Academy of Sciences of the United States of America, 103*(4), 1071–1076. doi: 0506305103 10.1073/pnas.0506305103.

Chklovskii, D. B., Mel, B. W., & Svoboda, K. (2004). Cortical rewiring and information storage. *Nature, 431*(7010), 782–788. doi: nature03012 10.1038/nature03012.

Chugani, H. T. (1998). A critical period of brain development: Studies of cerebral glucose utilization with PET. *Preventive Medicine, 27*(2), 184–188. doi: S0091-7435(98)90274-2 10.1006/pmed.1998.0274.

Chugani, H. T., Phelps, M. E., & Mazziotta, J. C. (1987). Positron emission tomography study of human brain functional development. *Annals of Neurology, 22*(4), 487–497.

Cirelli, C. (2009). The genetic and molecular regulation of sleep: From fruit flies to humans. *Nature Reviews Neuroscience, 10*(8), 549–560. doi: nrn2683 10.1038/nrn2683.

Cirelli, C., & Tononi, G. (2008). Is sleep essential? *PLoS Biology, 6*(8), e216. doi: 08-PLBI-E-1874 10.1371/journal.pbio.0060216.

Courchesne, E., Chisum, H. J., Townsend, J., Cowles, A., Covington, J., Egaas, B., et al. (2000). Normal brain development and aging: Quantitative analysis at in vivo MR imaging in healthy volunteers. *Radiology, 216*(3), 672–682.

Deoni, S. C., Mercure, E., Blasi, A., Gasston, D., Thomson, A., Johnson, M., et al. (2011). Mapping infant brain myelination with magnetic resonance imaging. *The Journal of Neuroscience, 31*(2), 784–791. doi: 31/2/784 10.1523/JNEUROSCI.2106-10.2011.

Diekelmann, S., & Born, J. (2010). The memory function of sleep. *Nature Reviews Neuroscience, 11*(2), 114–126. doi: nrn2762 10.1038/nrn2762.

Elston, G. N., Oga, T., & Fujita, I. (2009). Spinogenesis and Pruning Scales across Functional Hierarchies. *Journal of Neuroscience, 29*(10), 3271–3275. doi:10.1523/Jneurosci.5216-08.2009.

Esser, S. K., Hill, S. L., & Tononi, G. (2007). Sleep homeostasis and cortical synchronization: I. Modeling the effects of synaptic strength on sleep slow waves. *Sleep, 30*(12), 1617–1630.

Faraguna, U., Vyazovskiy, V. V., Nelson, A. B., Tononi, G., & Cirelli, C. (2008). A causal role for brain-derived neurotrophic factor in the homeostatic regulation of sleep. *The Journal of Neuroscience, 28*(15), 4088–4095.

Feinberg, I. (1982). Schizophrenia: Caused by a fault in programmed synaptic elimination during adolescence? *Journal of Psychiatric Research, 17*(4), 319–334.

Feinberg, I., de Bie, E., Davis, N. M., & Campbell, I. G. (2011). Topographic differences in the adolescent maturation of the slow wave EEG during NREM sleep. *Sleep, 34*(3), 325–333.

Feinberg, I., March, J. D., Flach, K., Maloney, T., Chern, W. -J., & Travis, F. (1990). Maturational Changes in Amplitude, Incidence and Cyclic Pattern of the 0 to 30 Hz (Delta) Electroencephalogram of Human Sleep. *British Dysfunction*, *3*, 183–192.

Finelli, L. A., Achermann, P., & Borbely, A. A. (2001). Individual 'fingerprints' in human sleep EEG topography. *Neuropsychopharmacology*, *25*(5 Suppl.), S57–62. doi: S0893133X01003207 10.1016/S0893-133X(01)00320-7.

Finelli, L. A., Borbely, A. A., & Achermann, P. (2001). Functional topography of the human nonREM sleep electroencephalogram. *The European Journal of Neuroscience*, *13*(12), 2282–2290.

Frank, M. G., Issa, N. P., & Stryker, M. P. (2001). Sleep enhances plasticity in the developing visual cortex. *Neuron*, *30*(1), 275–287.

Frank, M. G., Morrissette, R., & Heller, H. C. (1998). Effects of sleep deprivation in neonatal rats. *American Journal of Physiology-Regulatory Integrative and Comparative Physiology*, *44*(1), R148–R157.

Frankenhaeuser, B. (1952). Saltatory conduction in myelinated nerve fibres. *The Journal of Physiology*, *118*(1), 107–112.

Gao, P. P., Yue, Y., Cerretti, D. P., Dreyfus, C., & Zhou, R. (1999). Ephrin-dependent growth and pruning of hippocampal axons. *Proceedings of the National Academy of Sciences of the United States of America*, *96*(7), 4073–4077.

Gasser, T., Rousson, V., Caflisch, J., & Jenni, O. G. (2010). Development of motor speed and associated movements from 5 to 18 years. *Developmental Medicine and Child Neurology*, *52*(3), 256–263. doi: DMCN3391 10.1111/j.1469-8749.2009.03391.x.

Gasser, T., Verleger, R., Bacher, P., & Sroka, L. (1988). Development of the EEG of school-age children and adolescents. I. Analysis of band power. *Electroencephalography and Clinical Neurophysiology*, *69*(2), 91–99.

Gaudreau, H., Carrier, J., & Montplaisir, J. (2001). Age-related modifications of NREM sleep EEG: From childhood to middle age. *Journal of Sleep Research*, *10*(3), 165–172.

Genoud, C., Knott, G. W., Sakata, K., Lu, B., & Welker, E. (2004). Altered synapse formation in the adult somatosensory cortex of brain-derived neurotrophic factor heterozygote mice. *The Journal of Neuroscience*, *24*(10), 2394–2400. doi:10.1523/JNEUROSCI.4040-03.2004. (24/10/2394)

Giedd, J. N. (2004). Structural magnetic resonance imaging of the adolescent brain. *Annals of the New York Academy of Sciences*, *1021*, 77–85.

Giedd, J. N. (2008). The teen brain: Insights from neuroimaging. *The Journal of Adolescent Health*, *42*(4), 335–343. doi: S1054-139X(08)00075-X 10.1016/j.jadohealth.2008.01.007.

Giedd, J. N., Blumenthal, J., Jeffries, N. O., Castellanos, F. X., Liu, H., Zijdenbos, A., et al. (1999). Brain development during childhood and adolescence: A longitudinal MRI study. *Nature Neuroscience*, *2*(10), 861–863.

Gilestro, G. F., Tononi, G., & Cirelli, C. (2009). Widespread changes in synaptic markers as a function of sleep and wakefulness in Drosophila. *Science*, *324*(5923), 109–112. doi: 324/5923/109 10.1126/science.1166673.

Gogtay, N., Giedd, J. N., Lusk, L., Hayashi, K. M., Greenstein, D., Vaituzis, A. C., et al. (2004). Dynamic mapping of human cortical development during childhood through early adulthood. *Proceedings of the National Academy of Sciences of the United States of America*, *101*(21), 8174–8179. doi:10.1073/pnas.0402680101. (0402680101)

Gogtay, N., Lu, A., Leow, A. D., Klunder, A. D., Lee, A. D., Chavez, A., et al. (2008). Three-dimensional brain growth abnormalities in childhood-onset schizophrenia visualized by using tensor-based morphometry. *Proceedings of the National Academy of Sciences of the United States of America*, *105*(41), 15979–15984. doi: 0806485105 10.1073/pnas.0806485105.

Goldman, P. S., & Nauta, W. J. (1977). Columnar distribution of cortico-cortical fibers in the frontal association, limbic, and motor cortex of the developing rhesus monkey. *Brain Research*, *122*(3), 393–413. doi: 0006-8993(77)90453-X.

Good, C. D., Johnsrude, I. S., Ashburner, J., Henson, R. N., Friston, K. J., & Frackowiak, R. S. (2001). A voxel-based morphometric study of ageing in 465 normal adult human brains. *Neuroimage, 14*(1 Pt 1), 21–36. doi: S1053-8119(01)90786-4 10.1006/nimg.2001.0786.

Hairston, I. S., Peyron, C., Denning, D. P., Ruby, N. F., Flores, J., Sapolsky, R. M., et al. (2004). Sleep deprivation effects on growth factor expression in neonatal rats: A potential role for BDNF in the mediation of delta power. *Journal of Neurophysiology, 91*(4), 1586–1595. doi:10.1152/jn.00894.2003. (00894.2003)

Halter, J. A., & Clark, J. W., Jr. (1991). A distributed-parameter model of the myelinated nerve fiber. *Journal of Theoretical Biology, 148*(3), 345–382.

Hensch, T. K. (2004). Critical period regulation. *Annual Review of Neuroscience, 27*, 549–579. doi:10.1146/annurev.neuro.27.070203.144327.

Hensch, T. K. (2005). Critical period plasticity in local cortical circuits. *Nature Reviews Neuroscience, 6*(11), 877–888. doi: nrn1787 10.1038/nrn1787.

Heynen, A. J., & Bear, M. F. (2001). Long-term potentiation of thalamocortical transmission in the adult visual cortex in vivo. *The Journal of Neuroscience, 21*(24), 9801–9813. doi: 21/24/9801.

Hill, S., & Tononi, G. (2005). Modeling sleep and wakefulness in the thalamocortical system. *Journal of Neurophysiology, 93*(3), 1671–1698. doi: 00915.2004 10.1152/jn.00915.2004.

Hoffman, R. E., & McGlashan, T. H. (1993). Neurodynamics and schizophrenia research: Editors' introduction. *Schizophrenia Bulletin, 19*(1), 15–19.

Holmes, S. J. (1907). Regeneration as functional adjustment. *Journal of Experimental Zoology, 4*(3), 12.

Holtmaat, A., & Svoboda, K. (2009). Experience-dependent structural synaptic plasticity in the mammalian brain. *Nature Reviews Neuroscience, 10*(9), 647–658. doi: nrn2699 10.1038/nrn2699.

Hua, J. Y. Y., & Smith, S. J. (2004). Neural activity and the dynamics of central nervous system development. *Nature Neuroscience*, 7(4), 327–332.

Huber, R., Ghilardi, M. F., Massimini, M., Ferrarelli, F., Riedner, B. A., Peterson, M. J., et al. (2006). Arm immobilization causes cortical plastic changes and locally decreases sleep slow wave activity. *Nature Neuroscience, 9*(9), 1169–1176.

Huber, R., Ghilardi, M. F., Massimini, M., & Tononi, G. (2004). Local sleep and learning. *Nature, 430*(6995), 78–81.

Huber, R., Mäki, H., Rosanova, M., Casarotto, S., Canali, P., Casali, A., et al. (2012). Human cortical excitability increases with time awake. *Cerebral Cortex,* Epub 2012 Feb 7, PMID: 22314045.

Huber, R., Tononi, G., & Cirelli, C. (2007). Exploratory behavior, cortical BDNF expression, and sleep homeostasis. *Sleep, 30*(2), 129–139.

Huttenlocher, P. R. (1979). Synaptic density in human frontal cortex - developmental changes and effects of aging. *Brain Research, 163*(2), 195–205.

Huttenlocher, P. R. (1990). Morphometric study of human cerebral cortex development. *Neuropsychologia, 28*(6), 517–527.

Huttenlocher, P. R., & Dabholkar, A. S. (1997). Regional differences in synaptogenesis in human cerebral cortex. *The Journal of Comparative Neurology, 387*(2), 167–178.

Huxley, A. F., & Stampfli, R. (1949). Evidence for saltatory conduction in peripheral myelinated nerve fibres. *The Journal of Physiology, 108*(3), 315–339.

Jenni, O. G., Achermann, P., & Carskadon, M. A. (2005). Homeostatic sleep regulation in adolescents. *Sleep, 28*(11), 1446–1454.

Jenni, O. G., Borbely, A. A., & Achermann, P. (2004). Development of the nocturnal sleep electroencephalogram in human infants. *American Journal of Physiology Regulatory, Integrative and Comparative Physiology, 286*(3), R528–538. doi:10.1152/ajpregu.00503.2003. (00503.2003).

Jenni, O. G., & Carskadon, M. A. (2004). Spectral analysis of the sleep electroencephalogram during adolescence. *Sleep, 27*(4), 774–783.

Jennings, H. S. (1904). *Contributions to the study of the behaviour of lower organisms*. Washington: Carnegie Institution of Washington.

Jha, S. K., Jones, B. E., Coleman, T., Steinmetz, N., Law, C. T., Griffin, G., et al. (2005). Sleep-dependent plasticity requires cortical activity. *The Journal of Neuroscience, 25*(40), 9266–9274.

Jiang, B., Akaneya, Y., Hata, Y., & Tsumoto, T. (2003). Long-term depression is not induced by low-frequency stimulation in rat visual cortex in vivo: A possible preventing role of endogenous brain-derived neurotrophic factor. *The Journal of Neuroscience, 23*(9), 3761–3770. doi: 23/9/3761.

Kandel, E. R. (2001). The molecular biology of memory storage: A dialogue between genes and synapses. *Science, 294*(5544), 1030–1038. doi:10.1126/science.1067020. (294/5544/1030).

Kattler, H., Dijk, D. J., & Borbely, A. A. (1994). Effect of unilateral somatosensory stimulation prior to sleep on the sleep EEG in humans. *Journal of Sleep Research, 3*(3), 159–164. doi: jsr003003159.

Keshavan, M. S., Anderson, S., & Pettegrew, J. W. (1994). Is schizophrenia due to excessive synaptic pruning in the prefrontal cortex? The Feinberg hypothesis revisited. *Journal of Psychiatric Research, 28*(3), 239–265.

Kety, S. S. (1956). Human cerebral blood flow and oxygen consumption as related to aging. *Journal of Chronic Diseases, 3*(5), 478–486.

Kety, S. S., & Schmidt, C. F. (1948). The nitrous oxide method for the quantitative determination of cerebral blood flow in man; theory, procedure and normal values. *The Journal of Clinical Investigation, 27*(4), 476–483.

Kleim, J. A., Chan, S., Pringle, E., Schallert, K., Procaccio, V., Jimenez, R., et al. (2006). BDNF val-66met polymorphism is associated with modified experience-dependent plasticity in human motor cortex. *Nature Neuroscience, 9*(6), 735–737. doi: nn1699 10.1038/nn1699.

Knudsen, E. I. (2004). Sensitive periods in the development of the brain and behavior. *Journal of Cognitive Neuroscience, 16*(8), 1412–1425. doi:10.1162/0898929042304796.

Kurth, S., Jenni, O. G., Riedner, B. A., Tononi, G., Carskadon, M. A., & Huber, R. (2010). Characteristics of sleep slow-waves in children and adolescents. *Sleep, 33*(4), 475–480.

Kurth, S., Ringli, M., Geiger, A., LeBourgeois, M., Jenni, O. G., & Huber, R. (2010). Mapping of cortical activity in the first two decades of life: A high-density sleep electroencephalogram study. *The Journal of Neuroscience, 30*(40), 13211–13219. doi: 30/40/13211 10.1523/JNEUROSCI.2532-10.2010.

Kurth, S., Ringli, M., LeBourgeois, M., Geiger, A., Buchmann, A., Jenni, O. G., et al. (2012). Mapping the electrophysiological marker of sleep depth reveals skill maturation in children and adolescents. *NeuroImage,* Epub 2012 Mar 27, PMID: 22498654.

Landsness, E. C., Crupi, D., Hulse, B. K., Peterson, M. J., Huber, R., Ansari, H., et al. (2009). Sleep-dependent improvement in visuomotor learning: A causal role for slow waves. *Sleep, 32*(10), 1273–1284.

Lante, F., Toledo-Salas, J. C., Ondrejcak, T., Rowan, M. J., & Ulrich, D. (2011). Removal of synaptic Ca(2)+-permeable AMPA receptors during sleep. *The Journal of Neuroscience, 31*(11), 3953–3961. doi: 31/11/3953 10.1523/JNEUROSCI.3210-10.2011.

Largo, R. H., Caflisch, J. A., Hug, F., Muggli, K., Molnar, A. A., Molinari, L., et al. (2001). Neuromotor development from 5 to 18 years. Part 1: Timed performance. *Developmental Medicine and Child Neurology, 43*(7), 436–443.

Lebel, C., & Beaulieu, C. (2011). Longitudinal Development of human brain wiring continues from childhood into adulthood. *The Journal of Neuroscience, 31*(30), 10937–10947.

Lebel, C., Walker, L., Leemans, A., Phillips, L., & Beaulieu, C. (2008). Microstructural maturation of the human brain from childhood to adulthood. *Neuroimage, 40*(3), 1044–1055. doi: S1053-8119(07)01177-9 10.1016/j.neuroimage.2007.12.053.

Leemburg, S., Vyazovskiy, V. V., Olcese, U., Bassetti, C. L., Tononi, G., & Cirelli, C. (2010). Sleep homeostasis in the rat is preserved during chronic sleep restriction. *Proceedings of the National Academy of Sciences of the United States of America, 107*(36), 15939–15944. doi: 1002570107 10.1073/pnas.1002570107.

Lewis, D. A., & Lieberman, J. A. (2000). Catching up on schizophrenia: Natural history and neurobiology. *Neuron, 28*(2), 325–334. doi: S0896-6273(00)00111-2.

Lidow, M. S., Goldman-Rakic, P. S., & Rakic, P. (1991). Synchronized overproduction of neurotransmitter receptors in diverse regions of the primate cerebral cortex. *Proceedings of the National Academy of Sciences of the United States of America, 88*(22), 10218–10221.

Liu, Z. W., Faraguna, U., Cirelli, C., Tononi, G., & Gao, X. B. (2010). Direct evidence for wake-related increases and sleep-related decreases in synaptic strength in rodent cortex. *The Journal of Neuroscience, 30*(25), 8671–8675. doi: 30/25/8671 10.1523/JNEUROSCI.1409-10.2010.

Luna, B., & Sweeney, J. A. (2004). The emergence of collaborative brain function: FMRI studies of the development of response inhibition. *Annals of the New York Academy of Sciences, 1021*, 296–309. doi:10.1196/annals.1308.035. (1021/1/296).

Maatta, S., Landsness, E., Sarasso, S., Ferrarelli, F., Ferreri, F., Ghilardi, M. F., et al. (2010). The effects of morning training on night sleep: A behavioral and EEG study. *Brain Research Bulletin, 82*(1–2), 118–123. doi: S0361-9230(10)00023-7 10.1016/j.brainresbull.2010.01.006.

Mahowald, M. W., & Schenck, C. H. (2005). Insights from studying human sleep disorders. *Nature, 437*(7063), 1279–1285. doi: nature04287 10.1038/nature04287.

Majewska, A., & Sur, M. (2003). Motility of dendritic spines in visual cortex in vivo: Changes during the critical period and effects of visual deprivation. *Proceedings of the National Academy of Sciences of the United States of America, 100*(26), 16024–16029. doi:10.1073/pnas.2636949100. (2636949100).

Maret, S., Faraguna, U., Nelson, A. B., Cirelli, C., & Tononi, G. (2011). Sleep and waking modulate spine turnover in the adolescent mouse cortex. *Nature Neuroscience, 14*(11), 1418–1420. doi: nn.2934 10.1038/nn.2934.

Marshall, L., Helgadottir, H., Molle, M., & Born, J. (2006). Boosting slow oscillations during sleep potentiates memory. *Nature, 444*(7119), 610–613.

Massimini, M., Huber, R., Ferrarelli, F., Hill, S., & Tononi, G. (2004). The sleep slow oscillation as a traveling wave. *The Journal of Neuroscience, 24*(31), 6862–6870.

Mataga, N., Mizuguchi, Y., & Hensch, T. K. (2004). Experience-dependent pruning of dendritic spines in visual cortex by tissue plasminogen activator. *Neuron, 44*(6), 1031–1041. doi: S089662730400755X 10.1016/j.neuron.2004.11.028.

Miller, E. K., & Cohen, J. D. (2001). An integrative theory of prefrontal cortex function. *Annual Review of Neuroscience, 24*, 167–202. doi:10.1146/annurev.neuro.24.1.167. (24/1/167)

Mistlberger, R. E. (2005). Circadian regulation of sleep in mammals: Role of the suprachiasmatic nucleus. *Brain Research Brain Research Reviews, 49*(3), 429–454. doi: S0165-0173(05)00020-2 10.1016/j.brainresrev.2005.01.005.

Mistlberger, R. E., Bergmann, B. M., Waldenar, W., & Rechtschaffen, A. (1983). Recovery sleep following sleep deprivation in intact and suprachiasmatic nuclei-lesioned rats. *Sleep, 6*(3), 217–233.

Miyamoto, H., Katagiri, H., & Hensch, T. (2003). Experience-dependent slow-wave sleep development. *Nature Neuroscience, 6*(6), 553–554. doi:10.1038/nn1064. (nn1064)

Murphy, M., Riedner, B. A., Huber, R., Massimini, M., Ferrarelli, F., & Tononi, G. (2009). Source modeling sleep slow waves. *Proceedings of the National Academy of Sciences of the United States of America, 106*(5), 1608–1613. doi: 0807933106 10.1073/pnas.0807933106.

Nir, Y., Staba, R. J., Andrillon, T., Vyazovskiy, V. V., Cirelli, C., Fried, I., et al. (2011). Regional slow waves and spindles in human sleep. *Neuron, 70*(1), 153–169. doi: S0896-6273(11)00166-8 10.1016/j.neuron.2011.02.043.

Nobili, L., Ferrara, M., Moroni, F., De Gennaro, L., Russo, G. L., Campus, C., et al. (2011). Dissociated wake-like and sleep-like electro-cortical activity during sleep. *Neuroimage* doi: S1053-8119(11)00650-1 10.1016/j.neuroimage.2011.06.032.

Oleksenko, A. I., Mukhametov, L. M., Polyakova, I. G., Supin, A.Y., & Kovalzon, V. M. (1992). Unihemispheric sleep deprivation in bottlenose dolphins. *Journal of Sleep Research*, *1*(1), 40–44. doi: jsr001001040.

Park, M., Watanabe, S., Poon, V.V. N., Ou, C.Y., Jorgensen, E. M., & Shen, K. (2011). CYY-1/ Cyclin Y and CDK-5 Differentially Regulate Synapse Elimination and Formation for Rewiring Neural Circuits. *Neuron*, *70*(4), 742–757. doi:10.1016/j.neuron.2011.04.002.

Paus, T. (2005). Mapping brain maturation and cognitive development during adolescence. *Trends in Cognitive Sciences*, *9*(2), 60–68. doi: S1364-6613(04)00320-1 10.1016/j.tics.2004.12.008.

Paus, T., Collins, D. L., Evans, A. C., Leonard, G., Pike, B., & Zijdenbos, A. (2001). Maturation of white matter in the human brain: A review of magnetic resonance studies. *Brain Research Bulletin*, *54*(3), 255–266. doi: S0361-9230(00)00434-2.

Paus, T., Keshavan, M., & Giedd, J. N. (2008). Why do many psychiatric disorders emerge during adolescence? *Nature Reviews Neuroscience*, *9*(12), 947–957. doi: nrn2513 10.1038/nrn2513.

Petanjek, Z., Judas, M., Kostovic, I., & Uylings, H. B. (2008). Lifespan alterations of basal dendritic trees of pyramidal neurons in the human prefrontal cortex: A layer-specific pattern. *Cerebral Cortex*, *18*(4), 915–929. doi: bhm124 10.1093/cercor/bhm124.

Rabinowicz, T., Petetot, J. M., Khoury, J. C., & de Courten-Myers, G. M. (2009). Neocortical maturation during adolescence: Change in neuronal soma dimension. *Brain and Cognition*, *69*(2), 328–336. doi: S0278-2626(08)00251-0 10.1016/j.bandc.2008.08.005.

Rakic, P., Arellano, J. I., & Breunig, J. (2009). Development of the Primate Cerebral Cortex. In M. S. Gazzangia (Ed.), *The cognitive neurosciences* (4th ed.). Cambridge, Massachusetts: The MIT Press.

Rakic, P., Bourgeois, J. P., Eckenhoff, M. F., Zecevic, N., & Goldman-Rakic, P. S. (1986). Concurrent overproduction of synapses in diverse regions of the primate cerebral cortex. *Science*, *232*(4747), 232–235.

Rall, W. (1967). Distinguishing theoretical synaptic potentials computed for different soma-dendritic distributions of synaptic input. *Journal of Neurophysiology*, *30*(5), 1138–1168.

Rapoport, J. L., & Gogtay, N. (2008). Brain neuroplasticity in healthy, hyperactive and psychotic children: Insights from neuroimaging. *Neuropsychopharmacology*, *33*(1), 181–197. doi: 1301553 10.1038/sj.npp.1301553.

Rappelsberger, P., Pockberger, H., & Petsche, H. (1982). The contribution of the cortical layers to the generation of the EEG: Field potential and current source density analyses in the rabbit's visual cortex. *Electroencephalography and Clinical Neurophysiology*, *53*(3), 254–269.

Riedner, B. A., Vyazovskiy, V.V., Huber, R., Massimini, M., Esser, S., Murphy, M., et al. (2007). Sleep homeostasis and cortical synchronization: III. A high-density EEG study of sleep slow waves in humans. *Sleep*, *30*(12), 1643–1657.

Ringli, M., & Huber, R. (2011). Developmental aspects of sleep slow waves: Linking sleep, brain maturation and behavior. *Progress in Brain Research*, *193*, 63–82. doi: B978-0-444-53839-0.00005-3 10.1016/B978-0-444-53839-0.00005-3.

Ringli, M., Souissi, S., Kurth, S., Brandeis, D., Jenni, O. G., & Huber, R. (2011). *Topography of sleep slow wave activity in children with attention deficit hyperactivity disorder.* Paper presented at the Annual Conference of the SSSSC (Swiss Society of Sleep Research, Sleep Medicine and Chronobiology) and SNS (Society of Neurological Surgeons) 2011, St. Gallen, Switzerland.

Rusterholz, T., & Achermann, P. (2011). Topographical aspects in the dynamics of sleep homeostasis in young men: Individual patterns. *BMC Neuroscience*, *12*, 84. doi: 1471-2202-12-84 10.1186/1471-2202-12-84.

Sanchez-Vives, M. V., & McCormick, D. A. (2000). Cellular and network mechanisms of rhythmic recurrent activity in neocortex. *Nature Neuroscience*, *3*(10), 1027–1034. doi:10.1038/79848.

Saper, C. B., Lu, J., Chou, T. C., & Gooley, J. (2005). The hypothalamic integrator for circadian rhythms. *Trends in Neurosciences*, *28*(3), 152–157. doi: S0166-2236(04)00395-9 10.1016/j.tins.2004.12.009.

Shaw, P., Gogtay, N., & Rapoport, J. (2010). Childhood psychiatric disorders as anomalies in neurodevelopmental trajectories. *Human Brain Mapping*, *31*(6), 917–925. doi:10.1002/hbm.21028.

Shaw, P., Kabani, N. J., Lerch, J. P., Eckstrand, K., Lenroot, R., Gogtay, N., et al. (2008). Neurodevelopmental trajectories of the human cerebral cortex. *The Journal of Neuroscience*, *28*(14), 3586–3594. doi: 28/14/3586 10.1523/JNEUROSCI.5309-07.2008.

Siegel, J. M. (2009). Sleep viewed as a state of adaptive inactivity. *Nature Reviews Neuroscience*, *10*(10), 747–753. doi: nrn2697 10.1038/nrn2697.

Sowell, E. R., Thompson, P. M., Leonard, C. M., Welcome, S. E., Kan, E., & Toga, A. W. (2004). Longitudinal mapping of cortical thickness and brain growth in normal children. *The Journal of Neuroscience*, *24*(38), 8223–8231.

Steriade, M., Contreras, D., Curro Dossi, R., & Nunez, A. (1993). The slow (<1 Hz) oscillation in reticular thalamic and thalamocortical neurons: Scenario of sleep rhythm generation in interacting thalamic and neocortical networks. *The Journal of Neuroscience*, *13*(8), 3284–3299.

Steriade, M., Timofeev, I., & Grenier, F. (2001). Natural waking and sleep states: A view from inside neocortical neurons. *Journal of Neurophysiology*, *85*(5), 1969–1985.

Sur, M., & Rubenstein, J. L. (2005). Patterning and plasticity of the cerebral cortex. *Science*, *310*(5749), 805–810. doi: 310/5749/805 10.1126/science.1112070.

Tarokh, L., Carskadon, M. A., & Achermann, P. (2010). Developmental Changes in Brain Connectivity Assessed Using the Sleep Eeg. *Neuroscience*, *171*(2), 622–634. doi:10.1016/j.neuroscience.2010.08.071.

Tartar, J. L., Ward, C. P., McKenna, J. T., Thakkar, M., Arrigoni, E., McCarley, R. W., et al. (2006). Hippocampal synaptic plasticity and spatial learning are impaired in a rat model of sleep fragmentation. *The European Journal of Neuroscience*, *23*(10), 2739–2748. doi: EJN4808 10.1111/j.1460-9568.2006.04808.x.

Tassinari, C. A., Rubboli, G., Volpi, L., Billard, C., & Bureau, M. (2005). Electrical status epilepticus during slow sleep (ESES or CSWS) including acquired epileptic aplasia. (Landau-Kleffner sindrome). In J. Roger, M. Bureau, C. Dravet, P. Genton, C. A. Tassinari & P. Wolf (Eds.), *Epileptic syndromes in infancy, childhood and adolescence.* (pp. 295–314). Montrouge: John Libbey Eurotext.

Teller, D. Y. (1981). The Development of Visual-Acuity in Human and Monkey Infants. *Trends in Neurosciences*, *4*(1), 21–24.

Terzaghi, M., Sartori, I., Tassi, L., Didato, G., Rustioni, V., LoRusso, G., et al. (2009). Evidence of dissociated arousal states during NREM parasomnia from an intracerebral neurophysiological study. *Sleep*, *32*(3), 409–412.

Thomas, D. A., Poole, K., McArdle, E. K., Goodenough, P. C., Thompson, J., Beardsmore, C. S., et al. (1996). The effect of sleep deprivation on sleep states, breathing events, peripheral chemoresponsiveness and arousal propensity in healthy 3 month old infants. *The European Respiratory Journal*, *9*(5), 932–938.

Timofeev, I., Grenier, F., Bazhenov, M., Sejnowski, T. J., & Steriade, M. (2000). Origin of slow cortical oscillations in deafferented cortical slabs. *Cerebral Cortex*, *10*(12), 1185–1199.

Timofeev, I., Grenier, F., & Steriade, M. (2001). Disfacilitation and active inhibition in the neocortex during the natural sleep-wake cycle: An intracellular study. *Proceedings of the National Academy of Sciences of the United States of America*, *98*(4), 1924–1929. doi:10.1073/pnas.041430398. (041430398).

Tobler, I. (2000). Phylogeny of Sleep Regulation. In M. H. Kryger, T. Roth, & W. C. Dement (Eds.), *Principles and practice of sleep medicine* (pp. 72–81) (3rd ed.). Philadelphia: Saunders Co..

Tobler, I., Borbely, A. A., & Groos, G. (1983). The Effect of Sleep-Deprivation on Sleep in Rats with Suprachiasmatic Lesions. *Neuroscience Letters*, *42*(1), 49–54.

Tononi, G., & Cirelli, C. (2006). Sleep function and synaptic homeostasis. *Sleep Medicine Reviews*, *10*(1), 49–62.

Trachsel, L., Edgar, D. M., Seidel, W. F., Heller, H. C., & Dement, W. C. (1992). Sleep homeostasis in suprachiasmatic nuclei-lesioned rats: Effects of sleep deprivation and triazolam administration. *Brain Research*, *589*(2), 253–261.

Trachtenberg, J. T., Trepel, C., & Stryker, M. P. (2000). Rapid extragranular plasticity in the absence of thalamocortical plasticity in the developing primary visual cortex. *Science*, *287*(5460), 2029–2032. doi: 8345.

Turrigiano, G. G. (1999). Homeostatic plasticity in neuronal networks: The more things change, the more they stay the same. *Trends in Neurosciences*, *22*(5), 221–227. doi: S0166-2236(98)01341-1.

Vyazovskiy, V. V., Cirelli, C., Pfister-Genskow, M., Faraguna, U., & Tononi, G. (2008). Molecular and electrophysiological evidence for net synaptic potentiation in wake and depression in sleep. *Nature Neuroscience*, *11*(2), 200–208.

Vyazovskiy, V. V., Olcese, U., Hanlon, E. C., Nir, Y., Cirelli, C., & Tononi, G. (2011). Local sleep in awake rats. *Nature*, *472*(7344), 443–447. doi: nature10009 10.1038/nature10009.

Vyazovskiy, V. V., Olcese, U., Lazimy, Y. M., Faraguna, U., Esser, S. K., Williams, J. C., et al. (2009). Cortical firing and sleep homeostasis. *Neuron*, *63*(6), 865–878. doi: S0896-6273(09)00637-0 10.1016/j.neuron.2009.08.024.

Vyazovskiy, V. V., Riedner, B. A., Cirelli, C., & Tononi, G. (2007). Sleep homeostasis and cortical synchronization: II. A local field potential study of sleep slow waves in the rat. *Sleep*, *30*(12), 1631–1642.

Webster, M. J., Weickert, C. S., Herman, M. M., & Kleinman, J. E. (2002). BDNF mRNA expression during postnatal development, maturation and aging of the human prefrontal cortex. *Brain Research Developmental Brain Research*, *139*(2), 139–150. doi: S0165380602005400.

Wheatley, D. N. (1998). Diffusion theory, the cell and the synapse. *Biosystems*, *45*(2), 151–163. doi: S0303264797000737.

White, L. E., & Fitzpatrick, D. (2007). Vision and cortical map development. *Neuron*, *56*(2), 327–338. doi: S0896-6273(07)00773-8 10.1016/j.neuron.2007.10.011.

Whitlock, J. R., Heynen, A. J., Shuler, M. G., & Bear, M. F. (2006). Learning induces long-term potentiation in the hippocampus. *Science*, *313*(5790), 1093–1097. doi: 313/5790/1093 10.1126/science.1128134.

Woo, T. U., & Crowell, A. L. (2005). Targeting synapses and myelin in the prevention of schizophrenia. *Schizophrenia Research*, *73*(2–3), 193–207. doi: S0920-9964(04)00242-7 10.1016/j.schres.2004.07.022.

Woo, T. U., Pucak, M. L., Kye, C. H., Matus, C. V., & Lewis, D. A. (1997). Peripubertal refinement of the intrinsic and associational circuitry in monkey prefrontal cortex. *Neuroscience*, *80*(4), 1149–1158. doi: S0306452297000596.

Yuste, R., & Bonhoeffer, T. (2004). Genesis of dendritic spines: Insights from ultrastructural and imaging studies. *Nature Reviews Neuroscience*, *5*(1), 24–34.

Zee, P. C., & Manthena, P. (2007). The brain's master circadian clock: Implications and opportunities for therapy of sleep disorders. *Sleep Medicine Reviews*, *11*(1), 59–70. doi: S1087-0792(06)00053-0 10.1016/j.smrv.2006.06.001.

Zuo, Y., Lin, A., Chang, P., & Gan, W. B. (2005). Development of long-term dendritic spine stability in diverse regions of cerebral cortex. *Neuron*, *46*(2), 181–189.

INDEX

F

G

J

K

L

M

O

P

U

V

W

Z